Fatigue Damage

Special Issue Editor
Filippo Berto

MDPI • Basel • Beijing • Wuhan • Barcelona • Belgrade

MDPI

Special Issue Editor
Filippo Berto
Norwegian University of Science and Technology
Norway

Editorial Office
MDPI AG
St. Alban-Anlage 66
Basel, Switzerland

This edition is a reprint of the Special Issue published online in the open access journal *Metals* (ISSN 2075-4701) from 2016–2017 (available at: http://www.mdpi.com/journal/metals/special_issues/Fatigue_Damage).

For citation purposes, cite each article independently as indicated on the article page online and as indicated below:

Author 1; Author 2. Article title. *Journal Name* **Year**, *Article number*, page range.

First Edition 2017

ISBN 978-3-03842-624-0 (Pbk)
ISBN 978-3-03842-625-7 (PDF)

Table of Contents

About the Special Issue Editor

Filippo Berto is Chair of Mechanics and Materials at the Department of Mechanical Engineering in the Norwegian University of Science and Technology. He is author of more than 400 papers dedicated to the fracture and fatigue assessment of different materials, including additive manufactured materials. The main areas of expertise are fatigue and fracture of structural materials, advanced approaches for the design, additive materials, notch mechanics.

metals

MDPI

Editorial

Fatigue Damage

Filippo Berto

Department of Mechanical and Industrial Engineering, Norwegian University of Science and Technology, Richard Birkelands vei 2b, 7491 Trondheim, Norway; Filippo.berto@ntnu.no; Tel.: +47-7359-3831

Academic Editor: Hugo F. Lopez
Received: 22 September 2017; Accepted: 25 September 2017; Published: 26 September 2017

1. Introduction and Scope

Prevention of unexpected failures is a fundamental design objective in any engineering structure or system subjected to fatigue. Nevertheless, the complexity of modern structures and the interactivity among engineering systems, coupled with human fallibility, means that failure and its consequences can only be avoided to a statistical probability. Hence, occasional catastrophic failures will occur, with some of them involving the loss of human lives. Within the last few decades, a dramatic advancement has been achieved in many of the necessary technologies to either avoid or mitigate the consequences of failure. This advancement is associated with ever-increasing performance objectives for materials, structures, and machines, an increased complexity of engineered products and processes. Alongside this, catastrophic failure and its consequences are considered less tolerable in society as a whole; this ensures that efforts to prevent unexpected failures are now a cornerstone in modern engineering design and, simultaneously, a technological and scientific challenge. There is an increasing acknowledgement in the engineering community that the response to this challenge, that is, prevention of catastrophic failure, generally requires a systems approach and necessitates engagement of a large pool of multidisciplinary expertise and the deployment of tools for systems analysis. This multidisciplinary pool includes materials science, structural analysis, manufacturing technologies, quality control and evaluation, mathematics, physics, and probability and reliability. Furthermore, from the scientific point of view, there is also an increasing acknowledgement that addressing the complex engineering problems of today requires the use of concepts and approaches that can account for size- and time-scaling effects. The Special Issue scope embraces interdisciplinary work aimed at understanding and deploying physics of fatigue and failure techniques, advancing experimental and theoretical failure analysis, modelling of the structural response with respect to both local and global failures, and structural design that accounts for scale and time effects in preventing engineering failures.

2. Contributions

Fourteen articles have been published in the present Special Issue of *Metals*, encompassing the fields of fatigue damage, high-cycle fatigue, fatigue and creep interaction, and fatigue in aggressive corrosive media. This grouping is not hard-bound, and thematic links can be established beyond them, which shall be emphasized in the following presentation.

2.1. High-Cycle Fatigue

High-cycle fatigue is a traditional topic, but still offers many challenges, as is well described in some of the present contributions. In particular, it becomes increasingly critical if large-scale effects occur in a full-scale component, increasing the presence of defects [1]. Aspects related to microstructure and interaction between fatigue properties and metallurgical properties are also very important for the fatigue design [2]. The special issue treats this topic in a comprehensive way, providing an overview of the state of the art of recent developments useful for the readers of *Metals*.

2.2. Fatigue and Creep Interaction

The interest in fatigue assessment of steels and different alloys at high temperature has increased continuously in recent years. The applications in which the fatigue phenomenon is affected by high temperature are of considerable interest, and involve various industrial sectors, such as transportation, energy, and metal-manufacturing (e.g., jet engine components, nuclear power plant, pressure vessel, hot rolling of metal). To provide as optimal a performance as possible in these highly demanding conditions, it is necessary to be aware of the application and of the proper tools for performing the fatigue assessment, including at high temperatures. Interaction between creep and fatigue is a crucial point for the design, and it is well treated in some contributions to the present Special Issue (see, for example Ref. [3]).

2.3. Fatigue in Corrosive Media

An aggressive environment can be extremely critical for the fatigue life of a structure working in an aggressive environment, and protection against corrosion is necessary to maintain adequate fatigue properties and warrant the safety of the component. Designers must consider corrosion in service for a proper design against fatigue loadings. Corrosion is also undesirable for reasons related to a safe and economic use of a structure during its service life. Some recent advances in this area are well presented in [4], providing a useful and up-to-date overview of the problem in connection with fatigue damage of structural materials.

3. Conclusions and Outlook

A variety of connected topics have been compiled in the present Special Issue of *Metals*, providing a wide overview of recent developments on different aspects of fatigue damage. Hopefully, this special issue will be a starting point for future discussions and scientific debate on challenging topics related to fatigue damage and fatigue design. The topic, in fact, remains current, with a high and relevant impact for many applications. The selected papers touch on a variety of important topics related to fatigue and fracture in structural materials. Scale effect and multiscaling approaches are a fundamental part of these topics, and allow a better understanding of the fatigue damage at different scale levels.

As guest editor of this special issue, I am very happy with the final result, and hope that the present papers will be useful to researchers and designers, working towards the demanding objective of failure prevention in presence cyclic loadings. I would like to warmly thank all the authors for their contributions, and all of the reviewers for their efforts in ensuring a high-quality publication. At the same time, I would like to thank the many anonymous reviewers who assisted me in the reviewing process. Sincere thanks also to Editors of *Metals* for their continuous help, and to the *Metals* Editorial Assistants for the valuable and inexhaustible engagement and support during the preparation of this volume. In particular, my sincere thanks to Natalie Sun for her help and support.

Conflicts of Interest: The author declares no conflict of interest.

References

1. Benedetti, M.; Torresani, E.; Fontanari, V.; Lusuardi, D. Fatigue and Fracture Resistance of Heavy-Section Ferritic Ductile Cast Iron. *Metals* **2017**, *7*, 88. [CrossRef]
2. Liu, C.; Liu, Y.; Ma, L.; Yi, J. Effects of Solution Treatment on Microstructure and High-Cycle Fatigue Properties of 7075 Aluminum Alloy. *Metals* **2017**, *7*, 193. [CrossRef]

Metals **2017**, *7*, 394

3. Liu, D.; Pons, D.J. Physical-Mechanism Exploration of the Low-Cycle Unified Creep-Fatigue Formulation. *Metals* **2017**, *7*, 379. [CrossRef]

4. Yang, H.; Wang, Y.; Wang, X.; Pan, P.; Jia, D. The Effects of Corrosive Media on Fatigue Performance of Structural Aluminum Alloys. *Metals* **2016**, *6*, 160. [CrossRef]

metals

MDPI

Article

The Effects of Corrosive Media on Fatigue Performance of Structural Aluminum Alloys

Huihui Yang [1], Yanling Wang [1], Xishu Wang [1,*], Pan Pan [1] and Dawei Jia [2]

[1] Department of Engineering Mechanics, School of Aerospace Engineering, AML, Tsinghua University, Beijing 100084, China; yanghh14@mails.tsinghua.edu.cn (H.Y.); 1362135868@126.com (Y.W.); pp13@mails.tsinghua.edu.cn (P.P.)

[2] Engineering Research & Development Center, AVIC SAC Commercial Aircraft Company LTD, Shenyang 110850, China; jia.dawei@sacc.com.cn

* Correspondence: xshwang@tsinghua.edu.cn; Tel.: +86-10-6279-2972

Academic Editor: Filippo Berto
Received: 3 June 2016; Accepted: 7 July 2016; Published: 13 July 2016

Abstract: The effects of corrosive media on rotating bending fatigue lives (the cyclic numbers from 10^4 to 10^8) of different aluminum alloys were investigated, which involved the corrosion fatigue lives of five kinds of aluminum alloys in air, at 3.5 wt. % and 5.0 wt. % NaCl aqueous solutions. Experimental results indicate that corrosive media have different harmful influences on fatigue lives of different aluminum alloys, in which the differences of corrosion fatigue lives depend strongly on the plastic property (such as the elongation parameter) of aluminum alloys and whether to exist with and without fracture mode II. The other various influence factors (such as the dropping corrosive liquid rate, the loading style, and the nondimensionalization of strength) of corrosion fatigue lives in three media were also discussed in detail by using the typical cases. Furthermore, fracture morphologies and characteristics of samples, which showed the different fatigue cracking behaviors of aluminum alloys in three media, were investigated by scanning electron microscopy (SEM) in this paper.

Keywords: aluminum alloy; corrosive environment; fatigue performance; rotating bending test; relative strength

1. Introduction

As important types of light metals (Al, Mg, and Ti alloys) [1], aluminum alloys have received much attention in the past several decades due to their excellent properties such as high strength-to-weight ratio, high corrosion resistance and recyclability, which have been widely applied to aeronautic and automotive fields. However, components and structures made of aluminum alloys are usually directly exposed to the corrosive environment, and their fatigue strengths are highly susceptible in the environmental conditions. More data, especially fatigue properties in corrosive environments, is crucial for guaranteeing the service safety and reliability. Therefore, many studies have focused on fatigue behaviors and damage mechanisms of aluminum alloys in corrosive environments in recent years [2–17]. Yi et al. [1] explored the effect of temperature, humidity and environment (air, humidity and 3.5 wt. % NaCl salt spray) on the fatigue life and fracture mechanisms of 2524 aluminum alloys by scanning electron microscopy (SEM), transmission electron microscopy (TEM) and fatigue property testing. The results showed that temperature has a detrimental influence on corrosion fatigue life, and the increased crack growth rate was attributed to a combination of hydrogen embrittlement and anodic dissolution at the growing fatigue crack tip. Wang [2] attempted to extend the concept of material element fracture ahead of a crack tip, during fatigue crack propagation (FCP) to corrosion fatigue crack propagation (CFCP) of aluminum alloys in corrosive environment. He derived a new expression for the CFCP rate according to the principle of fracture mechanics, which exhibited the important feature of correlation between the CFCP rate, corrosion damage, and mechanical parameters. Na et al. [3] investigated the susceptibility to pitting corrosion of AA2024-T4, AA7075-T651, and AA7475-T761 in aqueous neutral chloride solutions for the purpose of comparison using electrochemical noise measurement. They concluded that the susceptibility was decreased in the following order: AA2024-T4 (the naturally aged condition), AA7475-T761 (the overaged condition), and AA7075-T651 (the near-peak-aged condition). Warner [4] demonstrated the effective inhibition of environmental fatigue crack propagation in age-hardenable aluminum alloys: the addition of the molybdate effectively inhibited the fatigue crack propagation behavior. With regard to the fatigue strength of high strength aluminum alloys in corrosive environments, it has been reported that the degradation of fatigue strength was mainly caused by an acceleration of the fatigue crack growth due to the anodic dissolution and hydrogen embrittlement mechanism [5,6]. Giacomo et al. [7] reported on the effect of fatigue life reduction of 2024 Al alloy for aerospace components due to the corrosive (exfoliation) environment. Both standard fatigue tests on prior corroded samples and fatigue tests conducted with the samples in corrosive solution are developed to define some guidelines for the inclusion of such effects in design and to improve aircraft life management [7]. In addition, as one of the basic effect factors of environments to the hydrogen embrittlement fracture of metals, Murakami [5], Takahashi et al. [8], Wang et al. [6,12], and Ricker et al. [17], etc. also found that corrosive fatigue summed up the hydrogen embrittlement (HE) mechanism for aluminum alloys. In the corrosive fatigue damage process, the hydrogen evolution process based on electrochemical reacting equations can be expressed as follows:

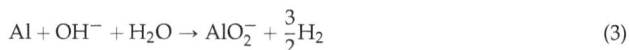

$$H_2O + e \rightarrow OH^- + \frac{1}{2}H_2 \tag{1}$$

$$Al_2O_3 + 2OH^- \rightarrow 2AlO_2^- + H_2O \tag{2}$$

$$Al + OH^- + H_2O \rightarrow AlO_2^- + \frac{3}{2}H_2 \tag{3}$$

However, the presence of corrosive environment greatly increased the complexity of the corrosion fatigue performance evaluation [7], in which there is a need to account for both mechanical and environmental driving forces, so that the fatigue properties in corrosive environments are very complex and are not fully understood.

In the present study, the effects of corrosive media on fatigue behaviors and fatigue lives of aluminum alloys were investigated focusing on fracture characteristics of AA 6000 and AA 7000 series aluminum alloys and S-N curves in three types of media, namely air, 3.5 wt. % NaCl and 5.0 wt. % NaCl aqueous solutions. As fracture characteristics of AA 2000 series aluminum alloys, please see the reference [14]. In addition, the fatigue life of AA2024-CZ aluminum alloy in 3.5 wt. % NaCl aqueous solution under different cyclic loading types (rotation bending cyclic loading and axial cyclic loading, $R = -1$) and under different corrosion modes (stress corrosion and pre-corrosion) were also compared, respectively.

2. The Experimental Method

2.1. Materials and Specimens

Materials used in this work were the commercial AA7475-T7351, AA7075-T651, AA2024-CZ, AA2024-T4, and AA6063-T4 aluminum alloys. The major mechanical properties of those aluminum alloys were listed in Table 1 and the chemical compositions in Table 2, respectively.

Table 1. Major mechanical properties of representative aluminum alloys.

Al Alloys	E (GPa)	$\sigma_{0.2}$ (MPa)	σ_b (MPa)	δ (%)
AA7475-T7351	70.5	434	503	10.2
AA7075-T651	72.0	505	570	11.0
AA2024-CZ	71.0	260	290	17.0
AA2024-T4	71.0	325	470	20.0
AA6063-T4	69.0	90	170	22.0

E: Young's modulus; $\sigma_{0.2}$: the material's offset yield strength; σ_b: the tensile strength; δ: the percentage elongation.

Table 2. Chemical compositions (wt. %) of representative aluminum alloys.

Al Alloys	Zn	Mg	Cu	Mn	Cr	Ti	Fe	Si	Al
AA7475-T7351	5.89	2.48	1.59	<0.01	0.22	0.02	0.06	<0.03	Bal.
AA7075-T651	5.60	2.50	1.60	0.32	0.40	0.06	0.50	0.40	Bal.
AA2024-CZ	0.07	1.49	4.36	0.46	<0.01	0.01	0.25	0.14	Bal.
AA2024-T4	0.25	1.60	4.50	0.80	0.10	<0.01	0.50	0.50	Bal.
AA6063-T4	0.10	0.70	0.10	0.10	0.10	0.10	0.35	0.40	Bal.

All plate or round bars of aluminum alloys were machined as experimental samples based on the demands of shape and size as shown in Figure 1. The specimen (total length = 100 mm and diameter = 10 mm) is the minimum diameter of φ4 mm, which formed a circumferential notch to localize the crack initiation site. After machining, all of the specimens were prepared by an electro-polishing method after hand grinding at −20 °C at a voltage of 70 V for approximately 5–10 s. The polishing etchant was a solution of 5% perchloric acid and 95% glacial acetic acid. This treatment can remove the work-hardened or oxidized layer of the sample and then reduce the residue stress. The surface roughness (Re) and the stress concentration factor of each specimen were controlled approximately 0.80 µm by using the different abrasive papers (1000, 1200, 2000 grits, respectively) and 1.08 by using the radius of curvature of notch, respectively. The applied stress level in the rotation bending fatigue tests is estimated as follows [6,12,14]:

$$\sigma = \frac{32g\alpha LW}{\pi d^3} \tag{4}$$

where d is the diameter of the critical section (i.e., 4 mm as shown in Figure 1a), g is the acceleration of gravity (9.8 m/s^2), α is the stress concentration factor (1.08), L is the distance from the critical section of specimen to the applied loading end (about 40.5 mm for a standard sample) and W is the applied loading (kgf), respectively.

2.2. Testing Procedures

All rotation bending fatigue tests were controlled by the load at a stress ratio $R = -1$ and a rotating frequency about 58 Hz referring to literature [18]. The temperature of the testing environment is room temperature of about 25 °C and the relative humidity of the air environment is about 25%. Three kinds of corrosion media were air, 3.5 wt. % and 5.0 wt. % NaCl aqueous solutions, respectively. The mechanical rotating bending loading was applied on the sample as shown in Figure 1b and the corrosion liquid was dropped on the free surface of samples as shown in Figure 1c. In addition, considering the effect of corrosion solution sorption time or amount on the fatigue damage of samples, the different dropping rates of corrosive liquid, 1.6 mL/min and 2.5 mL/min, were controlled in the fatigue tests. In order to obtain fracture morphologies, the typical fracture cross-sections of samples were carefully cleaned by the ultrasonic then measured in the vacuum chamber (10^{-4} Pa) of Quanta FEG 450, and scanning electron microscopy (SEM) was conducted.

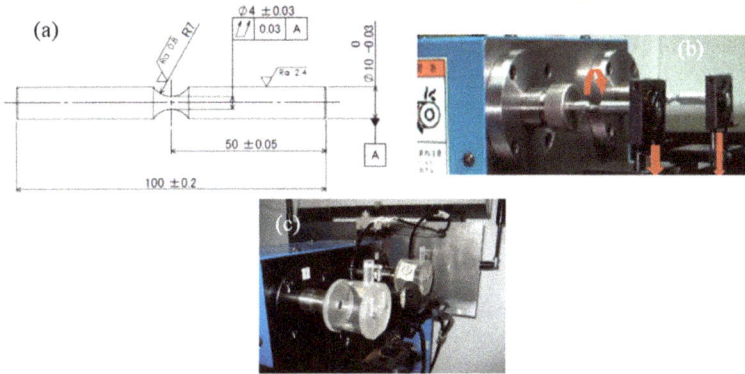

Figure 1. Schematic drawing of rotating bending fatigue specimen. (**a**) Size and shape of sample; (**b**) rotation bending applied to schematic drawing in the air case; (**c**) environmental box for the corrosion liquid dropping.

3. Results and Discussions

3.1. Fatigue Life Characteristics

Figure 2 indicates S-N curves of representative AA6063-T4 aluminum alloy in the air case, at 3.5 wt. % and 5.0 wt. % NaCl aqueous solutions. High-cyclic fatigue (HCF) lives of AA6063-T4 aluminum alloy decreased along with the concentration percentage of NaCl aqueous solution increasing, in which the corrosion fatigue limits (10^7 cycles) at the two corrosion media were not clear, especially at 5.0 wt. % NaCl aqueous solution. If we define the fatigue limit of 10^7, the fatigue limit stress at the air case is about three times that at 3.5 wt. % NaCl aqueous solution. These curves indicated that the effect of the corrosive environment on the HCF damage of AA6063-T4 aluminum alloy was not ignored even though the material exposed time in NaCl aqueous solution did not exceed 48 h. According to the descended trends of S-N curves, the fatigue resistance of AA6063-T4 aluminum alloy at 3.5 wt. % and 5.0 wt. % NaCl aqueous solutions is much lower than that in the air case within 5×10^5 cycles. In addition, the larger the cyclic number, the smaller the slope difference of S-N curves. It means that the coupled effect of corrosive environment and applied stress on the high-cycle fatigue (HCF) behavior becomes more and more important.

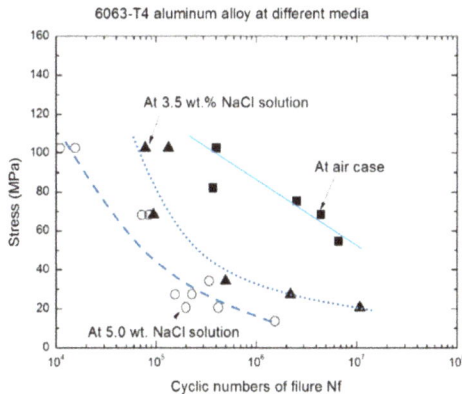

Figure 2. S-N curves of AA6063-T4 in the different media at $R = -1$ and $f = 58$ Hz.

Figure 3, in which fatigue data marked as Δ (AA2024-CZ) is from Chang et al. [19], indicated the influence of the relative strength ($\sigma/\sigma_{0.2}$ by the non-dimension of applied stress and offset yield strength) on HCF performances of aluminum alloys in different media. AA7475-T7351 and AA7075-T651 are high strength aluminum alloys (the tensile strength is more than 500 MPa), and AA2024-CZ and AA2024-T4 are high ductility aluminum alloys (the elongation ratio (δ%) is about 20%). The difference of fatigue behaviors between high strength aluminum alloys (AA7475-T7351 and AA7075-T651) can be ignored based on the slopes and the difference of stress ratio for S-N curves in the different media, in which the slopes in S-N curves of high strength aluminum alloys (AA7475-T7351 and AA7075-T651) are approximately the same and the differences of fatigue lives are approximately the same at the different stress ratios, especially in corrosion media. However, like the high ductility aluminum alloys (AA2024-CZ and AA2024-T4), there are not small differences of fatigue performances at the same stress ratios in the different media, although the slopes in S-N curves of AA2024-CZ and AA2024-T4 have only slight differences in either the air case (Figure 3a) or NaCl aqueous solutions (Figure 3b,c). It was found that the effect of deformed capacity on the fatigue damage resistance in the different media is much more obvious. On the other hand, the fatigue lives of high strength (AA7000 series) aluminum alloys are shorter than that of high ductility (AA2000 series) aluminum alloys in the different media, especially in NaCl aqueous solutions. This reflects a fact that the effect of corrosive environment is much stronger on the fatigue performance of high strength aluminum alloys than that of high ductility aluminum alloys under the stress ratio. For example, when the stress ratio is 0.5, the corrosion fatigue lives of AA7475-T7351 and AA7075-T651 at 5.0 wt. % NaCl aqueous solution are the same approximate values of about 2.5×10^4, and the corrosion fatigue lives of AA2024-CZ and AA2024-T4 at 5.0 wt. % NaCl aqueous solution are 2×10^5, 6×10^5, respectively. With the increasing of cyclic number to failure or the reducing of stress ratio, the influencing trend becomes smaller. One of the reasons is that slight strain change-enhanced anodic dissolution and concomitant cathodic hydrogen generation might play an important role in accelerating fatigue damage of high strength aluminum alloys compared with high ductility aluminum alloys [20]. However, the difference of corrosion fatigue behavior between high strength aluminum alloys (such as AA7475-T7351 and AA7075-T651) at 3.5 wt. % NaCl aqueous solution is close to that at 5.0 wt. % NaCl aqueous solution as shown in Figure 3b,c, which will be beneficial to predicating fatigue lives of high strength aluminum alloys.

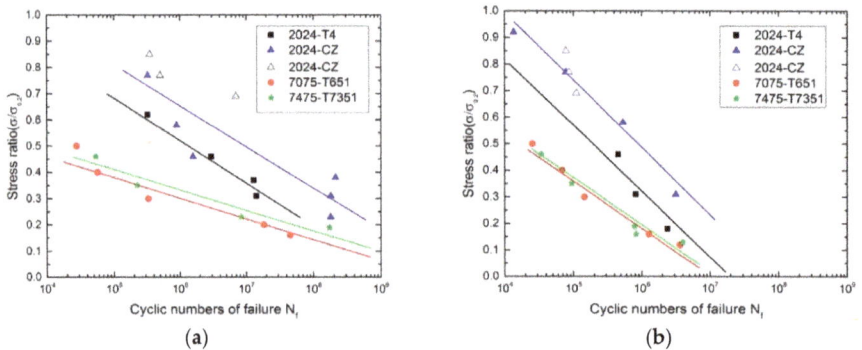

(a)

(b)

Figure 3. *Cont.*

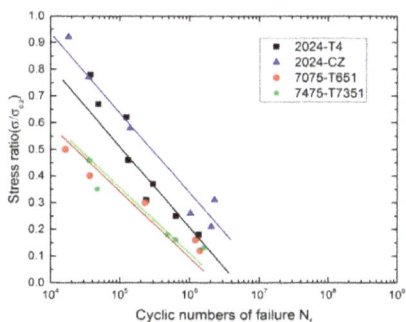

(c)

Figure 3. Effect of relative strength on the high-cyclic fatigue performance of representative aluminum alloys in the different environmental media. (**a**) In the air case; (**b**) at 3.5 wt. % NaCl solution; (**c**) at 5.0 wt. % NaCl solution.

To explore the effect of corrosion solution sorption time or amount on the fatigue damage of samples' free surface, Figure 4 shows the S-N curves of AA2024-T4 aluminum alloy under the different rates of dropping corrosion liquid (at 5.0 wt. % NaCl aqueous solution). It was clearly found that the rate of dropping corrosion liquid directly affects the fatigue life of AA2024-T4 aluminum alloy. The faster the rate of dropping corrosion liquid, the shorter the fatigue life of AA2024-T4 aluminum alloy. One of the effective reasons may be because the corrosive NaCl liquid with the 2.5 mL/min is much easier to accumulate on the specimen surface than that with the 1.6 mL/min, in which the corrosion solution always adheres to the surface of samples at the 2.5 mL/min state, but its difference is not greater than that at 1.6 mL/min during the HCF damage process. That is, the effect of corrosion solution sorption time or amount on the surface damage of sample can be ignored when the rate of dropping corrosion liquid is beyond 1.6 mL/min. As the rate of dropping corrosion liquid is less than 1.6 mL/min, it is worth further study.

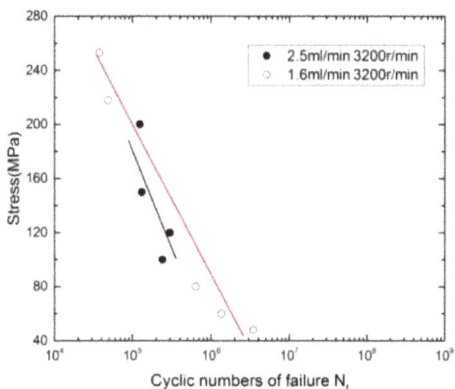

Figure 4. The S-N curves of AA2024-T4 at the different dropping corrosive liquid rates at 5.0 wt. % NaCl aqueous solution.

Figure 5 shows the comparative results about the effect of applied cyclic loading styles on the fatigue life of AA2024-CZ aluminum alloy at 3.5 wt. % NaCl aqueous solution. Fatigue data of AA2024-CZ aluminum alloy under the axial cyclic loading are from Wu [21], and fatigue data of

AA2024-CZ under the rotational bending cyclic loading style are based on our experimental results. As the same AA2024-CZ aluminum alloy, it was found that the effect of applied cyclic loading style on the fatigue life of AA2024-CZ aluminum alloy is obvious. This is because the stress (strain) gradient exists in the cross-section of sample under rotation bending cyclic loading style, but it does not exist in that under push-pull cyclic loading style although the stress ratio ($R = -1$) and the stress levels reach the same conclusions [22]. Therefore, for the same material and stress level, the fatigue life of the former is obviously longer than that of the latter as shown in Figure 5.

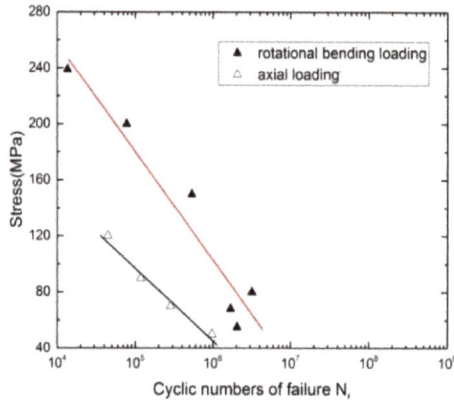

Figure 5. The S-N curves of AA2024-CZ under the different cyclic loading styles at 3.5 wt. % NaCl aqueous solution, all data of mark Δ come from literature [21].

Most researchers have previously reported fatigue data of aluminum alloys under pre-corrosion mode based on the equivalent life concept according to airplane structure subjected to two main factors of environment corrosion and fatigue loading [9,23–25]. These results indicated that the differences between the cyclic stress corrosion and prior-corrosion HCF damage behaviors of aluminum alloys are always existent. Figure 6 shows quantificationally the difference of fatigue behaviors under different corrosion modes for AA2024-CZ aluminum alloys. The fatigue life under the prior-corrosion mode is obviously longer than that under the stress corrosion fatigue mode. However, with the decreasing of the applied stress level or increasing of the cyclic number, the difference of fatigue life becomes insignificant. It hints that the fatigue damage model under the prior-corrosion mode can be divided into two parts including low-cyclic fatigue (LCF) and HCF models in the different media. Typical results indicated that for AA2024-CZ aluminum alloys, the fatigue crack initiation and propagation under stress corrosion mode is easier than that under prior-corrosion in the same condition. One of the important reasons is that the size and shape of corrosion pits depend directly on the fatigue crack initiation of aluminum alloys, in which the maximum depth of corrosion pit (measured area is about 3 mm × 3 mm) is about 50–60 μm. The depth of corrosion pit was controlled by using the 20% EXCO (NaCl: 3.5 wt. %, KNO$_3$: 0.5 mol/L, HNO$_3$: 0.1 mol/L) solution at 25 ± 3 °C for holding 6 h in the salt fog cabinet to accelerate surface corrosion of AA2024-CZ aluminum alloy. With increasing of holding time, the depth of corrosion pits becomes bigger and bigger so that the fatigue crack initiation life of aluminum alloy becomes shorter [9].

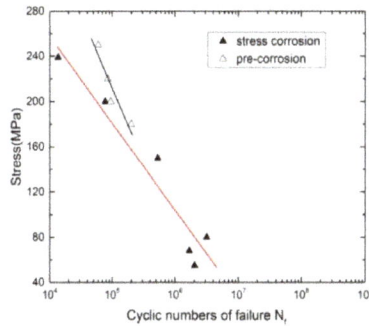

Figure 6. The S-N curves of AA2024-CZ under the different corrosion modes at 3.5 wt. % NaCl aqueous solution, all data of mark △ come from literature.

3.2. Fatigue Fracture Characteristics

Figure 7 illustrates the typical fatigue fracture morphologies and characteristics of AA6063-T4 and AA7475-T7351 aluminum alloys in the different media under the same stress level. According to the concave and convex degrees of fracture surface, it is clearly seen that the cross sections of samples can be divided into two different fracture regions including the fatigue crack initiation and propagation region (Mark A) and the instantaneous fracture region (Mark B). For the sample of AA6063-T4 aluminum alloy, the area of the instantaneous fracture region in the center of sample decreased along with the percentage of NaCl aqueous solution increasing as shown in Figure 7(a1,3,4). It reflects the fact that the instantaneous fracture region is relatively smooth, in which many micro plastic dimples can be still seen in the instantaneous fracture region as shown in Figure 7(a2). It means that the fatigue fracture of AA6063-T4 aluminum alloy is the typical ductility fracture behavior in the air case. In addition, the corrosion fatigue crack initiation and propagation regions present the different fatigue fracture characteristics in air, at 3.5 wt. % and at 5.0 wt. % NaCl solutions, in which there are different corrosion fatigue crack initiation positions and propagation directions just as the results showed some crack propagation sidesteps or transition angles of crack propagation paths due to the round bar geometry and applied loading style as shown in Figure 7a. It means that the shear stress vector and crack initiation position number is much greater along with the transition angle or sidestep increasing [6]. At the same time, the fatigue crack propagation behavior was inclined to mode II (i.e., the shear mode), with many sidesteps containing an angle of approximately 50° as shown in Figure 7(a1,3,4). Therefore, fatigue life of samples containing more sidesteps in cross-section is lower than that of samples containing less sidesteps. On the other hand, the corrosive action with more crack initiation positions is stronger than that with less crack initiation positions. More importantly, there is a significant difference in branching crack morphologies (or sidesteps characteristics) in both AA6063-T4 Al alloy as shown in Figure 7a and AA7475-T7351 Al alloy as shown in Figure 7b. For the typical high strength AA7475-T7351 Al alloy, corrosion fatigue fracture characteristics in air, at 3.5 wt. % and 5.0 wt. % NaCl aqueous solutions were almost similar to those shown in Figure 7b. In the air case, the corrosion fatigue crack propagation region (Mark A) is much less than the instantaneous fracture region as shown in Figure 7(b1) and the corrosion fatigue crack initiation position can be thought as not more than two. This is because the corrosion fatigue crack propagation region has a relatively concentrated area so that the crack propagation sidestep is relatively less. However, in the different percentage NaCl solutions, there are confused corrosion fatigue crack propagation paths and a spot of sidesteps or transition angles of crack propagation. It means that the fatigue crack initiation and propagation behavior in different percentage NaCl solutions are much more complex than those in air conditions. In addition, fracture surfaces of AA7475-T7351 aluminum alloy were much smoother than those of AA6063-T4 aluminum alloy, especially in the fatigue crack initiation and propagation regions.

This is because the shear stress vector component of high strength AA7475-T7351 aluminum alloy is smaller than that of AA6063-T4 aluminum alloy.

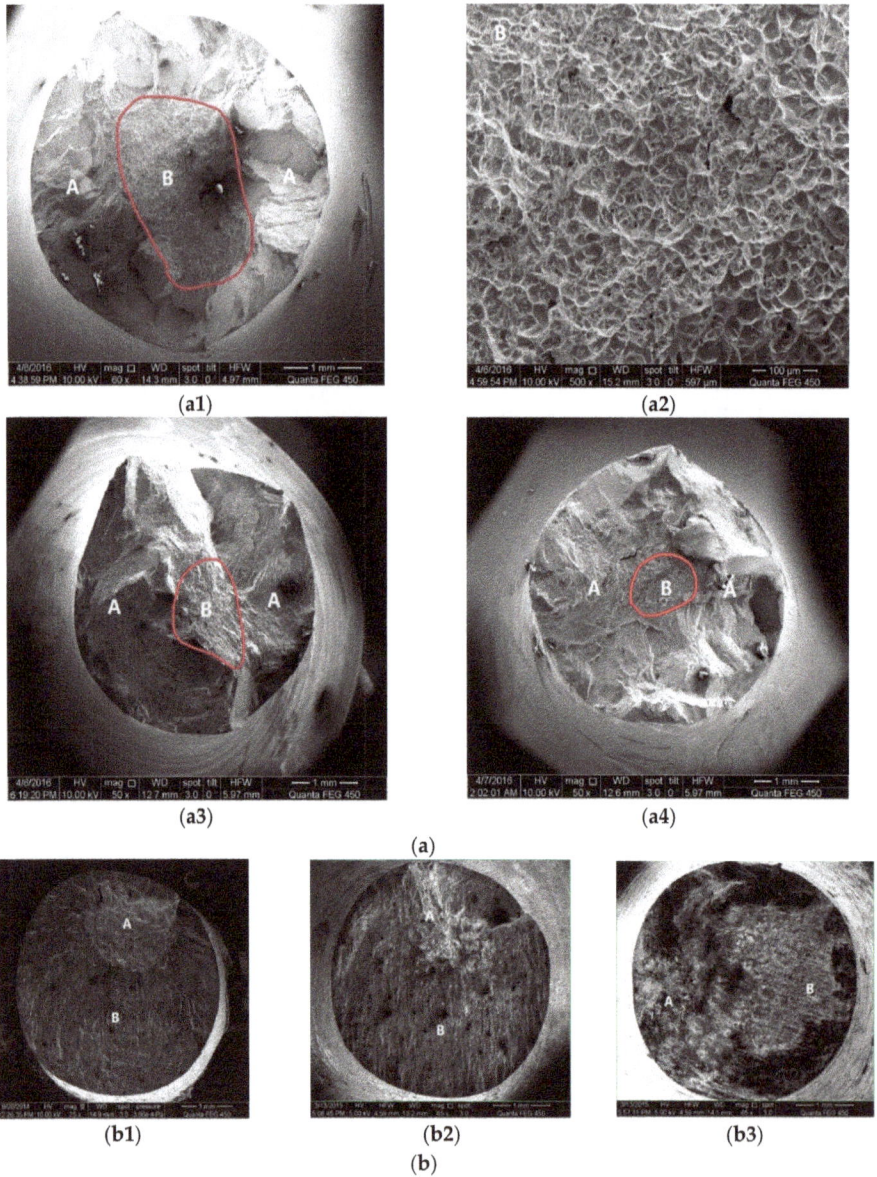

(a1)   (a2)

(a3)   (a4)

(a)

(b1)   (b2)   (b3)

(b)

Figure 7. Scanning electron microscopy (SEM) fracture morphologies for AA6063-T4 and AA7475-T7351 in the different media. (**a**) AA6063-T4; (**a1**) in the air case; (**a2**) in the air case; (**a3**) at 3.5% NaCl solution; (**a4**) at 5.0% NaCl solution; (**b**) AA7475-T7351; (**b1**) in the air case; (**b2**) at 3.5% NaCl solution; (**b3**) at 5.0% NaCl solution.

Figure 8 shows the amplification regions of fracture morphologies in fatigue crack propagation regions for AA6063-T4 and AA7475-T7351 aluminum alloys in the different media. For 6063-T4 aluminum alloy, a number of obvious slip sidesteps and transition angles of slip planes were clearly found in the fatigue crack initiation and propagation regions as shown in Figure 8a, which is attributed to the multiple different crack propagation directions, due to the multi crack initiation sites occurring at the free surfaces under the rotating bending loading. For AA7475-T7351 aluminum alloy, three fracture surfaces were occupied by fatigue crack propagation striations and brittle fracture characteristics as shown in Figure 8b. In addition, second fatigue cracks were clearly found so that their effects on the fatigue fracture could not be ignored. There is no or little difference in three fracture surfaces in the different media, while the crack propagation was accelerated by corrosive environment (seen the above section). This is because the aqueous NaCl solution, a hydrogen-induced environment, may result in localized deformation and accelerating of crack propagation due to the anodic dissolution and hydrogen embrittlement mechanisms, which are important factors for weakening the fatigue resistance or strength of AA7475-T7351 aluminum alloy.

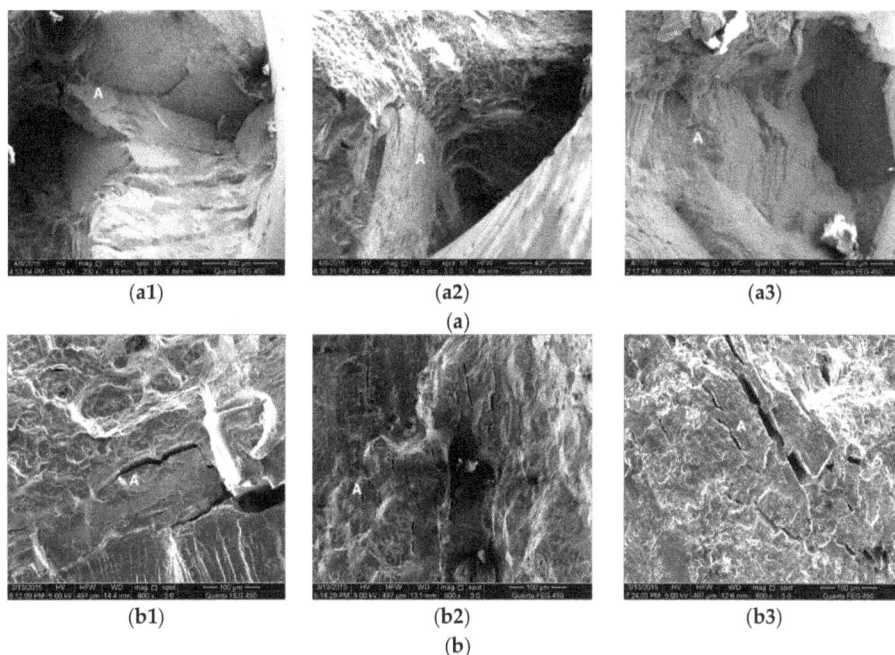

(a1)   (a2)   (a3)

(a)

(b1)   (b2)   (b3)

(b)

Figure 8. SEM fracture morphologies in fatigue crack initiation and propagation regions for AA6063-T4 and AA7475-T7351 in the different media. (**a**) AA6063-T4; (**a1**) in the air case; (**a2**) at 3.5 wt. % NaCl solution; (**a3**) at 5.0 wt. % NaCl solution; (**b**) AA7475-T7351; (**b1**) air, σ = 100 MPa; (**b2**) 3.5 wt. %, σ = 80 MPa; (**b3**) 5.0 wt. %, σ = 54.6 MPa.

4. Conclusions

Environment-induced corrosion fatigue behaviors of representative aluminum alloys depend strongly on the coupled action of corrosive media and applied stress level. Main conclusions obtained in this paper are as follows:

1. The effect of corrosive media is much stronger on the fatigue life of high strength aluminum alloys (such as AA7475-T7351 and AA7075-T651) than that of high ductility aluminum alloys (such as AA2024-CZ and AA2024-T4). For example, when the stress ratio is 0.5, the corrosion fatigue lives of AA7475-T7351 and AA7075-T651 at 5.0 wt. % NaCl aqueous solution are approximately the same values of about 2.5×10^4, and the corrosion fatigue lives of AA2024-CZ and AA2024-T4 at 5.0 wt. % NaCl aqueous solution are 2×10^5, 6×10^5, respectively.

2. With increasing of cycles to failure, the effect of mechanical properties of materials on the corrosion fatigue performance becomes relatively weak. For example, when the stress ratio is 0.2, the corrosion fatigue lives of AA7475-T7351 and AA7075-T651 at 5.0 wt. % NaCl aqueous solution are approximately the same values of about 6.5×10^5, and the corrosion fatigue lives of AA2024-CZ and AA2024-T4 at 5.0 wt. % NaCl aqueous solution are 1.6×10^6, 3×10^6, respectively.

3. When the dropping corrosive liquid rate is less than 1.6 mL/min, the effect of dropping liquid rate on the fatigue performance of AA2024-T4 aluminum alloy cannot be ignored. This is because the fatigue life at corrosive NaCl liquid rate with 1.6 mL/min is slightly smaller than that at corrosive NaCl liquid rate with 2.5 mL/min.

4. For the AA2024-CZ aluminum alloy, the crack initiation and propagation life under stress corrosion mode is much shorter than that under mechanical fatigue mode after prior-corrosion at the same stress level, in which in the former, the coupling effect of corrosion media and stress is stronger than that in the latter, even if there are some corrosion pits including the maximum depth of corrosion pit of about 50–60 μm.

Acknowledgments: The present project research was supported by the National Natural Science Foundation of China (Grant No.: 11272173, 11572170) and the State Key Lab. of Traction Power of Southwest Jiaotong University (Grant No.: TPL1503).

Author Contributions: Huihui Yang: one of main authors who finished the fatigue experiments and wrote a text of paper; Yanling Wang: one of main authors who finished the fracture analysis of paper; Xishu Wang: a corresponding author who wrote the paper and analyzed all data of experiments; Pan Pan: one of authors who participated in a part experiment; Dawei Jia: one of author who provided all materials for fatigue tests and checked a manuscript.

Conflicts of Interest: The authors declare no conflicts of interest.

Abbreviation

α	the stress concentration factor (1.08)
E	Young's modulus
HCF	high-cyclic fatigue
L	geometry sizes of sample
LCH	low-cyclic fatigue
g	acceleration of gravity (9.8 m/s^2)
R	stress ratio
Re	surface roughness
N_f	number of cycles to failure
σ	engineering stress amplitude (MPa)
$\sigma_{0.2}$	the material's offset yield strength
σ_b	the tensile strength
δ	the percentage elongation

References

1. Yi, D.Q.; Zhou, M.Z.; Liu, H.Q.; Wang, B.; Yang, S. Effect of Temperature and Corrosive Environment on Cyclic Fatigue and Final Fracture Behavior of 2524 Aluminum Alloy. *Int. J. Soc. Mater. Eng. Resour.* **2010**, *17*, 58–63. [CrossRef]

2. Wang, R. A fracture model of corrosion fatigue crack propagation of aluminum alloys based on the materials elements fracture ahead of a crack tip. *Int. J. Fatigue* **2008**, *30*, 1376–1386. [CrossRef]
3. Na, K.H.; Pyun, S.I. Comparison of susceptibility to pitting corrosion of AA2024-T4, AA7075-T651 and AA7475-T761 aluminium alloys in neutral chloride solutions using electrochemical noise analysis. *Corros. Sci.* **2007**, *50*, 248–258. [CrossRef]
4. Warner, J.S. *Inhibition of Environmental Fatigue Crack Propagation in Age-Hardenable Aluminum Alloys*; The School of Engineering and Applied Science University of Virginia: Charlottesville, VA, USA, 2010.
5. Murakami, Y.; Kanezaki, T.; Mine, Y.; Metall, S. Effect of Humidity on Fatigue Strength of Age-Hardened Al Alloy under Rotating Bending. *Mater. Trans. A* **2008**, *39*, 1327–1339. [CrossRef]
6. Wang, X.S.; Li, X.D.; Yang, H.H.; Kawagoishi, N.; Pan, P. Environment-induced fatigue cracking behavior of aluminum alloys and modification methods. *Corros. Rev.* **2015**, *33*, 119–137. [CrossRef]
7. Frulla, G.; Avalle, G.; Sapienza, V. Preliminary evaluation of the fatigue behavior of aluminum alloy in corrosive environment. *Aircr. Eng. Aerosp. Technol.* **2015**, *87*, 165–171.
8. Takahashi, H.; Kasahara, K.; Fujiwara, K.; Seo, M. The cathodic polarization of aluminum covered with anodic oxide-films in a neutral borate solution—I. The mechanism of rectification. *Corros. Sci.* **1994**, *36*, 677–688. [CrossRef]
9. Li, X.D.; Wang, X.S.; Ren, H.H.; Chen, Y.L.; Mu, Z.T. Effect of prior corrosion state on the fatigue small cracking behavior of 6151-T6 aluminum alloy. *Corros. Sci.* **2012**, *55*, 26–33. [CrossRef]
10. Stanzl, S.E.; Mayer, H.R.; Tschegg, E.K. The influence of air humidity on near-threshold fatigue crack growth of 2024-T3 aluminum alloy. *Mater. Sci. Eng. A* **1991**, *147*, 45–54. [CrossRef]
11. Wasekar, N.P.; Jyothirmayi, A.; Sundarajian, G. Influence of pre-corrosion on the high cycle fatigue behavior of microarc oxidation coated 6061-T6 aluminum alloy. *Int. J. Fatigue* **2011**, *33*, 1268–1276. [CrossRef]
12. Wang, X.S.; Guo, X.W.; Li, X.D.; Ge, D.Y. Improvement on the fatigue performance of 2024-T4 alloy by synergistic coating technology. *Materials* **2014**, *7*, 3533–3546. [CrossRef]
13. Ishihara, S.; Saka, S.; Nan, Z.Y.; Goshima, T.; Sunada, S. Prediction of corrosion fatigue lives of aluminum alloy on the basis of corrosion pit growth law. *Fatigue Fract. Eng. Mater. Struct.* **2006**, *29*, 472–480. [CrossRef]
14. Yang, H.H.; Wang, Y.L.; Wang, X.S.; Pan, P.; Jia, D.W. Synergistic effect of environmental media and stress on the fatigue fracture behavior of aluminum alloys. *Fatigue Fract. Eng. Mater. Struct.* **2016**. [CrossRef]
15. Magnin, T. Recent advances for corrosion fatigue mechanisms. *ISIJ Int.* **1995**, *35*, 223–233. [CrossRef]
16. Maitra, S.; English, G.C. Environmental Factors Affecting Localized Corrosion of 7075-T7351 Aluminum Alloy Plate. *Metall. Trans. A* **1982**, *13*, 161–166. [CrossRef]
17. Ricker, R.E.; Duquette, D.J. The role of hydrogen in corrosion fatigue of high purity Al-Zn-Mg exposed to water vapor. *Metall. Trans. A* **1988**, *19*, 1775–1783. [CrossRef]
18. *Standard Practice for Presentation of Constant Amplitude Fatigue Test Results for Metallic Materials*; ASTM International: West Conshohocken, PA, USA, 2004. [CrossRef]
19. Chang, H.; Han, E.H.; Wang, J.Q.; Ke, W. Effect of cathodic polarization on corrosion fatigue life of the LY12CZ aluminum alloy. *Acta Metall.* **2005**, *41*, 556–560. (In Chinese)
20. Hall, M.M., Jr. Effect of cyclic crack opening displacement rate on corrosion fatigue crack velocity and fracture mode transitions for Al-Zn-Mg-Cu alloys. *Corros. Sci.* **2014**, *81*, 132–143. [CrossRef]
21. Wu, X.R. *Handbook of Mechanical Properties*; Aviation Industry Press: Beijing, China, 1997.
22. Wang, X.S.; Kawagoishi, N. Effect of specimen configuration and loading types on fatigue life of an annealed 0.42% carbon steels. *Key Eng. Mater.* **2010**, *417*, 121–124.
23. Jones, K.; Hoeppner, D.W. Prior corrosion and fatigue of 2024-T3 aluminum alloy. *Corros. Sci.* **2006**, *48*, 3109–3122. [CrossRef]
24. Sundararajan, G.; Wasekar, N.P.; Ravi, N. The influence of the coating technique on the high cycle fatigue life of alumina coated Al 6061 alloy. *Trans. Indian Inst. Met.* **2010**, *63*, 203–208. [CrossRef]
25. Nickel, D.; Dietrich, D.; Mehner, T.; Frint, P.; Spieler, D.; Lampke, T. Effect of Strain Localization on Pitting Corrosion of an AlMgSi0.5 Alloy. *Metals* **2015**, *5*, 172–191. [CrossRef]

![metals logo] *metals*

MDPI

Article

Understanding Low Cycle Fatigue Behavior of Alloy 617 Base Metal and Weldments at 900 °C

Rando Tungga Dewa [1], Seon Jin Kim [1,*], Woo Gon Kim [2] and Eung Seon Kim [2]

[1] Department of Mechanical Design Engineering, Pukyong National University, Busan 608-739, Korea;
    rando.td@gmail.com
[2] Korea Atomic Energy Research Institute (KAERI), Daejeon 305-353, Korea; wgkim@kaeri.re.kr (W.G.K.);
    kimes@kaeri.re.kr (E.S.K.)
*   Correspondence: sjkim@pknu.ac.kr; Tel.: +82-51-629-6163; Fax: +82-51-629-6150

Academic Editor: Filippo Berto
Received: 4 July 2016; Accepted: 26 July 2016; Published: 2 August 2016

Abstract: In order to better understand the high temperature low cycle fatigue behavior of Alloy
617 weldments, this work focuses on the comparative study of the low cycle fatigue behavior of
Alloy 617 base metal and weldments, made from automated gas tungsten arc welding with Alloy
617 filler wire. Low cycle fatigue tests were carried out by a series of fully reversed strain-controls
(strain ratio, $R_\varepsilon = -1$), i.e., 0.6%, 0.9%, 1.2% and 1.5% at a high temperature of 900 °C and a constant
strain rate of 10^{-3}/s. At all the testing conditions, the weldment specimens showed lower fatigue
lives compared with the base metal due to their microstructural heterogeneities. The effect of very
high temperature deformation behavior regarding cyclic stress response varied as a complex function
of material property and total strain range. The Alloy 617 base weldments showed some cyclic
hardening as a function of total strain range. However, the Alloy 617 base metal showed some cyclic
softening induced by solute drag creep during low cycle fatigue. An analysis of the low cycle fatigue
data based on a Coffin-Manson relationship was carried out. Fracture surface characterizations were
performed on selected fractured specimens using standard metallographic techniques.

Keywords: Alloy 617; very high temperature gas-cooled reactor (VHTR); gas tungsten arc welding
(GTAW); weldments; low cycle fatigue (LCF); fatigue life; fracture surface characterization

1. Introduction

The Next Generation Nuclear Plant (NGNP) being developed in the Republic of Korea is the
Very High Temperature gas-cooled Reactor (VHTR). The VHTR merges the diversities of the baseline
design to allow eventual operation at gas outlet temperatures up to 950 °C. In the VHTR, some of
the major components such as the reactor internals, the reactor pressure vessel, the piping, the hot
gas ducts (HGD), and the intermediate heat exchangers (IHX) are classified as key components,
with helium as a primary and secondary coolant. The IHX performs the main purpose in the
operation of the NGNP, transferring heat from the primary reactor helium to an active working
fluid at a lower temperature. Leading materials of potential concern include nickel-base Alloy 800H,
Alloy 617, Alloy 230, and Hastelloy X for the high temperature components. In the high temperature
design, creep and fatigue resistance, oxidation resistance, and phase stability need to be satisfied [1,2].
Alloy 617, a nickel-base super alloy, is a leading candidate material for a VHTR because of its excellent
high-temperature mechanical properties, formability, and weldability. Alloy 617 is strengthened by
solid solution hardening precipitates provided by the alloy chemical compositions of chromium,
cobalt, and molybdenum, which are required for high temperature strength [3]. In Alloy 617, the high
temperature oxidation resistance is derived from the high nickel and chromium content. In addition,
grain boundary strengthening takes place during the solidification process, with the aid of carbide

precipitates. Primary carbides, M_6C, have a complex structure and they precipitate in a relatively high temperature process. A more complex secondary carbide, $M_{23}C_6$, is suspected and mainly grows along the grain boundary. These carbide precipitates are known to have a high content of chromium. They diffuse and form a depleted zone of a Cr-rich oxide layer on the outer surface, namely Cr_2O_3 [4]. Consequently, the Alloy 617 is expected to provide good thermal stability for components of power generating plants with a high temperature strength up to 950 °C [5]. The IHX have to be joined to piping or other components by welding technique. Very high temperature deformation is expected to be a predominant failure mechanism of the IHX, and thus, weldments used in its fabrication experience varying cyclic deformation and are a key element of all designs [6,7].

In an actual high temperature design evaluation, however, fatigue and creep damage are usually more critical than other design parameters. In this circumstance, the low cycle fatigue (LCF) loadings represent a predominant failure mode from the temperature gradient induced thermal strain during operation as well as in the startups and shutdowns and in power transients or with temperature change of the flowing coolant having a low loading rate [5–8]. Because of these shortcomings, significant consideration of LCF behavior is needed in the design and life assessment of such components working in high temperature conditions. The welded section material could be considerably affected by the welding process which is responsible for heterogeneities. As such, the weldments are critical considerations in the engineering design because they are the weakest links in the components and may have some original defects. Experience with nickel alloy weldments in structural applications suggests that most cases of high temperature fatigue failures occur at the weldments or in the heat affected zone (HAZ) [6]. Although Alloy 617 has many superior properties, numerous researchers have reported that the fatigue life varies widely at high temperature and it is generally found that the weldment specimens have a lower fatigue life compared to the base metal, although only limited data were available on the weldments material [5–11]. A draft Code Case was developed to qualify the Alloy 617 for nuclear service; the need for fatigue data, such as the influence of strain ranges, strain rate, and temperature at thermally induced strain rates in the IHX parent material and weldments material is necessary to predict the lifetime of the reactor components. However, the behavior of Alloy 617 weldments is not yet fully understood, and there remains a need for further experiments; a lot of data needs to be supplemented at very high temperatures due to the variability in the fatigue response of the element parts of the weldments (i.e., weld, HAZ, and base metal) to confirm the suitability of a baseline draft Code Case [6].

The aim of this work focuses on the understanding of the LCF behavior of Alloy 617 base metal and weldments, made from an automated gas tungsten arc welding (GTAW) process with Alloy 617 filler wire. LCF tests have been carried out through a series of fully reversed strain-controls (strain ratio, $R_\varepsilon = -1$) regarding to the four different total strain ranges, i.e., 0.6%, 0.9%, 1.2% and 1.5% at a high temperature of 900 °C in an air environment, in accordance with the ASTM Standard E606. The effect of very high temperature deformation behavior of Alloy 617 base metal and weldment specimens is comparatively investigated as a function of total strain range. The plastic deformation regarding stress response was reflected as damage accumulation in the structural material, and it could be correlated to the fatigue life of the material. An evaluation of LCF behavior data was performed using the well-known relationship based on a Coffin-Manson relationship, and the material constants were also determined. The LCF fracture surface microstructures were characterized on selected fractured specimens, and thus, the microstructural changes under various conditions are also reported quantitatively using standard metallographic techniques.

2. Materials and Experiments

A commercial grade Alloy 617 is approved for non-nuclear construction in the ASME Code. Thus the composition (wt %) of the Alloy 617 used for material chosen in this study is 53.11Ni, 22.2Cr, 12.3Co, 9.5Mo, 1.06Al, 0.08C, 0.949Fe, 0.4Ti, 0.084Si, 0.029Mn, 0.027Cu, 0.003P, <0.002S, and <0.002B. The initial microstructure analysis of Alloy 617 was revealed in a previous study [2,6]. Alloy 617 has a fully austenitic face centred cubic (fcc) structure which maintains superior mechanical properties at high temperature. The fcc matrix, known as, γ, mainly consists of nickel, cobalt, iron, chromium, and molybdenum. Figure 1 shows the microstructure of the cross-section of a weld, the HAZ, and the base region. However, the microstructure of the base metal with well-uniformed equiaxed grains is approximately 100 μm in diameter. The weld region is comprised of austenitic large columnar grains with a dendritic structure due to solidification during the welding process. As such, the HAZ of the weld was formed by carbide dissolution and a small amount of grain growth. Cylindrical specimens with 6.0 mm in diameter in the reduced section with a parallel length of 18 mm and gauge length of 12 mm were used for the LCF test specimens. Low stress grinding and polishing were applied in the final machining to avoid the formation of notches. LCF weldment specimens were machined from weld pad in the transverse direction to the welding direction. Alloy 617 filler wire was used with a diameter of 2.4 mm. After the welding process, the soundness of the weldments was qualified through an ultrasonic test (UT), a tensile test, and a bending test. The bending testing results coincide well with ASME specifications, which means the micro-crack is within 3.2 mm. It was also observed that the weldments exhibited acceptable ductility. Nevertheless, the soundness of the weldments gives no indication of welding defects. The shape of the weld pad has a single V-groove with an angle of 80° and 10 mm root gap from a 25 mm thick rolled plate. Figure 2 shows the shape and dimension of the weld pad configuration and the schematic of weldment specimens used in this investigation. The gauge section of the weldment specimen mainly covers the weld and HAZ materials only.

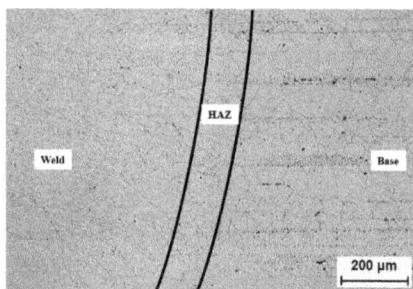

Figure 1. Microstructure of the cross-section of Weld-Heat affected zone (HAZ)-Base region.

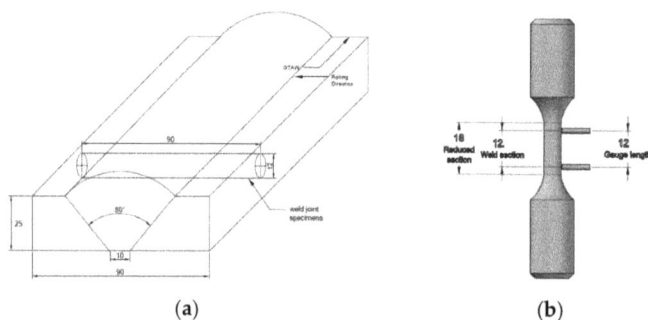

(a)   (b)

Figure 2. (a) The weld pad configuration; (b) The schematic of shape and dimension of the weldment specimen (all dimensions are in millimeters).

A closed loop 100 kN servo hydraulic testing machine (MTS 370 Landmark, Eden Prairie, MN, USA) was used and equipped with a tube furnace for heating the specimens of Alloy 617 base metal and weldment in dry air, which ensured that the temperature remained within ±2 °C of the nominal temperature throughout the test. The specimen was held at a target temperature with zero load for about 30 min to allow temperatures to stabilize before the commencement of the test. A full-scale photograph of the LCF testing apparatus is shown in Figure 3. We performed fully reversed (strain ratio, $R_\varepsilon = -1$) axial strain controlled LCF tests of Alloy 617 base metal and weldment specimens at 900 °C regarding to the four different total strain ranges, i.e., 0.6%, 0.9%, 1.2% and 1.5%. A triangular waveform and a constant strain rate of 10^{-3}/s were applied. The failure criterion was defined by the number of cycles which means a 20% reduction in the stress ratio (peak tensile over compressive stress ratio). In order to examine the LCF fracture morphologies, we characterized the post-fracture analysis of the selected specimens which split into two pieces in the preliminary observation using standard metallographic techniques. The fatigue specimens, cut around crack initiation sites, were chemical-polished with a mixed solution of ethanol, hydrochloric acid, and copper II chloride. The characterization was carried out using a scanning electron microscopy (SEM Hitach JEOL JSM 5610, JEOL Ltd., Tokyo, Japan) along with an energy dispersive X-ray (EDX, INCA Energy, Oxford Instruments Analytical, Halifax Road, UK) facility to determine the position of the crack initiation site as well as an optical microscope (OM JP/GX51, Olympus Corp., Tokyo, Japan) to provide an explanation of the LCF failure mechanism.

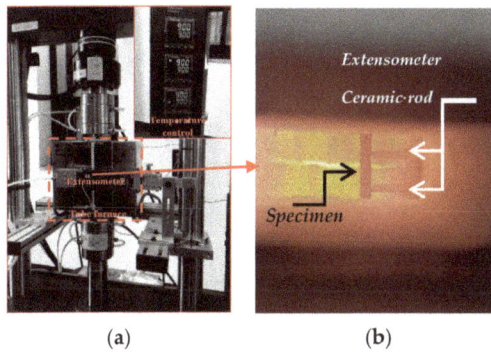

(a)   (b)

Figure 3. (a) Full-scale view of the low cycle fatigue (LCF) testing apparatus; (b) Photograph of the view port during LCF testing.

3. Results and Analysis

3.1. Fatigue Life and Cyclic Stress Response Behavior

Figure 4 shows the tensile test results of the Alloy 617 base metal and weldment specimens. Alloy 617 base metal and weldment specimens showed different characteristics of mechanical properties as microstructural differences: As revealed in the previous study [2], the hardness value of the HAZ and weldments had a higher value due to the austenitic phase with fine equiaxed dendrites. It is noticeable that the Young's modulus was similar for the Alloy 617 base metal and weldment specimens, while the strength of the weldments was higher than the base metal, but with a lower percentage of elongation. The higher strength of the weldment specimens could be attributed to the austenitic nickel-chromium morphology with a dendritic structure, and formation of precipitates of the weldments in the solidification grain boundaries [4].

At all the testing conditions, the weldment specimens showed lower fatigue lives compared with the base metal, and also the fatigue life of both base metal and weldment specimens decreased with increase in the total strain range, as shown in Figure 5. Figure 6 shows the peak tensile and compressive stresses of the cyclic stress response under four different total strain ranges during LCF testing. The peak tensile and compressive stress response were of the same magnitude, which means the strain accumulation is practically similar during cyclic deformation. The peak tensile stress, as a function of the number of cycles, attained a stable value within less than 10 cycles. The peak stress value of the half-life cycle increases with increasing total strain range of the weldment specimens, otherwise, the lower total strain ranges showed higher peak stresses for the base metal. These results could suggest that the reduction in fatigue life is not strongly dependent on cyclic deformation. At all the testing conditions, the cyclic stress response behavior of Alloy 617 at the high temperature of 900 °C showed a cyclic softening region for the major portion of the time in each total strain range. At the end of the test, the stress amplitude decreased rapidly at the formation of macro-crack initiation or just before fracture. Under the lowest total strain range, i.e., 0.6%, saturation phase was also observed. A short period of cyclic initial hardening was observed below 10 cycles for all weldment specimens at any given total strain ranges, the saturation phase also appeared at lower total strain range, and remained in the softening phase until failure.

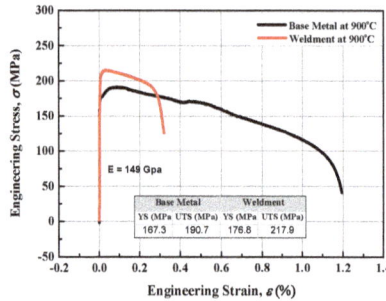

Figure 4. Tensile stress-strain curves of the Alloy 617 base metal and weldment specimens.

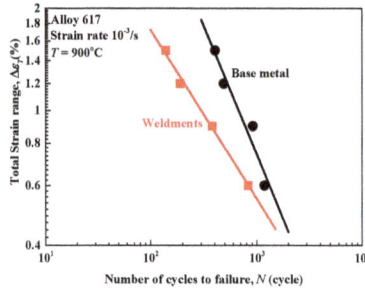

Figure 5. Comparison of low cycle fatigue resistance of Alloy 617 base metal and weldment specimens.

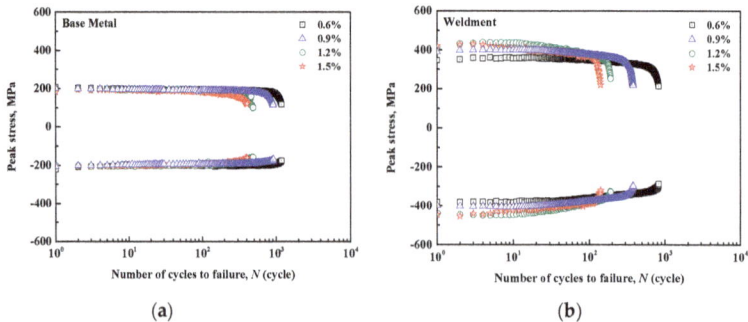

Figure 6. Peak tensile and compressive stress response of the Alloy 617: (**a**) base metal; (**b**) weldment specimen, as a function of the number of cycles.

The cyclic stress-strain response behavior could be determined through the well-known Ramberg-Osgood relationship (Equation (1)), and also linear regressions could be taken which provide a measure to cyclic straining.

$$\frac{\triangle\sigma}{2} = K' \left(\frac{\triangle\varepsilon_p}{2}\right)^{n'} \tag{1}$$

where $\triangle\sigma$, $\triangle\varepsilon_p$, K', and n' represent the stress amplitude, the plastic strain magnitude, the cyclic strength coefficient and cyclic strain hardening, or softening exponent, respectively. Figure 7 shows the cyclic stress amplitude versus plastic strain response at half-life. The material constants, K' and n', were obtained through the cyclic stress-strain curves by the least square fit method.

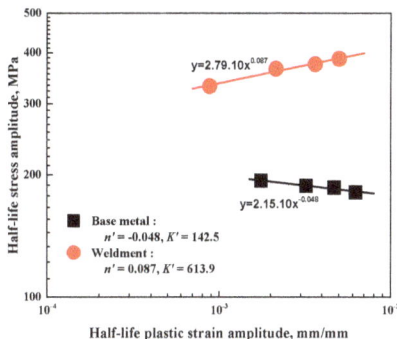

Figure 7. The cyclic stress-strain behavior of Alloy 617 base metal and weldment specimens at 900 °C, determined through the Ramberg-Osgood relationship.

On the Alloy 617 weldment specimens, the plastic strain increased continuously under LCF loading as the cyclic stress amplitude at the half-life increased. Alloy 617 weldment specimens were deformed by the plastic flow mechanism and show a cyclic hardening mechanism. On the contrary, the Alloy 617 base metal, showed a decrease in cyclic stress amplitude when the total strain range was elevated, which is a definition of the cyclic softening mechanism. This finding on the base metal was similar to a literature review [8], Wright et al. confirmed that under fatigue loading at very high temperature, the Alloy 617 exhibits an exponential decrease of peak stress amplitude with increasing strain amplitude induced by the solute drag creep mechanism. These stress drops with increasing total strain ranges have been attributed to solute-drag creep. Under high total strain ranges (beyond the flow stress peak) the solute drag flow peak only occurs during the first cycle, and the lower flow stress is maintained for each cycle, after the first cycle has established enough plastic flow to reach this steady state value. Figure 8 shows that a visible pattern of stress-strain behavior is noticeable under high-temperature fatigue, featured by an initial stress drop in the first cycle. Solute drag creep occurs due to the dynamic interactions between dislocations and the solute atoms. The solutes hinder dislocation motion resulting in relatively high resistance to plastic flow, and during cyclic loading the rearrangement and annihilation of the dislocation substructures overcome the energy and begin dragging the solutes. This causes the overall dislocation density to decrease and reduces the flow stress. In the case of the Alloy 617 the base metal experiences a small total strain range, the cyclic stress response remains at the higher value. The results are in good agreement with that of the empirical rule ($\sigma_{uts}/\sigma_{ys} < 1.2$), in which metallic materials cyclically soften; in this study we found that ($\sigma_{uts}/\sigma_{ys} = 1.13$) for the base metal and ($\sigma_{uts}/\sigma_{ys} = 1.23$) for the weldment specimens.

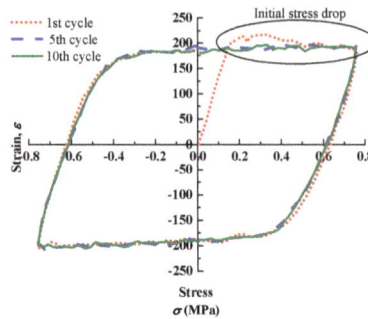

Figure 8. First, fifth, and tenth cycle of a 1.5% total strain range at 900 °C, showing initial stress drop in the stress-strain curve.

3.2. Stress-Strain Hysteresis Loops

Figure 9 shows the stress-strain hysteresis loops of the Alloy 617 base metal and weldment specimens for cycles 1, 5, 10, stable cycle (half-life), and fractured cycle, at a selected total strain range of 0.9%. The Alloy 617 base metal showed a continuous decrease in the peak stress as a function of the cycles, as shown in Figure 9a. Meanwhile, in the case of the weldment specimens as shown in Figure 9b, initial hardening in the peak stress was observed below 10 cycles, and continued to decrease until failure.

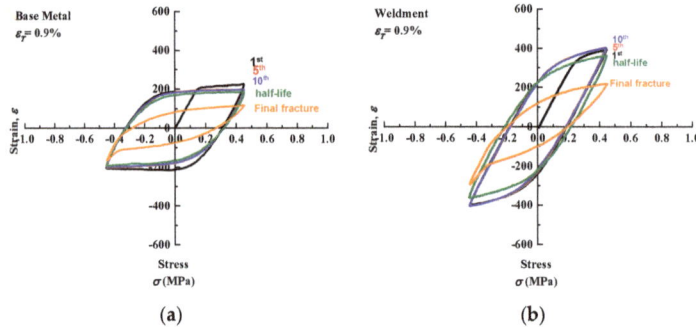

Figure 9. Stress-strain hysteresis plots of the Alloy 617: (**a**) base metal; (**b**) weldment specimens for selected cycles; total strain range of 0.9%, at 900 °C.

Figure 10 shows comparative stress-strain hysteresis loops for the Alloy 617 base metal and weldment specimens for the half-life cycle at each total strain range. The plastic strain accumulation (represented by the shape of the hysteresis loops) shows more significantly an increase as a function of total strain range, although, we found a relatively higher plastic strain magnitude for the base metal (Figure 10a) compared with those of the weldment specimens (Figure 10b). At each total strain range, the hysteresis loops of the weldment specimens were relatively narrower than those of the base metals. The overall cyclic stress response curve of weldment specimens is higher than that of the base metal. The inner area of the hysteresis loop represents the plastic energy per unit volume dissipated as plastic work during a cycle (J/m^3), and decreases as the stress amplitude decreases. As previously described, Figure 10a shows the cyclic strain softening. From Figure 10b some cyclic strain hardening is obvious but depends on the total strain ranges. This behavior conforms well to a cyclic hardening characteristic, and is related to the accumulation of dislocation density within the matrix, γ, and the configuration of materials. As the LCF testing proceeds, the total dislocation density depends on the plastic strain accumulation (in accordance with the strain range), and the material resistance against the dislocation increases, while the stress amplitude continues to be slightly increasing [12].

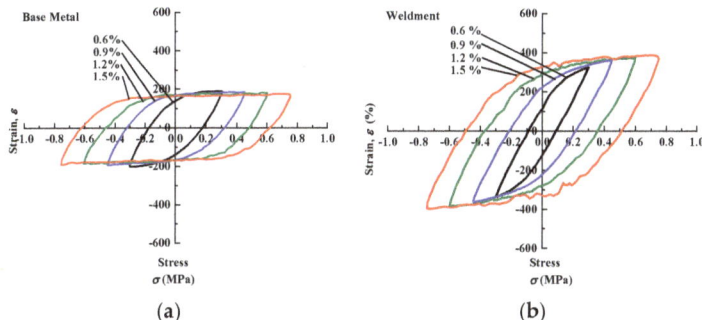

Figure 10. Comparative stress-strain hysteresis loop plots at the half-life cycle of the Alloy 617: (**a**) base metal; (**b**) weldment specimens regarding different total strain ranges at 900 °C.

3.3. Strain-Life Data Analysis

The axial strain controlled LCF test provides crucial information on the strain-life relationship, and could be used to improve the fatigue performance. It is obvious that the LCF life is sensitive to the total strain range and the material property (namely, base metal and weldments). Therefore, the Alloy 617 base metal and weldments represent a typical Coffin-Manson life dependence on total strain range. The well-known, Coffin-Manson fatigue design method correlates the number of cycles to failure, N_f, with the presented strain history of the specimen under LCF loadings. This model could be used to model the LCF problem, when the plastic strain range, $\Delta \varepsilon_P$, is the same or even larger than the elastic strain range, $\Delta \varepsilon_e$. The Coffin-Manson method could be worked out with the stress-strain (log scale) relationship, as well as the Ramberg-Osgood relationship, to describe the hysteresis loop, considering cyclic softening or hardening of the material. Equation (2) can be introduced:

$$\frac{\Delta \varepsilon_T}{2} = \frac{\Delta \varepsilon_P}{2} + \frac{\Delta \varepsilon_e}{2} = \frac{\sigma_f'}{E} (2N_f)^b + \varepsilon_f' (2N_f)^c \qquad (2)$$

where, $\Delta \varepsilon_T/2$ is the total strain amplitude, $\Delta \varepsilon_e/2$ is the elastic strain amplitude, $\Delta \varepsilon_P/2$ is the plastic strain amplitude, $2N_f$ is the number of reversals to failure, σ_f' is the fatigue strength coefficient, b is the fatigue strength exponent, ε_f' is the fatigue ductility coefficient, c is the fatigue ductility exponent, and E is the elastic modulus, respectively.

Equation (2) mathematically represents the Coffin-Manson curves of the strain amplitude and the number of reversals to failure shown in Figure 11. The results show the success of this approach through experimental testing. The material constants were calculated through Equation (2), and simply by fitting the curves using a least square fit method. The derived fatigue life coefficients in this work are also listed in Table 1. In this study, the c slopes of -1.04 and -0.96 of the Alloy 617 base metal and weldments, respectively, were obtained according to the experimental data. At all total strain ranges, the base metals exhibited the plastic regime controlling the fatigue deformation. Over a large strain range, it could be seen that the plastic regime controlled the fatigue deformation, and furthermore, the small strain range was induced by a larger elastic deformation rather than plastic deformation of the weldments. The domain in the intersection between plastic and elastic strain line, is called the transition of the fatigue life, N_t. From Figure 11, we obtained $2N_t$ of 3766 cycles and 628 cycles for base metal and weldment specimens, respectively.

(a)   (b)

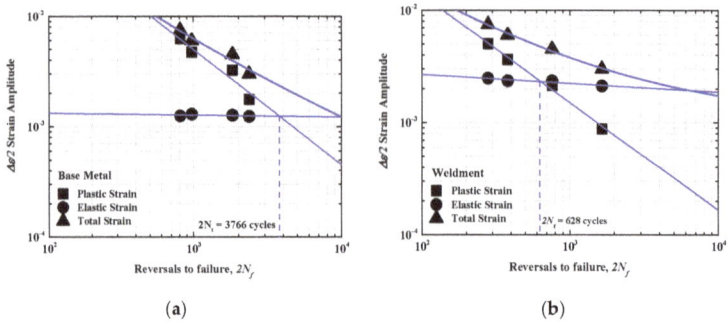

Figure 11. The Coffin-Manson curves of the Alloy 617: (**a**) base metal; (**b**) weldment specimens at 900 °C condition.

Table 1. The Coffin-Manson coefficients of Alloy 617 completed at 900 °C and a strain rate of 10^{-3}/s.

Reference	ε_f'	c	σ_f' (MPa)	b	E (GPa)	n'	K' (MPa)
Base Metal	6.527	−1.040	212.4	−0.017	149	−0.048	142.5
Weldments	1.120	−0.958	572.2	−0.078	149	0.087	613.9

3.4. Fracture Surface Characterization

In order to investigate the failure location and mechanism of Alloy 617, the macro- and micro-structural analysis of selected specimens, were investigated using SEM along with an EDX facility, and OM. Figure 12 shows the comprehensive photographs and micrographs of the cracking and failure location of the Alloy 617 base metal and weldment specimens at 0.6% total strain range. In this investigation, we observed that the LCF crack initiation sites of Alloy 617 base metal and weldment fractured specimens illustrate a flat- and a shear-type crack in the free surface, and that LCF cracking occurred within the gauge section.

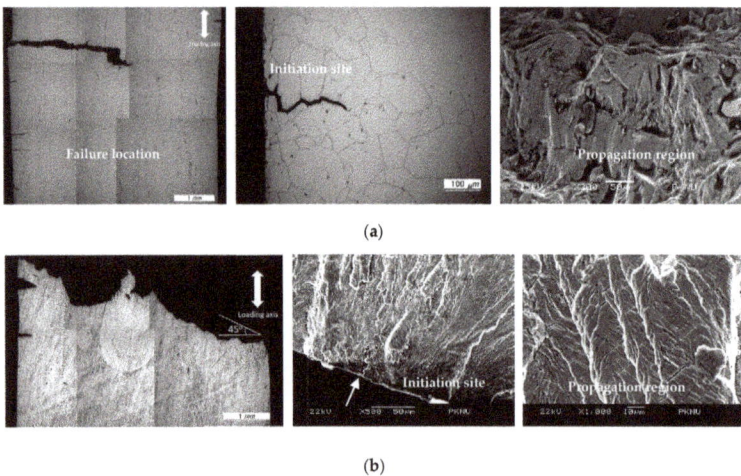

(a)

(b)

Figure 12. Scanning electron microscopy (SEM) and cross-view images of the selected specimens of Alloy 617 at 900 °C condition under LCF testing for (**a**) base metal; and (**b**) weldment specimen, total strain range of 0.6%. (Left to right: Failure location, crack initiation, and crack propagation regions).

From Figure 12a, it can be seen that the crack initiation site proceeded in the free surface through intergranular crack initiation resulting from oxidation of surface connected grain boundaries. It was found that the major role of reduction in the fatigue life of Alloy 617 base metal was probably generated from surface carbide oxidation. The LCF failure mechanism on the crack propagation as a result of the slip band containing extrusions and intrusions led to the propagation of a transgranular nature within the grain on the specimen surface, and finally at the critical stage, the primary crack is just about to fail. The LCF resistance of Alloy 617 is not affected by the intergranular creep damage mechanism. The strengthening of carbides creates a barrier to prevent intergranular creep crack by hindering grain boundary sliding and avoiding the concentration of stress. All of the base metal primary crack mechanism propagated in a transgranular nature with some quasi cleavage-facets and secondary cracks due to higher ductility of grain boundary carbides.

Figure 12b demonstrates that the main crack for weldment specimens was presented in the weld metal region. However, since the weld metal has a fully austenitic and dendritic structure, these implications resulted in a strength and hardness much larger than the base, but lower than the HAZ due to grain boundary strengthening [2]. Especially in the fusion line, or migrated boundary, it is coupled with their bonding strength and interface connection which prevents the occurrence of cracking. In fact, the possible weakness of the weld metal as well as the presence of microstructural heterogeneity, including the residual stress effect due to the welding process, cause a strain localization in the weld region. Figure 12b also shows the crack propagation of the weldments was followed by some quasi-cleavage like facets with striations due to the higher strength and lower ductility aside from the base metal. The formation of quasi-cleavage facets may be induced from the high stress concentration in the slip planes, thus, generating some cracks through the grains. We found that the crack initiation of the weldments emerged from a multi-site on the specimen surface, and was oriented about 45 or 135 degrees according to the loading direction due to the maximum shear stress.

Figure 13 shows the failure mechanism of the weldment specimens occurred due to the slip band process on the free surface, leading to the formation of a micro-crack within the grain on the specimen surface, and finally at the critical stage, coalescence of the micro-cracks develop into the macro-crack and the crack is propagated in transgranular nature across the dendritic structure. Again, as is seen from Figure 12, the fatigue striations on the weldment specimens are more obvious than on the base metal. The occurrence of fatigue striations was due to a repeated plastic blunting sharpening process arising from the slip of dislocations in the plastic zone of the fatigue crack tip [12]. However, the weldments could easily form fatigue striations and undergo quasi-cleavage facets due to their higher stress-strain interaction and being more brittle than that of the base metal [13].

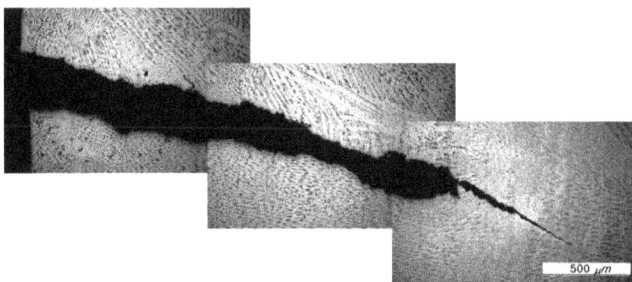

Figure 13. Typical optical microscopy (OM) images of the Alloy 617 weldment specimen tested at 900 °C, total strain range of 1.2% showing the LCF failure mechanism propagated across the dendritic structure.

Figure 14 shows typical micrographs of the Alloy 617 base metal crack initiation site. It is interesting to note that under higher total strain ranges, the reduction in fatigue life of the base metal is more clearly acquired, partly from homogenization of the slip process and an increase in plastic strain generated in the cycle, evidenced by several micro-cracks on the surface. At higher plastic deformation, this may induce a weak stress response because of more secondary cracks occurring in the grain boundary. In Figure 14 also indicates the LCF failures are often attended by intergranular crack initiation resulting from oxidation of surface connected grain boundaries.

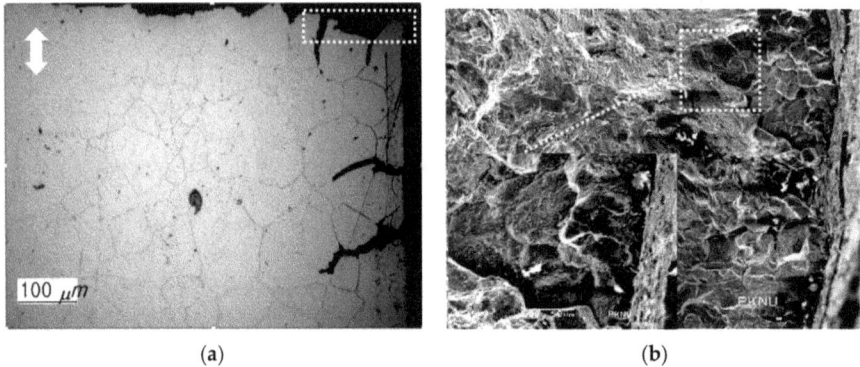

(a) (b)

Figure 14. Typical micrographs of the Alloy 617 base metal tested at 900 °C, total strain range of 1.2% shows critical damage induced by large plastic deformation, (**a**) micro-cracks on the surface; and (**b**) intergranular crack initiation due to the oxidation process.

The coarse carbide precipitations, Cr-rich $M_{23}C_6$, evolved mainly on the grain boundaries. From in Figure 15, EDX analysis shows the carbide precipitations from Cr metallic particles. The primary carbides known as, M_6C, with large precipitations were formed in the body of the grain structure. Also, the secondary carbides ($M_{23}C_6$) were found at the grain boundary. These carbide precipitations exist from chromium activities maintaining oxide formation on the surface.

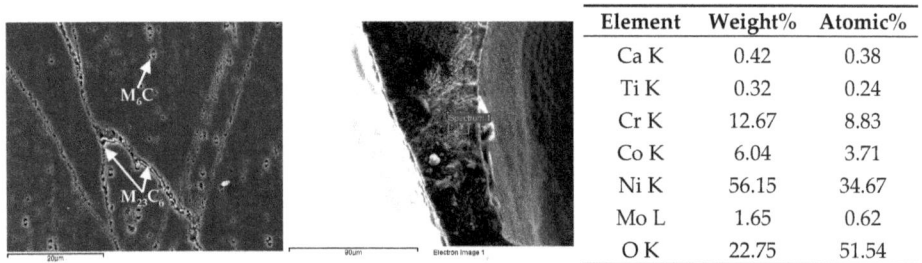

Element	Weight%	Atomic%
Ca K	0.42	0.38
Ti K	0.32	0.24
Cr K	12.67	8.83
Co K	6.04	3.71
Ni K	56.15	34.67
Mo L	1.65	0.62
O K	22.75	51.54

Figure 15. SEM and energy dispersive X-ray (EDX) examination of the Alloy 617 weldment specimen tested at 900 °C, total strain range of 0.6% demonstrating precipitations formed on the grain boundaries.

On the outer surface, a Cr-rich (mainly Cr_2O_3) oxide layer was formed owing to Cr diffusion from the matrix. Surface oxidation is evident owing to prolonged exposure time at very high temperature. In a previous work [15], it was stated that internal sub layers consisting of (Al-rich and carbides) were formed by the diffusion of oxygen through the surface oxide and matrix just beneath the Cr_2O_3 layer. The Cr_2O_3 layer is of the anion diffusion type, which prevents further diffusion of oxygen. We believe that transgranular crack propagation is associated with the excellent oxidation resistance of the Alloy 617, although, the intergranular cracking is initiated and developed by incorporation of oxidation of the surface connected grain boundaries together with environmentally generated mixed mode propagation.

At the end, the final stage where the material no longer holds the applied stress, namely the failure region, both base metal and weldment specimens exhibit shear, or dimple features. In Figure 16, SEM and EDX spectrum examination show a failure region of the weldments. However, the weldment specimens showed less dimple features due to lower ductility properties compared to the base metal. Furthermore, the EDX examinations show that the dimple features were observed with the occurrence of dominating metallic precipitations by Cr, Mo, Co, and Ni.

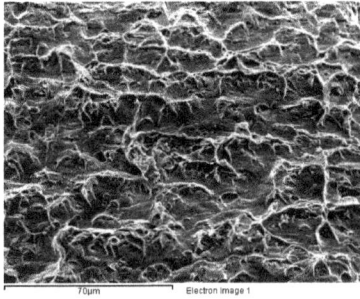

Element	Weight%	Atomic%
Al K	1.40	3.03
Si K	0.68	1.40
Ti K	0.80	0.97
Cr K	23.65	26.48
Fe K	0.59	0.62
Co K	11.05	10.92
Ni K	49.51	49.11
Mo L	12.32	7.48

Figure 16. SEM and EDX examination of the Alloy 617 weldment specimen tested at 900 °C, total strain range of 0.6% refers to the final failure region with some dimples features together with its composition.

4. Conclusions

The LCF tests of Alloy 617 base metal and weldments were conducted at a high temperature of 900 °C under four different total strain ranges in air. Under all the testing conditions, the weldment specimens showed lower fatigue lives compared with the base metal, and also the fatigue lives both of base metal and weldment specimens were decreased on increasing the total strain range. The overall cyclic stress response curve of the weldment specimens was higher than that of the base metal. The weldment specimens exhibited a cyclic hardening behavior during cyclic loadings, whereas the base metal experienced a cyclic softening induced by solute drag creep as a function of total strain range. In addition, the Coffin-Manson relationship analysis revealed that the *c* slope value is equal to -1.04 and -0.96 of the Alloy 617 base metal and weldment specimens, respectively.

The failure mechanism for Alloy 617 base metal occurring in the intergranular cracking is initiated by incorporation of oxidation of surface connected grain boundaries, and propagated in a transgranular nature., The main crack for weldment specimens presented itself in the weld metal region, whereas crack initiation emerged from multi-sites and was also oriented at about 45 or 135 degrees according to the loading direction. The main crack for weldment specimens occurred of a transgranular nature across the dendritic structure while the propagation area with clear striations was also observed.

Acknowledgments: The authors would like to recognize KAERI, and acknowledge that this research was supported by Nuclear Research & Development Program through the National Research Foundation of Korea (NRF) funded by the Ministry of Science, ICT & Future Planning (NRF-2016M2A8A2902895).

Author Contributions: Seon Jin Kim formulated this research with cooperation from Woo Gon Kim and Eung Seon Kim. Rando Tungga Dewa performed the experiment works, with the help of Seon Jin Kim, interpreted the results and prepared the manuscript. All co-authors contributed to the manuscript proof and submissions.

Conflicts of Interest: The authors declare no conflict of interest.

References

1. Dewa, R.T.; Kim, S.J.; Kim, W.G.; Kim, E.S. Low cycle fatigue behaviors of Alloy 617 (INCONEL 617) weldments for high temperature applications. *Metals* **2016**, *6*, 100. [CrossRef]
2. Kim, S.J.; Dewa, R.T.; Kim, W.G.; Kim, M.H. Cyclic stress response and fracture behaviors of Alloy 617 base metal and weldments under LCF loading. *Adv. Mater. Sci. Eng.* **2015**, *2015*. [CrossRef]

3. Lee, H.Y.; Kim, Y.W.; Song, K.N. Preliminary application of the draft code case for alloy 617 for a high temperature component. *J. Mech. Sci. Technol.* **2008**, *22*, 856–863. [CrossRef]

4. Lee, G.G.; Jung, S.; Park, J.Y.; Kim, W.G.; Hong, S.D.; Kim, Y.W. Microstructural investigation of Alloy 617 creep-ruptured at high temperature in a helium environment. *J. Mater. Sci. Technol.* **2013**, *29*, 1177–1183. [CrossRef]

5. Wright, J.K.; Carroll, L.J.; Cabet, C.; Lillo, T.M.; Benz, J.K.; Simpson, J.A.; Lloyd, W.R.; Chapman, J.A.; Wright, R.N. Characterization of elevated temperature properties of heat exchanger and steam generator alloys. *Nucl. Eng. Des.* **2012**, *251*, 252–260. [CrossRef]

6. Wright, J.K.; Carroll, L.J.; Wright, R.N. Creep and creep-fatigue of Alloy 617 weldments. Available online: http://www.osti.gov/scitech/biblio/1168621 (accessed on 26 July 2016).

7. Totemeier, T.C.; Tian, H.; Clark, D.E.; Simpson, J.A. Microstructure and strength characteristics of Alloy 617 welds. Available online: https://inldigitallibrary.inl.gov/sti/3310959.pdf (accessed on 26 July 2016).

8. Wright, J.K.; Carroll, L.J.; Simpson, J.A.; Wright, R.N. Low cycle fatigue of Alloy 617 at 850 °C and 950 °C. *J. Eng. Mater. Technol.* **2013**, *135*, 1–8. [CrossRef]

9. Rao, K.B.S.; Meurer, H.P.; Schuster, H. Creep-fatigue interaction of Inconel 617 at 950 °C in simulated nuclear reactor helium. *Mater. Sci. Eng. A* **1988**, *104*, 37–51. [CrossRef]

10. Ren, W.; Swindeman, R. A review on current status of Alloys 617 and 230 for Gen IV nuclear reactor internals and heat exchangers. *J. Press. Vessel Technol.* **2009**, *131*. [CrossRef]

11. Rahman, M.S.; Priyadarshan, G.; Raja, K.S.; Nesbitt, C.; Misra, M. Characterization of high temperature deformation behavior of INCONEL 617. *Mech. Mater.* **2009**, *41*, 261–270. [CrossRef]

12. Tian, D.D.; Liu, X.S.; He, G.Q.; Shen, Y.; Lv, S.Q.; Wang, Q.G. Low cycle fatigue behavior of casting A319 alloy under two different aging conditions. *Mater. Sci. Eng. A* **2016**, *654*, 60–68. [CrossRef]

13. Zhang, Q.; Zhang, J.; Zhao, P.; Huang, Y.; Yu, Z.; Fang, X. Low-cycle fatigue behaviors of a new type of 10% Cr martensitic steel and welded joint with Ni-based weld metal. *Int. J. Fatigue* **2016**, *88*, 78–87. [CrossRef]

14. Jang, C.; Lee, D.; Kim, D. Oxidation behaviour of an Alloy 617 in very high-temperature air and helium environments. *Int. J. Press. Vessels Pip.* **2008**, *85*, 368–377. [CrossRef]

15. Kim, W.G.; Park, J.Y.; Lee, G.G.; Hong, S.D.; Kim, Y.W. Temperature effect on the creep behavior of alloy 617 in air and helium environments. *Nucl. Eng. Des.* **2014**, *271*, 291–300. [CrossRef]

metals

MDPI

Article

The Unified Creep-Fatigue Equation for Stainless Steel 316

Dan Liu [1], Dirk John Pons [1],* and Ee-hua Wong [2]

[1]   Department of Mechanical Engineering, University of Canterbury, Christchurch 8140, New Zealand;
      dan.liu@pg.canterbury.ac.nz
[2]   Energy Research Institute, Nanyang Technological University, Singapore 637553, Singapore;
      ehwong@ntu.edu.sg
*   Correspondence: dirk.pons@canterbury.ac.nz; Tel.: +64-021-069-0900

Academic Editor: Filippo Berto
Received: 1 August 2016; Accepted: 2 September 2016; Published: 10 September 2016

Abstract: Background—The creep-fatigue properties of stainless steel 316 are of interest because of the wide use of this material in demanding service environments, such as the nuclear industry. Need—A number of models exist to describe creep-fatigue behaviours, but they are limited by the need to obtain specialized coefficients from a large number of experiments, which are time-consuming and expensive. Also, they do not generalise to other situations of temperature and frequency. There is a need for improved formulations for creep-fatigue, with coefficients that determinable directly from the existing and simple creep-fatigue tests and creep rupture tests. Outcomes—A unified creep-fatigue equation is proposed, based on an extension of the Coffin-Manson equation, to introduce dependencies on temperature and frequency. The equation may be formulated for strain as $\varepsilon_p = C_{0C}(T, t, \varepsilon_p) N^{-\beta_0}$, or as a power-law $\varepsilon_p = C_{0C}(T, t) N^{-\beta_0 b(T,t)}$. These were then validated against existing experimental data. The equations provide an excellent fit to data ($r^2 = 0.97$ or better). Originality—This work develops a novel formulation for creep-fatigue that accommodates temperature and frequency. The coefficients can be obtained with minimum experimental effort, being based on standard rather than specialized tests.

Keywords: creep-fatigue; creep-rupture; unified equation; fatigue model

---

1. Introduction

   The life of nuclear power plants has been a major issue because it strongly relates to safety and economy [1]. Stainless steel 316 is widely used for the making of components, such as turbine blades and piping, because of its excellent corrosion resistance.

   As the two main fatigue evaluation and design methods in the nuclear industry, the linear damage rule and the crack growth law have been used for many years. However, microstructural characteristics lead to imperfect prediction of fatigue failure. The coefficients in these models are obtained from a large number of experiments, which are time-consuming and expensive for industry to perform. When these models are employed, the coefficients are normally obtained from specific temperature and frequency, so these coefficients cannot be used to predict fatigue life in other situations. Therefore, there is a need to develop a creep-fatigue model that can largely avoid the influence of microstructure, present the influence of creep effects on fatigue behaviour, and be generalised to other situations of temperature and frequency. Ideally, the parameters in this model should be easy to obtain from empirical tests with minimum effort.

   In this paper, the strain-form unified creep-fatigue equation and power-law form will be introduced and will be verified on stainless steel 316. As part of the validation, the simple experimental methods of extracting coefficients will be presented.

2. Existing Approaches

In the case of pure fatigue, three general fatigue models are used to predict the fatigue life of this material: the Basquin equation (Equation (1)) [2], the Coffin-Manson equation (Equation (2)) [3,4] and Morrow's energy-based equation (Equation (3)) [5,6].

$$\frac{\Delta\sigma}{2} = \sigma'_f \left(2N_f\right)^b \tag{1}$$

$$\frac{\Delta\varepsilon_p}{2} = \varepsilon'_f \left(2N_f\right)^c \tag{2}$$

$$\Delta W_P = W'_f \left(2N_f\right)^\beta \tag{3}$$

where $\Delta\sigma$ is the stress amplitude, σ'_f is the fatigue strength coefficient, b is the fatigue strength exponent, $\Delta\varepsilon_p$ is the plastic strain amplitude, ε'_f is the fatigue ductility coefficient, c is the fatigue ductility exponent, ΔW_p is the plastic energy, W'_f is the energy coefficient, β is the energy exponent and N_f is the number of cycles. The coefficients in Equations (1) and (2) can be related through the compatibility (Equation (4)) [7] and cyclic stress-strain relation (Equation (5)) [8].

$$K' = \frac{\sigma'_f}{\left(\varepsilon'_f\right)^{n'}} \quad n' = \frac{b}{c} \tag{4}$$

$$\varepsilon_p = \left(\frac{\sigma_a}{K'}\right)^{\frac{1}{n'}} \tag{5}$$

where σ_a is the stress amplitude, K' is the strain hardening coefficient and n' is the strain hardening exponent.

In nuclear power plants, some components which are made of stainless steel 316 are subjected to fatigue at elevated temperature, at which the mechanism of creep is active. The failure of these components is caused by the combination of fatigue damage and creep damage. Two major rules are used to evaluate creep-fatigue life: the linear damage rule and the crack growth law.

2.1. The Linear Damage Rule

The linear damage theory was proposed by Palmgren [9] in 1924; it was further developed by Miner [10] in 1945 and called the Palmgren-Miner rule. This rule is widely used in the nuclear industry to design and evaluate the life of nuclear power plants [11–14]. According to this rule, damage can be calculated through using Equation (6), and the engineering structure fails when D equals 1.

$$D = \sum_{i=1}^{k} \frac{n_i}{N_i} \tag{6}$$

where D is the accumulated fatigue damage, k is the number of block loading, n_i is the number of constant amplitude cycles under the ith strain/stress range, and N_i is the number of cycles to fatigue failure under the ith strain/stress range.

Combined with the creep effects, the total damage is divided into fatigue damage and creep damage at the elevated temperature (Equation (7)) [15], which shows that the accumulation of fatigue and creep damage happens at different stages.

$$D = D_f + D_c \tag{7}$$

where D_f is the fatigue damage and D_c is the creep damage.

However, as one of the simplified methods which are used to predict life in the nuclear industry, the linear damage rule can lead to inaccurate results because of the neglect of loading sequences [13].

This problem was also realized in other industries. Therefore, many studies were conducted to improve the accuracy of this rule. For example, Richard and Newmark [16] proposed a power-law damage rule. Manson [17] demonstrated that failure can still happen when D is less than 1 and the linear damage rule was modified to double linear damage rule. Although these models can improve the results, these modifications do not change the character of linear accumulation of damage. Therefore, when creep is active, the inaccuracy which comes from the linear accumulation of creep damage and fatigue damage still cannot be solved, because the linear addition of damage is inconsistent with the microstructural characteristics. To be specific, cyclic strain/stress causes the slips between lattices, which can lead to the persistent slip bands. These deformations then lead to cracks. Meanwhile, the damage caused by creep comes from diffusion and dislocation along the grain boundaries and within the lattice, which leads to the accumulation of voids.

2.2. The Linear Damage Rule

The crack growth law is also used in the nuclear industry to predict fatigue life of some components [18,19], such as piping and tanks. The crack growth law shows that the fatigue life is the number of loading cycles which is required to achieve the final crack size, and this process is divided into two stages: initiation and propagation. Therefore, the total crack size can be presented as the linear addition of initial crack size and propagative crack size. Normally, the initial crack is identified as the real crack size in the structures before loading, and the propagative crack (Equation (8)) [19,20] can be obtained through Paris's model [21]:

$$\frac{da}{dN} = C \left(\Delta J_{\text{eff}} \right)^l \tag{8}$$

where $\frac{da}{dN}$ is the total crack growth per cycle, ΔJ_{eff} is the effective range of J-integral, and C is a material constant obtained from experiments. When creep is active, the total damage is calculated through the sum of fatigue and creep crack growth (Equations (9a) and (9b)) [21]:

$$\frac{da}{dN} = \left(\frac{da}{dN} \right)_f + \left(\frac{da}{dN} \right)_c \tag{9a}$$

$$\left(\frac{da}{dN} \right)_f = C \left(\Delta J_{\text{eff}} \right)^l \text{ and } \left(\frac{da}{dN} \right)_c = \int_0^{t_h} AC^{*q} dt \tag{9b}$$

where $\left(\frac{da}{dN} \right)_f$ is the crack growth per cycle due to cyclic load changes, $\left(\frac{da}{dN} \right)_c$ is the crack growth per cycle due to hold time, t_h is the hold time, C^* is the time-dependent fracture parameter, and l, A and q are material constants obtained from experiments.

The crack growth law provides a good physical explanation of damage. However, the quantitative summation between the cracks caused by fatigue and the cracks caused by creep does not consider the directions of these cracks. This means that two parallel cracks can cause the same damage as two non-parallel cracks, which is inconsistent with the microstructure, because the angle between two cracks plays an important role in the total damage.

2.3. Recent Developments towards a Unified Creep-Fatigue Equation

As identified above, the limitations of these methods are that they do not fully accommodate the observed microstructural characteristics, they require extensive testing to determine the coefficients, and the results cannot be generalised to other situations of temperature and frequency.

In an attempt to address these problems, Wong and Mai [22] proposed a formalism to accommodate fatigue and creep-fatigue, which they called a unified creep-fatigue equation (hereafter WM equation); see Equation (10). The unified creep-fatigue equation takes into account the influence of temperature and frequency on fatigue life. This equation was developed by extension of the Coffin-Manson equation.

$$\varepsilon_p = C_0 s\,(\sigma)\,c\,(T,t)\,N^{-\beta_0 b(T,t)} \tag{10}$$

where σ is the stress, T is the temperature, t is the cyclic time (1/frequency), C_0 is the fatigue ductility coefficient, β_0 is the fatigue ductility exponent, ε_p is the plastic strain and N is the fatigue life. The basic premise of the WM formulation is that 'all fatigue phenomenon are indeed creep-fatigue, and "pure fatigue" is just a special case of creep-fatigue' [22]. They reasoned that the Coffin-Manson equation was "a special case of a unified creep-fatigue equation".

The general principles of this were shown for the case of SnPb solder [22]. However there are several issues with the WM formulation. They did not provide the method to get the stress function $s\,(\sigma)$. This means that this unified equation still cannot be used to predict fatigue life. A related issues is that they assumed that functions $c\,(T,t)$ and $b\,(T,t)$ share the same pattern and characteristics, such that internal coefficients $c_1/c_2 = b_1/b_2$, but no reason was provided for this assumption. Also, the method of extracting the coefficients of function $c\,(T,t)$ and $b\,(T,t)$ was not proposed.

The WM equation has potential, but the concept needs further development. It has not been applied other than to solder, so its universal applicability is uncertain. There is a need to further validate or modify and the relationships between the coefficients, or improve the formulation.

3. Methods

3.1. Research Question

The purpose of this paper was to extend and modify the WM unified creep-fatigue equation. The WM equation provided some helpful initial starting points for the present work. Firstly, they showed that the creep-fatigue behaviour is negatively influenced by temperature, frequency and stress. Secondly, the unified creep-fatigue equation could be deduced from the Coffin-Manson equation and the experimental data of Shi [23] on solder, and the reference condition could be introduced into this unified equation. Thirdly, function $c\,(T,t)$ and $b\,(T,t)$ could be related to Manson-Haferd parameter, at least numerically.

3.2. Approach

Work in progress towards a further conceptual development has been to show how plastic strain (ε_p) may be theoretically related, in the creep-fatigue situation, to conditions at the reference temperature (T_{ref}) and reference cycle time (t_{ref}) at which tests are performed [24]. This line of thinking results in two forms of the unified creep-fatigue equation. The first form (strain form) represents the plastic strain as a function of cycle time, number of cycles, and temperature: $\varepsilon_p = C_0 c\,(T,\,t,\varepsilon_p)\,N^{-\beta_0}$. The second form (power-law form) does the same, but is simplified to a power-law relationship: $\varepsilon_p = C_0 c\,(T,t)\,N^{-\beta_0 b(T,t)}$. Definitions of variables are provided below.

In this paper both forms are verified by application to stainless steel 316. As part of the validation, the simple experimental methods of extracting coefficients are presented. The approach is to take published empirical data for creep rupture tests and creep-fatigue tests. The experimental data for creep-fatigue are from [25] and the creep rupture data are from [26] for stainless steel 316. From the data which were obtained at an arbitrary temperature and reference cycle time, we extract coefficients for the strain-form unified creep-fatigue equation (Section 4.3.1), and validate them through the empirical data at other conditions (Section 4.3.2). Then, the empirical data at two temperatures and reference cycle time are used to extract the coefficients for the power-law form (Sections 4.4.1.1 and 4.4.2.1), and validate it through the empirical data at other conditions (Sections 4.4.1.2 and 4.4.2.2).

The unified creep-fatigue equations were originally developed for solder, a material that is very different to stainless steel. Results show that the two forms of the unified creep-fatigue equation provide excellent representation of the stainless steel 316 creep fatigue data. A temperature modified

Coffin-Manson equation was derived from the combination of the power-law unified creep-fatigue equation and the frequency modified Coffin-Manson equation.

4. Results: Theory and Calculation

4.1. Introduction to the Unified Creep-Fatigue Equation

At this point, the term "fatigue capacity" needs to be introduced to describe the relation between pure fatigue and creep. For creep-fatigue, the fatigue capacity is reduced because of the increasing influence of creep damage. This can be seen in Figure 1: the creep-fatigue curves between pure fatigue curve and pure creep curve show the residual fatigue capacities, and the reductions are caused by creeps at different plastic strains/stresses.

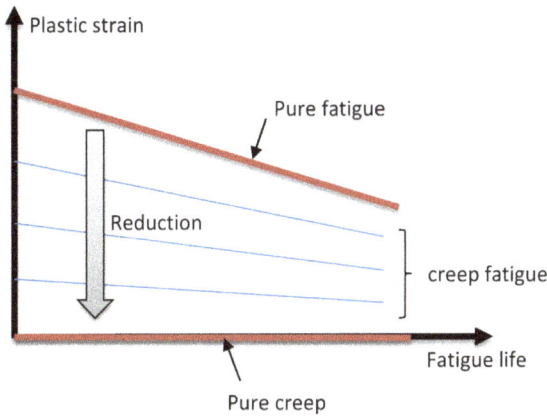

Figure 1. The relation between pure fatigue and pure creep.

Therefore, given the contribution of creep to the reduction of fatigue capacity, it is reasonable that the time-temperature-stress relationship, such as the Sherby-Dorn Parameter (Equation (11)) [27], the Larson-Miller Parameter (Equation (12)) [28] and the Manson-Haferd Parameter (Equation (13)) [29], is involved in the unified creep-fatigue equation.

$$P_{SD}(\sigma) = t e^{-Q/RT} \tag{11}$$

$$P_{LM}(\sigma) = T(\log t + C) \tag{12}$$

$$P_{MH}(\sigma) = \frac{T - T_a}{\log t - \log t_a} \tag{13}$$

where P_{SD} is the Sherby-Dorn Parameter, P_{LM} is the Larson-Miller Parameter, P_{MH} is the Manson-Haferd Parameter, Q is the activation energy of the creep mechanism, R is the Boltzmann's constant, t is the time, T is the absolute temperature, C is a constant and $(\log t_a, T_a)$ is the point of convergence of the $\log t$–T lines. Among these three parameters, the Manson-Haferd Parameter is regarded as the best description of stress-time-temperature relation [30].

It can be found that the unified creep-fatigue equation (Equations (14) and (15)) [24] shows a linear relationship between T and $\log t$:

$$\varepsilon_p = C_0 c(T, t, \varepsilon_p) N^{-\beta_0} \tag{14}$$

with

$$c(T, t, \varepsilon_p) = 1 - c_1(\varepsilon_p)(T - T_{ref}) - c_2 \log(t/t_{ref}) \quad (T > T_{ref}, t > t_{ref}) \tag{15}$$

Note that this relationship is consistent with the form of the P_{MH} parameter.

The *reference condition* refers to the threshold temperature T_{ref} at which creep first occurs (below this temperature no creep occurs), and cycle time (period) t_{ref} (arbitrary set as 1 second for comparing different data sets). The reference temperature and cyclic time for stainless steel 316 are identified as T_{ref} = 670 K and t_{ref} = 1 s. This temperature is 0.4 of the melting temperature and corresponds to the widely held assumption that below this temperature no creep occurs.

We then determine the plastic strain at reference condition, by transforming Equation (14):

$$\varepsilon_{p,ref} = C_0 N^{-\beta_0} = \frac{\varepsilon_p}{c\left(T, t, \varepsilon_p\right)} \tag{16}$$

This transformation would result in all ε_p: N data reducing into one single $\varepsilon_{p,ref}$: N curve if the creep function (Equation (15)) could describe the influence of creep on fatigue well.

However, the strain form returns an equation that is not power-law. Because power-law relation is expressed as a straight line on a log-log plot, it can provide an easy and clear way to present the creep-fatigue behaviours between different temperatures and cyclic times through translation and rotation. For this reason, the unified creep-fatigue equation is represented as a power-law form (Equation (17)) [24] through fitting the ε_p-N data with a power-law relation.

$$\varepsilon_p = C_0 c\left(T, t\right) N^{-\beta_0 b(T,t)} \tag{17}$$

with

$$c\left(T, t\right) = 1 - c_1\left(T - T_{ref}\right) - c_2 \log\left(t/t_{ref}\right) \ (T > T_{ref}, \ t > t_{ref}) \tag{18}$$

$$b\left(T, t\right) = 1 - b_1\left(T - T_{ref}\right) - b_2 \log\left(t/t_{ref}\right) \ (T > T_{ref}, \ t > t_{ref}) \tag{19}$$

Then, the plastic strain at reference condition is determined by transforming Equation (17):

$$\varepsilon_{p,ref} = C_0 N^{-\beta_0} = C_0 \left[\frac{\varepsilon_p}{C_0 c\left(T, t\right)}\right]^{1/b(T,t)} \tag{20}$$

This transformation would cause all ε_p:N data to collapse into one $\varepsilon_{p,ref}$:N curve if the creep function (Equation (18)) and stress function (Equation (19)) could present the influence of creep on fatigue well.

It can be seen that the unified creep-fatigue equation is restored to the Coffin-Manson equation at reference condition (T_{ref}, t_{ref}), which builds a bridge between pure fatigue and creep fatigue. The deduction of coefficients (Equations (A1)–(A13)) in the strain-form unified creep-fatigue equation and power-law form is shown in Appendix A.

Next, the relatively simple methods of obtaining the coefficients will be verified on stainless steel 316.

4.2. Extracting the Creep-Rupture Properties of Stainless Steel 316

The Manson–Haferd parameter was extracted through plotting creep-rupture data [26] as log *(time)* vs. *temperature*. According to the Manson-Haferd parameter, all log$t$–T lines at different stresses should converge to one point, where the temperature is regarded as the Creep Initiation Temperature, below which no creep occurs. We determine this temperature as 40% of the melting temperature. This corresponds to the reference temperature. In this case the reference temperature is found to be T_{ref} = 670 K. At this temperature, the rupture times at different stresses were found. According to Manson-Haferd parameter, these rupture times should be the same. However, because the accuracy of experiments is influenced by many factors, such as facilities and environment, the points (rupture time, reference temperature) at different stresses cannot be collected into one point. Therefore, the average of these rupture times is employed to define the rupture time at the point of convergence, and the point of convergence is identified to be (670, 9.54) (shown in Figure 2). The Manson-Haferd parameter is stress dependent, and the values are obtained as the inverse of the slope from the fitted lines. Then, the relationship between Manson-Haferd parameter and stress can be extracted from Figure 2, and Equation (21) is given through curve-fitting:

$$-\frac{1}{P_{MH}(\sigma)} = 4.50 \times 10^{-3} + 1.26 \times 10^{-5}\sigma + 1.70 \times 10^{-7}\sigma^2 \tag{21}$$

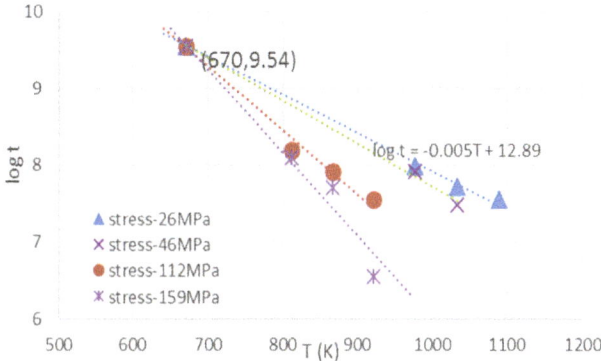

Figure 2. The $P_{MH}(\sigma, T, t)$ parameter of AISI stainless steel 316. Data extracted from [26].

To convert $P_{MH}(\sigma)$ to $P_{MH}(\varepsilon_p)$, the data of stress, total strain, temperature and Young's modulus are extracted from [31]. Because the time of tension tests is much less than the creep-rupture time, the influence of time (creep) on tension tests can be neglected. The temperature-dependent plastic strain-stress relationship (Equations (22)–(24)) can be developed through curve-fitting.

$$\varepsilon_p = \left(\frac{\sigma}{K(T)}\right)^{1/n(T)} \tag{22}$$

$$K(T) = -0.4875T + 769.79 \tag{23}$$

$$n(T) = 7.04 \times 10^{-2} + 1.94 \times 10^{-4}T - 1.22 \times 10^{-7}T^2 \tag{24}$$

Then, substituting Equations (22)–(24) into Equation (21) gives Equation (25):

$$-\frac{1}{P_{MH}(\varepsilon_p, T)} = 4.50 \times 10^{-3} + 1.26 \times 10^{-5}K(T)\ \varepsilon_p^{n(T)} + 1.70 \times 10^{-7}K(T)\ \varepsilon_p^{2n(T)} \tag{25}$$

4.3. Evaluation of the Coefficients of the Strain-Form Unified Creep-Fatigue Equation $\varepsilon_p = C_{0C}(T, t, \varepsilon_p)\ N^{-\beta_0}$ for the Stainless Steel 316 and Validation

4.3.1. Evaluation of Creep-Fatigue Coefficients

Based on the Equations (A1)–(A6) shown in Appendix A, the coefficients of the creep function, $c(T, t, \varepsilon_p)$, for the stainless steel 316 are established as follows.

Substituting $\log t_a = 9.54$ and $t_{ref} = 1$ s into Equation (A2), the coefficient c_2 is evaluated as:

$$c_2 = \frac{1}{\log(t_a/t_{ref})} = \frac{1}{\log(10^{9.54}/1)} = 0.105 \tag{26}$$

Substituting Equation (26) into Equation (A4), coefficient c_1 (σ) and c_1 (ε_p) are developed respectively:

$$c_1\left(\sigma, T\right) = 4.73 \times 10^{-4} + 1.32 \times 10^{-6}\sigma\left(T\right) + 1.79 \times 10^{-8}\sigma^2\left(T\right) \tag{27}$$

$$c_1\left(\varepsilon_p, T\right) = 4.73 \times 10^{-4} + 1.32 \times 10^{-6}K\left(T\right)\varepsilon_p^{n(T)} + 1.79 \times 10^{-8}K^2\left(T\right)\varepsilon_p^{2n(T)} \tag{28}$$

The fatigue coefficients C_0 and β_0 of the stainless steel 316 can be extracted from fatigue test at an arbitrary temperature at the reference cycle time. Selecting the data point (T = 873 K, t_{ref}), at which ε_p (T = 873 K, t_{ref}, N = 1) = 2.1705, C_0 is extracted through using Equation (A6), which is 2.95, and β_0 is given as 0.663. However, a big error ($\sum\left(N_{f-exp} - N_{f-equ}\right)^2$ = 49.85) is given through comparing the fatigue life obtained from experiments with the fatigue life obtained from unified creep-fatigue. Therefore, the C_0 is resolved as 0.959 through minimizing the error (0.312).

4.3.2. Validations

Using the fatigue and the creep coefficients evaluated in Section 4.3.1, namely, C_0 = 0.959, β_0 = 0.663, c_2 = 0.105, and c_1 (ε_p, T) as described by Equation (28), the generated raw fatigue data (ε_p–N) obtained from [25] (T = 723 K, 873 K and 973 K) are transformed to the reference condition ($\varepsilon_{p,ref}$–N) through Equation (16). The transformed data are plotted in Figure 3.

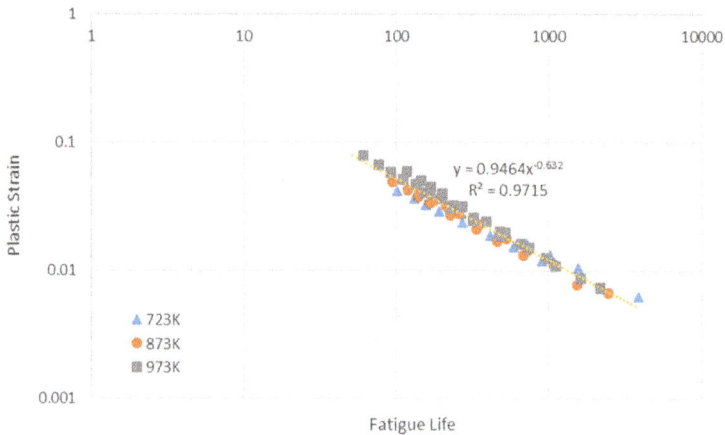

Figure 3. Transformed $\varepsilon_{p,ref}$-N data of stainless steel 316 using Equation (14) with creep function c (T, t, ε_p).

Significantly, these data are collected into a power-law curve of $\varepsilon_{p,ref}$ = $0.9464N^{-0.632}$ with the quality of fit as R^2 = 0.9715. This good transformation verifies the unified equation, $\varepsilon_p = C_0 c\left(T, t, \varepsilon_p\right)N^{-\beta_0}$, the form of creep function c (T, t, ε_p), and the methods of extracting the fatigue and the creep coefficients. The error between fatigue life from experiments and creep-fatigue equation is 0.312.

4.3.3. Application

This shows that the mathematical representation provided by the strain-form unified creep-fatigue equation accommodates the data for multiple temperatures, fatigue life, and plastic strain.

$$\varepsilon_p = 0.959\left[1 - c_1\left(\varepsilon_p, T\right)\left(T - 670\right) - 0.105\log\left(t\right)\right]N^{-0.663} \tag{29}$$

with

$$c_1\left(\varepsilon_p, T\right) = 4.73 \quad \times 10^{-4} + 8.712 \times 10^{-7}$$
$$\times \left(769.79 - 0.4875T\right) \; \varepsilon_p^{7.04 \times 10^{-2} + 1.94 \times 10^{-4}T - 1.22 \times 10^{-7}T^2} + 7.797 \quad (30)$$
$$\times 10^{-9} \times \left(769.79 - 0.4875T\right)^2 \; \varepsilon_p^{0.1408 + 3.88 \times 10^{-4}T - 2.44 \times 10^{-7}T^2}$$

This could be used to determine fatigue life for given plastic strain, temperature and cycle time. Alternatively, to determine plastic strain by numerical solution of the equation.

4.4. Evaluation of the Coefficients of the Power-Law Unified Creep-Fatigue Equation $\varepsilon_p = C_0 c\left(T, t\right) N^{-\beta_0 b(T,t)}$ for the Stainless Steel 316 and Validation

The fatigue behaviour of stainless steel 316 presents an inflection point at the temperature of 873 K [25]. Below this temperature, the fatigue life decreases with the increasing temperature, while, increases above this temperature. Given that the power-law unified creep-fatigue equation is built on the assumption of continual increasing/decreasing fatigue life with increasing temperature, the evaluation and validation of the power-law unified creep-fatigue equation are conducted at two temperature regimes (below 873 K and above 873 K).

4.4.1. Evaluation of Creep-Fatigue Coefficients and Validation below 873 K

4.4.1.1. Evaluation of Creep-Fatigue Coefficients

Based on the Equations (A7)–(A13) shown in Appendix A, the coefficients of c function and b function for the stainless steel 316 are established as follows.

The creep coefficient c_2 is identical as 0.105.

Substituting Equation (18) with the data points ($T = 873$ K, $t = t_{ref}$) and ($T = 873$ K, $t = 10$ s), where ε_p ($T = 873$ K, $t = t_{ref}$, $N = 1$) = 1.0296 (Equation (A8)) and ε_p ($T = 873$ K, $t = 10$ s, $N = 1$) = 0.5987 (Equation (A9)), gives $C_0 c_2 = 0.4309$. Then, C_0 is solved as 4.1038. In addition, substituting Equation (19) with these two data points, where $\beta_0 b$ (873 K, $t = t_{ref}$) = 0.663 (Equation (A10)) and $\beta_0 b$ ($T = 873$ K, $t = 10$ s) = 0.651 (Equation (A11)), gives $\beta_0 b_2 = 0.012$.

Substituting Equation (18) with the data points ($T = 723$ K, $t = t_{ref}$) and ($T = 873$ K, $t = t_{ref}$), where ε_p ($T = 723$ K, $t = t_{ref}$, $N = 1$) = 2.1705 (Equation (A12)) and ε_p ($T = 873$ K, $t = t_{ref}$, $N = 1$) = 1.0296 (Equation (A8)), gives $c_1 = 0.001853$. Then, substituting Equation (19) with these two data points, where $\beta_0 b$ (723 K, $t = t_{ref}$) = 0.634 (Equation (A13)) and $\beta_0 b$ (873 K, $t = t_{ref}$) = 0.663 (Equation (A10)), gives $\beta_0 = 0.624$ and $b_1 = -0.00031$, then b_2 is solved as 0.01924.

The error between the fatigue life from experiments and creep-fatigue equation is 52.88. As shown in the evaluation of coefficients for strain-form unified creep-fatigue equation, this poor prediction is caused by the inaccuracy of C_0. Therefore, the C_0 is resolved as 0.876 through minimizing the error (0.792).

4.4.1.2. Validations

Using the fatigue and creep coefficients found in Section 4.4.1.1, namely, $C_0 = 0.876$, $\beta_0 = 0.624$, $c_1 = 0.001853$, $c_2 = 0.105$, $b_1 = -0.0003094$ and $b_2 = 0.01924$, the raw fatigue data (ε_p–N) obtained from [25] ($T = 723$ K and 873 K) are transformed to the reference condition ($\varepsilon_{p,ref}$–N) through Equation (20). The transformed data are plotted in Figure 4, which shows that these data can be collapsed into a power-law curve of $\varepsilon_{p,ref} = 0.5959N^{-0.556}$ with the quality of fit as $R^2 = 0.9583$. This transformation has verified the unified equation, $\varepsilon_p = C_0 c\left(T, t\right) N^{-\beta_0 b(T,t)}$, the form of creep function $c\left(T, t\right)$ and stress function $b\left(T, t\right)$, and the methods of extracting the coefficients. The error between fatigue life from experiments and creep-fatigue equation is 0.792.

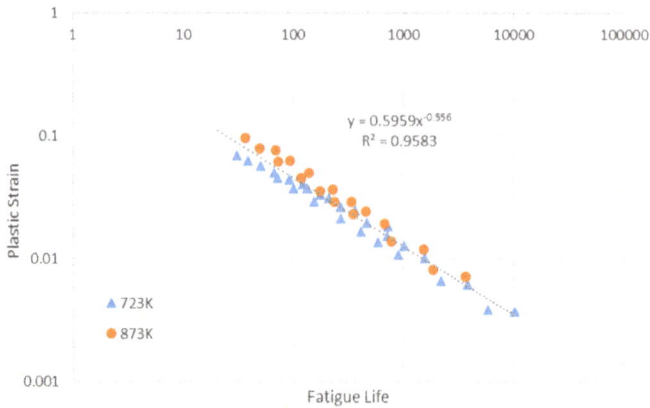

Figure 4. Transformed $\varepsilon_{p,ref}$–N data (below 873 K) of stainless steel 316 using Equation (17) with functions c (T, t) and b (T, t).

4.4.2. Evaluation of Creep-Fatigue Coefficients and Validation above 873 K

4.4.2.1. Evaluation of Creep-Fatigue Coefficients

The data points (T = 873 K, t = t_{ref}), (T = 873 K, t = 10 s) and (T = 973 K, t = t_{ref}) are selected to evaluate the coefficients. These coefficients are obtained through the same method shown in Section 4.4.1.1: C_0 = 0.879, β_0 = 0.807, c_1 = 0.00146, c_2 = 0.105, b_1 = 0.00088 and b_2 = 0.01487.

4.4.2.2. Validations

Using the fatigue and creep coefficients found in Section 4.4.2.1, the raw fatigue data (ε_p–N) obtained from [25] (T = 873 K and 973 K) are transformed to the reference condition ($\varepsilon_{p,ref}$–N) through Equation (20). The transformed data plotted in Figure 5, which shows that these data can be collapsed into a power-law curve of $\varepsilon_{p,ref} = 0.9439N^{-0.82}$ with a quality of fit of $R^2 = 0.9797$. This transformation has verified the unified equation, $\varepsilon_p = C_0 c\,(T, t)\,N^{-\beta_0 b(T,t)}$, the form of creep function c (T, t) and stress function b (T, t), and the methods of extracting the coefficients. The error between fatigue life from experiments and creep-fatigue equation is 0.300.

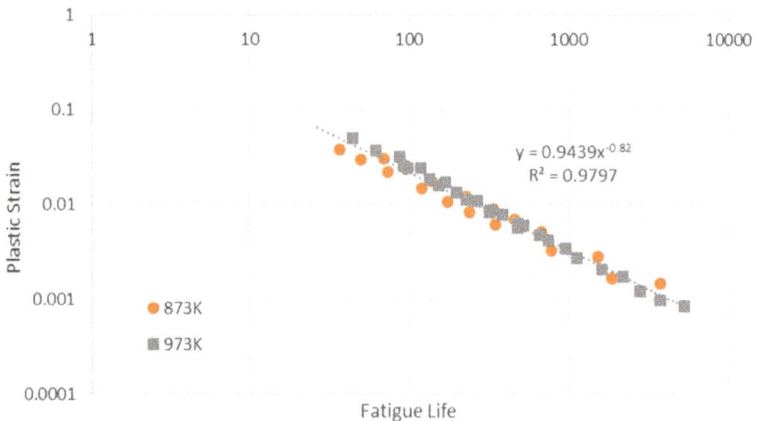

Figure 5. Transformed $\varepsilon_{p,ref}$–N data (above 873 K) of stainless steel 316 using Equation (17) with functions c (T, t) and b (T, t).

4.4.3. Application

The mathematical representation (Equation (31) for below 873 K and Equation (32) for above 873 K) provided by the unified creep-fatigue equation accommodates the data for multiple temperatures, fatigue life, and plastic strain. This could be used to determine fatigue life for given plastic strain, temperature and cycle time. Alternatively, to determine plastic strain by numerical solution of the equation:

$$\varepsilon_p = 0.876[2.2415 - 0.001853T - 0.105\log{(t)}]N^{-0.624[0.7927+0.0003094T-0.01924\log(t)]}$$
$$(T < 873\text{K}) \tag{31}$$

$$\varepsilon_p = 0.879[1.9782 - 0.00146T - 0.105\log{(t)}]N^{-0.807[1.5896+0.00088T-0.01487\log(t)]}$$
$$(T \geq 873\text{K}) \tag{32}$$

5. Discussion

5.1. The Moderating Factor

As shown above, C_0 in the strain-form unified creep-fatigue equation was obtained from data point ($T = 723$ K, t_{ref}). However, this result is different from C_0 obtained from data point ($T = 873$ K, t_{ref}) and ($T = 973$ K, t_{ref}). To improve the accuracy of C_0, these three data points are used to regress the magnitude of C_0, which is 2.517 (Equation (33)).

$$\varepsilon_p\left(T, t_{\text{ref}}, N = 1\right) = C_0' = C_0\left[1 - c_1\left(C_0'\right)\left(T - T_{\text{ref}}\right)\right] = 2.517\left[1 - 0.00278\left(T - T_{\text{ref}}\right)\right] \tag{33}$$

At data point ($T = 723$ K, t_{ref}), substituting C_0' into the Equation (28) cannot yield to 2.1705, and the magnitude of c_1 (2.1705) is bigger than 0.00278. Thus, according to the Equation (27), it appears that the big contribution of stress leads to higher magnitude of c_1 function. Therefore, mathematically, the amplitude of stress should be compressed in order to reduce the magnitude of c_1 (C_0') into the result of regression (0.00278). Then, a moderating factor, f, is introduced into Equation 27 to modify stress, and Equations (27) and (28) can be expressed as:

$$c_1\left(\sigma, f\right) = 4.73 \times 10^{-4} + 1.32 \times 10^{-6}f\sigma + 1.79 \times 10^{-8}f^2\sigma^2 \tag{34}$$

$$c_1\left(\varepsilon_p, T, f\right) = 4.73 \times 10^{-4} + 1.32 \times 10^{-6}fK\left(T\right)\varepsilon_p^{n(T)} + 1.79 \times 10^{-8}f^2K\left(T\right)\varepsilon_p^{2n(T)} \tag{35}$$

This moderating factor is solved as 0.69, and C_0 is given as 0.846 through minimizing the error (0.276).

Using the fatigue and the creep coefficients, namely, $C_0 = 0.846$, $\beta_0 = 0.663$, $c_2 = 0.105$, $c_1\left(\varepsilon_p, T, f\right)$ as described by Equation (35) and $f = 0.69$, the generated raw fatigue data (ε_p–N) obtained from [25] ($T = 723$ K, 873 K and 973 K) are transformed to the reference condition ($\varepsilon_{p,\text{ref}}$–$N$) through Equation (16). The transformed data are plotted in Figure 6.

Figure 6 illustrates how the transformed data are collected into a power-law curve of $\varepsilon_{p,\text{ref}} = 0.6633N^{-0.592}$ with the quality of fit $R^2 = 0.9759$. The error between fatigue life from experiments and creep-fatigue equation is 0.276. Comparing this result with the transformation in Section 4.3.2 shows that the introduction of moderating factor can provide a better description of creep effect through c_1 function and prediction of creep-fatigue behaviour.

The research conducted by Gary shows that the stress vs. creep rupture time curves under cyclic loading lie above the curves under constant loading [32]. This means that cyclic stress is higher than constant stress at the same rupture time. Because c_1 function is based on the time-temperature relation under constant stress (Manson-Haferd parameter), the creep effect is enlarged when the cyclic stress is imposed. Therefore, it is reasonable to introduce a moderating factor to compress the cyclic stress to an equivalent constant stress.

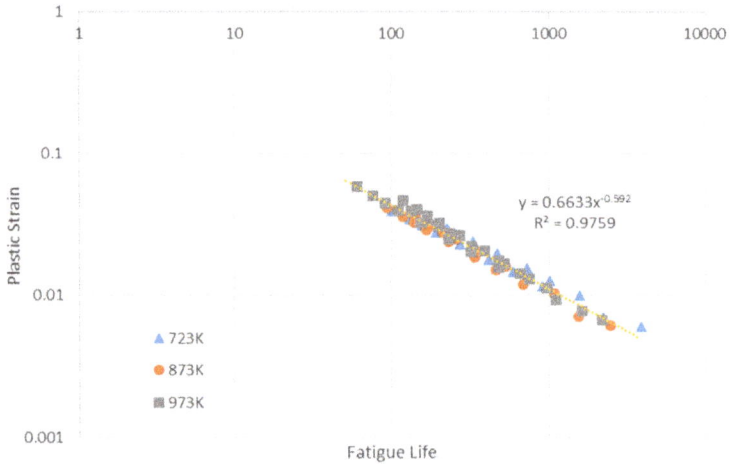

Figure 6. Transformed $\varepsilon_{p,ref}$–N data of stainless steel 316 using Equation (14) with creep function c (T, t, ε_p) and moderating factor.

5.2. The Heat Treatment

Heat treatment can change fatigue behaviour through hardening or softening. However, stainless steel 316 cannot be hardened by heat treatment, and this is proved by [33], where fatigue life changes slightly between aged condition and annealed condition [33]. This makes the unified creep-fatigue equation more universal for stainless steel 316 under different heat treatments. For example, the creep-fatigue data of aged stainless steel obtained from [33] (T = 839 K and 922 K) and quenched stainless steel 316 obtained from [25] (T = 723 K, 873 K and 973 K) can be collected into one power-law curve $\varepsilon_{p,ref} = 0.6533N^{-0.588}$ with a quality of fit of R^2 = 0.9827 (Figure 7).

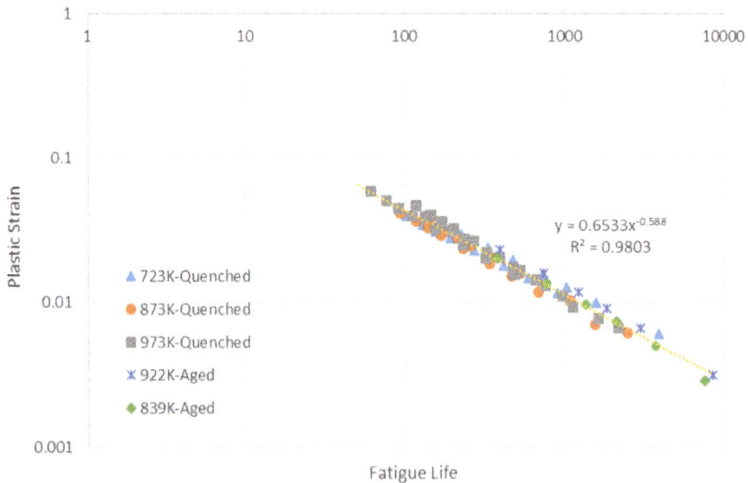

Figure 7. Transformed $\varepsilon_{p,ref}$–N data of quenched and aged stainless steel 316 using Equation (14) with creep function c (T, t, ε_p).

Although the unified creep-fatigue equation shows a good universality for stainless steel 316 under different heat treatments, the potential limitation for this equation still cannot be ignored. For example, this strong influence of heat treatment on fatigue life is shown on Inconel 718, and the aged condition can provide better fatigue strength than annealed condition [33]. Therefore, if the unified creep-fatigue equation is imposed on Inconel 718, the coefficients obtained at one specified heat treatment condition cannot be used to predict the fatigue life of this material under other heat treatments. This is an opportunity for future development of the formulation.

5.3. Reliability

The proposed new equations were validated against existing experimental data in the literature. The equations provide an excellent fit to data ($r^2 = 0.97$ or better). This demonstrates that the equations provide the desired level of fidelity to the original experimental data.

The validation only can show these selected data follow the unified creep-fatigue equation, but not all data. Hence it is important to consider the degree of reliability. In this section, the strain-form unified creep-fatigue equation was used to explore the reliability.

The fatigue life, temperature and strain rate are defined as random variables, and then 100 random creep-fatigue data points (*plastic strain* vs. *fatigue life*) were derived from [25] and the Coffin-Manson equation. These random data then were transformed into a reference condition through Equation (16). The transformation of 10 sets of random data (1000 data points) shows that these creep-fatigue data can collapse into almost one straight line at log-log scale (two of them are shown in Figure 8) with the quality of fit $R^2 = 0.975$–0.985. This transformation based on random data has further verified the unified equation, $\varepsilon_p = C_{0}c\left(T, t, \varepsilon_p\right) N^{-\beta_0}$ and the form of creep function $c\left(T, t, \varepsilon_p\right)$.

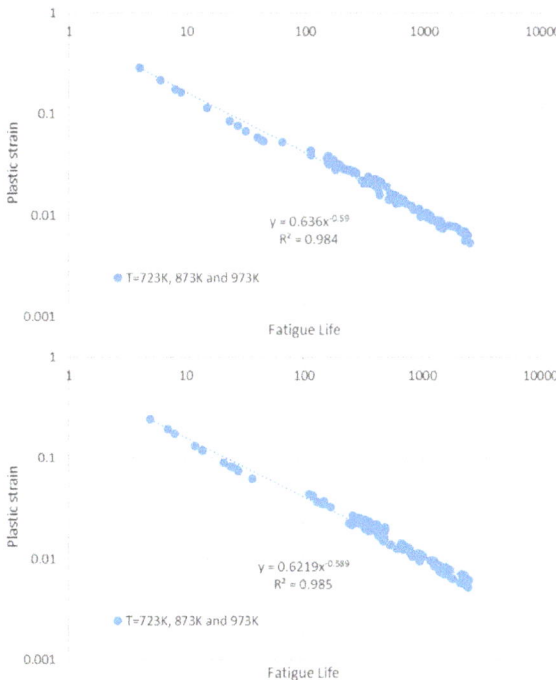

Figure 8. Transformed $\varepsilon_{p,\text{ref}}$–$N$ data of stainless steel 316 using Equation (14) with creep function $c\left(T, t, \varepsilon_p\right)$ and random variables. Data shown are for two of ten sets, as evidence of a consistent transformation process.

5.4. The Initial Proposal of Creep-Fatigue-Equation-Based Temperature Modified Coffin-Manson Equation

Frequency modified Coffin-Manson equation can be expressed as Equation (36) [34]:

$$\Delta \varepsilon_p = C_0 \left[N_f v^{k-1} \right]^m \tag{36}$$

where v is the frequency, and k and m are the constant determined by experiments. Comparing this equation with the conventional Coffin-Manson equation (Equation (2)), the $N_f v^{k-1}$ can be defined as the frequency modified fatigue life. Similarly, comparing the unified creep-fatigue equations (Equations (14) and (17)) with the conventional Coffin-Manson equation (Equation (2)), $[c\,(T,\,t,\varepsilon_p)]^{-1/\beta_0}\,N$ and $[c\,(T,t)]^{-1/\beta_0 b(T,t)}\,N$ can be defined as the temperature-frequency modified fatigue life (creep-fatigue life). If we get rid of frequency effect from the unified creep-fatigue equations, the creep-fatigue life ($N_{f-\mathrm{creep+fatigue}}$) can be transformed to temperature modified fatigue life ($N_{f-\mathrm{temp}}$) through Equation (37), then the temperature modified Coffin-Manson equation could be developed through this transformation.

$$N_{f-\mathrm{temp}} = N_{f-\mathrm{creep+fatigue}} \cdot v^{k-1} \tag{37}$$

Because c_1 function in the strain-form unified creep-fatigue equation related to time-temperature relation, it is difficult to remove the frequency effect from this function. Therefore, the power-law unified creep-fatigue equation is used to develop the temperature modified Coffin-Manson equation (Equation (38)) through removing the frequency-related items and transforming creep-fatigue life to temperature modified fatigue life.

$$\varepsilon_p = C_0 \left[1 - c_1 \left(T - T_{\mathrm{ref}} \right) \right] N_{f-\mathrm{temp}}^{-\beta_0 [1 - b_1 (T - T_{\mathrm{ref}})]} \tag{38}$$

Based on the creep-fatigue data [25], the coefficients of this equation are evaluated (shown in Table 1). Then, using the coefficient shown in Table 1, the generated raw fatigue data (ε_p–N) obtained from [25] (T = 723 K, 873 K and 973 K) are transformed to the reference condition ($\varepsilon_{p,\mathrm{ref}}$–N) through Equation (16). The transformed data at "723 K and 873 K", and "873 K and 973 K" are plotted in Figures 9 and 10 respectively.

Table 1. The coefficients of Equation (38).

Temperature Regimes	C_0	c_1	β_0	b_1	k
723 K–873 K	1.997	0.002955	0.62375	−0.000309	723 K: 0.728 873 K: 0.758
873 K–973 K	2.452	0.002668	0.80713	0.00088	873 K: 0.758 973 K: 0.873

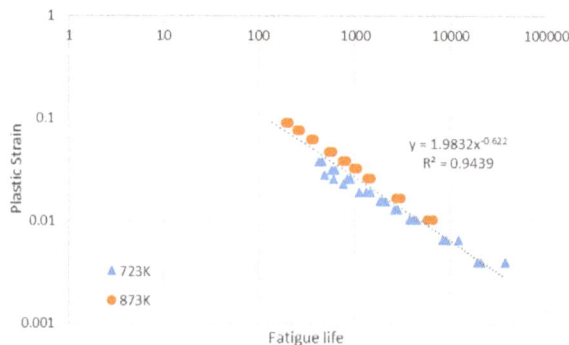

Figure 9. Transformed $\varepsilon_{p,\mathrm{ref}}$–N data (below 873 K) of stainless steel 316 using Equation (17).

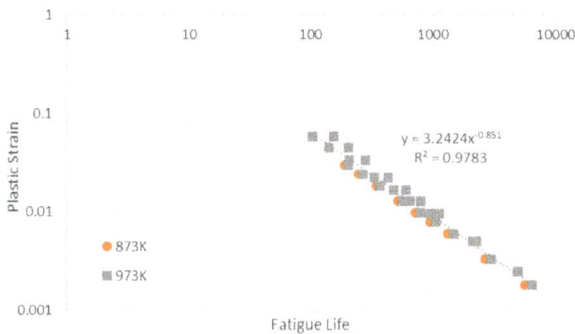

Figure 10. Transformed $\varepsilon_{p,\text{ref}}$–$N$ data (above 873 K) of stainless steel 316 using Equation (17).

Figures 9 and 10 show that the creep-fatigue data below 873 K and above 873 K can be collapsed into power-law curves of $\varepsilon_{p,\text{ref}} = 1.9832N^{-0.622}$ with the quality of fit as $R^2 = 0.9439$ and $\varepsilon_{p,\text{ref}} = 3.2424N^{-0.851}$ with the quality of fit as $R^2 = 0.9783$ respectively. The errors between fatigue life from experiments and creep-fatigue equation for these two temperature regimes are 0.814 and 0.294. This transformation has verified the creep-fatigue-equation-based temperature modified Coffin-Manson equation.

5.5. Application and Future Research

The results have provided a derivation of the equations and the methods for determining the coefficients. While this may appear mathematically complex, it is a necessary feature of the work as it provides reproducibility and assists other researchers in applying and adapting the work.

The implications for practitioners, e.g., nuclear industry, are given in Sections 4.3.3 and 4.4.3, using the two forms of the unified creep-fatigue equation. The equations are provided for stainless steel 316 and encapsulate the data for multiple temperatures, fatigue life, and plastic strain. In application, the equation might be used to determine fatigue life for given conditions (plastic strain, temperature and cycle time), or to determine acceptable plastic strain for a given life and other conditions.

6. Conclusions

The following major results were obtained:

(1) The strain-form unified creep-fatigue equation: $\varepsilon_p = C_0c\left(T,\ t,\ \varepsilon_p\right)N^{-\beta_0}$ and power-law form: $\varepsilon_p = C_0c\left(T,t\right)N^{-\beta_0 b(T,t)}$ have been verified on stainless steel 316, and the methods of extracting the fatigue and the creep coefficients by limited creep-fatigue tests and creep rupture tests have been presented. The equations are efficient as they represent a whole range of conditions, in contrast to some other formulations which have to be calibrated separately for each set of environmental conditions.

(2) A moderating factor is introduced into the unified creep-fatigue equation to compress the amplitude of cyclic stress, and leads to a more reasonable result than the situation without moderating factor.

(3) The reliability of this unified creep-fatigue equation is verified through random variables.

(4) The creep-fatigue equation-based, temperature-modified Coffin-Manson equation is proposed, and this equation is validated through transforming raw data to the reference condition.

(5) However, there is a potential limitation. For the materials whose fatigue behaviour is strongly influenced by heat treatment, the coefficients obtained from one specific material condition cannot be used to predict the fatigue life of the same material at material conditions. This is identified as an area for future development of the theory.

Author Contributions: The work was conducted by D.L. and supervised by D.J.P. and E.W. The initial form of the unified creep-fatigue equation was proposed by E.W., and the method of extracting the coefficients was then developed by D.L. and E.W. The discussion on the moderating factor, heat treatment and reliability was developed by D.L. and D.J.P. The review of creep-fatigue models in the nuclear industry and the evaluation and validation of the unified creep-fatigue equation for stainless steel 316 were conducted by D.L. The initial form of the creep-fatigue equation-based, temperature-modified Coffin-Manson equation was proposed by D.L.

Conflicts of Interest: The authors declare no conflict of interest. The research was conducted without personal financial benefit from any funding body, and no such body influenced the execution of the work.

Appendix A

A1. The Coefficients in the Strain-Form Unified Creep-Fatigue Equation

At pure creep-rupture condition, plastic strain $\varepsilon_p = 0$, then c function can be presented as:

$$c(T, t, \sigma) = 1 - c_1(\sigma)(T_R - T_{ref}) - c_2(\sigma) \log(t_R/t_{ref}) = 0 \tag{A1}$$

where T_R is rupture temperature and t_R is rupture time. According to Manson-Haferd Parameter (Equation (8)), the creep occurs at the point of convergence. At this point, $T_R = T_a = T_{ref}$ and $t_R = t_a$, then Equation (A1) can be deduced to Equation (A2):

$$c_2 = \frac{1}{\log(t_a/t_{ref})} \tag{A2}$$

This equation shows that c_2 is independent of stress/strain, and it can be obtained through creep rupture test.

In addition, creep rupture also occurs at the reference cyclic time, where $t = t_{ref}$, then Equation (A1) can be deduced to Equation (A3):

$$c_1(\sigma) = \frac{1}{T_R - T_{ref}} \tag{A3}$$

The Manson-Haferd Parameter shows the gradients of T vs. $\log t$ curves at different stresses, and keeps constant at one specific stress. The Manson–Haferd Parameter can be regarded as a function of stress $P_{MH}(\sigma)$. When the stress is converted into plastic strain through Ramberg-Osgood relation, the T vs. $\log t$ curves at different strains become nonlinear, and the Manson-Haferd Parameter is presented as the gradient of tangents at one specific strain and temperature. In this case, the Manson-Haferd Parameter can be shown as a function of strain and temperature $P_{MH}(\varepsilon_p, T)$.

Invoking the Manson–Haferd Parameter (Equation (8)) and Equations (A2) and (A3) can be reduced to Equation (A5):

$$c_1(\sigma) = -\frac{c_2}{P_{MH}(\sigma)} = -\frac{c_2}{P_{MH}(\varepsilon_p, T)} \tag{A4}$$

Then, the c function can be defined through Equations (A2) and (A4):

$$c(T, t, \varepsilon_p) = 1 + \frac{T - T_{ref}}{P_{MH}(\varepsilon_p, T)\Delta\log(t_a/t_{ref})} - \frac{\log(t/t_{ref})}{\log(t_a/t_{ref})} \tag{A5}$$

If plastic strain is expressed as C_0' when $N = 1$ and $t = t_{ref}$, Equation (14) can be presented as:

$$C_0[1 - c_1(C_0')(T - T_{ref})] = \varepsilon_p(T, t_{ref}, N = 1) = C_0' \tag{A6}$$

Because C_0' can be obtained from a single fatigue test performed at an arbitrary temperature and reference cyclic time, C_0 can be solved numerically. Meanwhile, β_0 also can be evaluated from this single fatigue test.

A2. The Coefficients in the Power-Law Unified Creep-Fatigue Equation

In Equation (18), the coefficient c_2 can be evaluated through Equation (A2).

When $N = 1$, Equation (17) can be represented as:

$$\varepsilon_p\left(T, t,\ N = 1\right) = C_0\left[1 - c_1\left(T - T_{\text{ref}}\right) - c_2\log\left(t/t_{\text{ref}}\right)\right] \tag{A7}$$

The coefficient C_0 can be extracted through performing fatigue tests at one arbitrary temperature (T_1) at t_{ref} and one arbitrary cyclic time (t_1). Substituting these fatigue data into Equation (A7) gives Equations (A8) and (A9), then C_0 can be solved.

$$\varepsilon_p\left(T, t_{\text{ref}},\ N = 1\right) = C_0\left[1 - c_1\left(T_1 - T_{\text{ref}}\right)\right] \tag{A8}$$

$$\varepsilon_p\left(T, t_1,\ N = 1\right) = C_0\left[1 - c_1\left(T - T_{\text{ref}}\right) - c_2\log\left(t_1/t_{\text{ref}}\right)\right] \tag{A9}$$

The same fatigue experiments performed for the magnitude of C_0 can be used to extract $\beta_0 b_2$ through Equations (A10) and (A11)

$$\beta_0 b\left(T_1,\ t_{\text{ref}}\right) = \beta_0\left[1 - b_1\left(T_1 - T_{\text{ref}}\right)\right] \tag{A10}$$

$$\beta_0 b\left(T_1,\ t_1\right) = \beta_0\left[1 - b_1\left(T_1 - T_{\text{ref}}\right) - b_2\log\left(t_1/t_{\text{ref}}\right)\right] \tag{A11}$$

Next, another set of fatigue testes at another arbitrary temperature (T_2) at t_{ref} is performed to obtain coefficient c_1. When $N = 1$ and $t = t_{\text{ref}}$, Equation (17) can be represented as:

$$\varepsilon_p\left(T, t_{\text{ref}},\ N = 1\right) = C_0\left[1 - c_1\left(T_2 - T_{\text{ref}}\right)\right] \tag{A12}$$

The combination of Equation (A8) and (A12) gives $-C_0c_1$, then c_1 can be solved.

In addition, when $T = T_2$ and $t = t_{\text{ref}}$, the exponent of the power-law unified creep-fatigue equation can be represented as:

$$\beta_0 b\left(T,\ t\right) = \beta_0\left[1 - b_1\left(T_2 - T_{\text{ref}}\right)\right] \tag{A13}$$

The combination of Equations (A10) and (A13) gives β_0 and b_1. Then β_0 can be used to solve b_2.

References

1. Holmström, S.; Pohja, R.; Nurmela, A.; Moilanen, P.; Auerkari, P. Creep and creep-fatigue behaviour of 316 stainless steel. *Procedia Eng.* **2013**, *55*, 160–164. [CrossRef]
2. Basquin, O. The Exponential Law of Endurance Tests. *Am. Soc. Test. Mater. Proc.* **1910**, *10*, 625–630.
3. Coffin, L.F., Jr. *A Study of the Effects of Cyclic Thermal Stresses on a Ductile Metal*; Knolls Atomic Power Lab.: Niskayuna, NY, USA, 1953.
4. Manson, S.S. *Behavior of Materials under Conditions of Thermal Stress*; National Advisory Committee for Aeronautics: Cleveland, OH, USA, 1954.
5. Feltner, C.E.; Morrow, J.D. Microplastic strain hysteresis energy as a criterion for fatigue fracture. *J. Basic Eng.* **1961**, *83*, 15–22. [CrossRef]
6. Morrow, J. Cyclic plastic strain energy and fatigue of metals. In *Internal Friction, Damping, and Cyclic Plasticity*; ASTM International: West Conshohocken, PA, USA, 1965.
7. Dowling, N.E. *Mechanical Behavior of Materials: Engineering Methods for Deformation, Fracture, and Fatigue*; Prentice Hall: Upper Saddle River, NJ, USA, 1993.
8. Ramberg, W.; Osgood, W.R. *Description of Stress-Strain Curves by Three Parameters*; National Advisory Committee for Aeronautics: Cleveland, OH, USA, 1943.
9. Palmgren, A. Die lebensdauer von kugellagern. *ZVDI* **1924**, *68*, 339–341.
10. Miner, M.A. Cumulative damage in fatigue. *J. Appl. Mech.* **1945**, *12*, 159–164.
11. Chopra, O.K. *Environmental Effects on Fatigue Crack Initiation in Piping and Pressure Vessel Steels*; Argonne National Lab.: Argonne, IL, USA, 2000.

12. Gosselin, S.R.; Deardorff, A.F.; Peltola, D.W. Fatigue assessments in operating nuclear power plants. In *Changing Priorities of Codes and Standards: Failure, Fatigue, and Creep. Pvp-vol. 286*; American Society of Mechanical Engineers: New York, NY, USA, 1994.

13. Rodabaugh, E. *Comparisons of Asme-Code Fatigue-Evaluation Methods for Nuclear Class 1 Piping with Class 2 or 3 Piping*; Rodabaugh (EC) and Associates: Hilliard, OH, USA, 1983.

14. Rudolph, J.; Heinz, B.; Jouan, B.; Bergholz, S. *Areva Fatigue Concept—A Three Stage Approach to the Fatigue Assessment of Power Plant Components*; INTECH Open Access Publisher: Rijeka, Croatia, 2012.

15. Zhu, Y.; Li, X.; Wang, C.; Gao, R. A new creep-fatigue life model of lead-free solder joint. *Microelectron. Reliab.* **2015**, *55*, 1097–1100. [CrossRef]

16. Richart, F.; Newmark, N. *An Hypothesis for the Determination of Cumulative Damage in Fatigue*; Selected Papers By Nathan M. Newmark@ sCivil Engineering Classics; ASCE: Reston, VA, USA, 1948; pp. 279–312.

17. Manson, S. Interfaces between fatigue, creep, and fracture. *Int. J. Fract. Mech.* **1966**, *2*, 327–363. [CrossRef]

18. Gosselin, S. *Fatigue Crack Flaw Tolerance in Nuclear Power Plant Piping: A Basis for Improvements to ASME Code Section XI Appendix L*; US Nuclear Regulatory Commission, Office of Nuclear Regulatory Research, Division of Fuel, Engineering and Radiological Research: Washington, DC, USA, 2007.

19. Rudolph, J.; Bergholz, S.; Willuweit, A.; Vormwald, M.; Bauerbach, K. Methods of detailed thermal fatigue evaluation of nuclear power plant components. *Mater. Werkst.* **2011**, *42*, 1082–1092. [CrossRef]

20. Ainsworth, R.; Ruggles, M.; Takahashi, Y. Flaw assessment procedure for high-temperature reactor components. *J. Press. Vessel Technol.* **1992**, *114*, 166–170. [CrossRef]

21. Paris, P.; Erdogan, F. A critical analysis of crack propagation laws. *J. Basic Eng.* **1963**, *85*, 528–533. [CrossRef]

22. Wong, E.; Mai, Y.-W. A unified equation for creep-fatigue. *Int. J. Fatigue* **2014**, *68*, 186–194. [CrossRef]

23. Shi, X.; Pang, H.; Zhou, W.; Wang, Z. Low cycle fatigue analysis of temperature and frequency effects in eutectic solder alloy. *Int. J. Fatigue* **2000**, *22*, 217–228. [CrossRef]

24. Wong, E.H.; Liu, D. The unified equations for creep-fatigue—Deriving creep function from creep-rupture parameters. *Int. J. Fatigue* submitted for publication. **2016**.

25. Kanazawa, K.; Yoshida, S. Effect of Temperature and Strain Rate on the High Temperature Low-Cycle Fatigue Behavior of Austenitic Stainless Steels. In Proceedings of the International Conference on Creep and Fatigue in Elevated Temperature Applications, Philadelphia, PA, USA, 23–27 September 1973; Institution of Mechanical Engineers: London, UK, 1975; Volume 1.

26. High Temperature Characteristics of Stainless Steels. Available online: https://www.nickelinstitute.org/~/Media/Files/TechnicalLiterature/High_TemperatureCharacteristicsofStainlessSteel_9004_.pdf (accessed on 8 September 2016).

27. Orr, R.L.; Sherby, O.D.; Dorn, J.E. *Correlations of Rupture Data for Metals at Elevated Temperatures*; DTIC Document: Fort Belvoir, VA, USA, 1953.

28. Larson, F.R.; Miller, J. *A Time-Temperature Relationship for Rupture and Creep Stresses*; Trans ASME: New York, NY, USA, 1952; pp. 765–771.

29. Manson, S.; Haferd, A. *A Linear Time-Temperature Relation for Extrapolation of Creep and Stress Rupture Data*; NaCA TN 2890; Lewis Flight Propulsion Laboratory Cleveland: Cleveland, OH, USA, 1953.

30. Penny, R.K.; Mariott, D.L. *Design for Creep*; Chapman & Hall: London, UK, 1995.

31. Engineering Virtual Organization for CyberDesign. 316 Stainless Steel. Available online: https://icme.hpc.msstate.edu/mediawiki/index.php/316_Stainless_Steel (accessed on 6 September 2016).

32. Halford, G. Cyclic creep-rupture behavior of three high-temperature alloys. *Metall. Trans.* **1972**, *3*, 2247–2256. [CrossRef]

33. Jaske, C.; Mindlin, H.; Perrin, J. *Development of Elevated Temperature Fatigue Design Information for Type 316 Stainless Steel*; Battelle Columbus Labs.: Columbus, OH, USA, 1975.

34. Coffin, L., Jr. *Predictive Parameters and Their Application to High Temperature, Low Cycle Fatigue*; ICF2: Brighton, UK, 1969.

metals

MDPI

Article

Effect of Pre-Fatigue on the Monotonic Deformation Behavior of a Coplanar Double-Slip-Oriented Cu Single Crystal

Xiao-Wu Li [1,2,*], Xiao-Meng Wang [1], Ying Yan [1], Wei-Wei Guo [1] and Cheng-Jun Qi [1]

[1]   Department of Materials Physics and Chemistry, School of Materials Science and Engineering, Northeastern University, No. 3-11, Wenhua Road, Shenyang 110819, China; wxm_1314@yahoo.com (X.-M.W.); yingyan@imp.neu.edu.cn (Y.Y.); gww_2016@sina.com (W.-W.G.); kakaqi@yahoo.com (C.-J.Q.)

[2]   Key Laboratory for Anisotropy and Texture of Materials, Ministry of Education, Northeastern University, No. 3-11, Wenhua Road, Shenyang 110819, China

*   Correspondence: xwli@mail.neu.edu.cn; Tel.: +86-24-8367-8479

Academic Editor: Filippo Berto
Received: 4 October 2016; Accepted: 17 November 2016; Published: 22 November 2016

Abstract: The $[\bar{2}33]$ coplanar double-slip-oriented Cu single crystals were pre-fatigued up to a saturation stage and then uniaxially tensioned or compressed to fracture. The results show that for the specimen pre-fatigued at a plastic strain amplitude γ_{pl} of 9.2×10^{-4}, which is located within the quasi-plateau of the cyclic stress-strain (CSS) curve, its tensile strength and elongation are coincidently improved, showing an obvious strengthening effect by low-cycle fatigue (LCF) training. However, for the crystal specimens pre-fatigued at a γ_{pl} lower or higher than the quasi-plateau region, due to a low pre-cyclic hardening or the pre-induction of fatigue damage, no marked strengthening effect by LCF training occurs, and even a weakening effect by LCF damage takes place instead. In contrast, the effect of pre-fatigue deformation on the uniaxial compressive behavior is not so significant, since the compressive deformation is in a stress state more beneficial to the ongoing plastic deformation and it is insensitive to the damage induced by pre-cycling. Based on the observations and comparisons of deformation features and dislocation structures in the uniaxially deformed $[\bar{2}33]$ crystal specimens which have been pre-fatigued at different γ_{pl}, the micro-mechanisms for the effect of pre-fatigue on the static mechanical behavior are discussed.

Keywords: Cu single crystal; coplanar double slip; pre-fatigue deformation; plastic strain amplitude; tensile behavior; compressive behavior; dislocation structure

1. Introduction

The mechanical degradation of materials will gradually occur during the process of their service due to fatigue deformation and damage accumulation. Therefore, it is extremely significant to explore the influencing mechanisms of pre-fatigue deformation on the mechanical properties of materials. Recently, there have been some investigations on the effect of pre-fatigue on the uniaxial deformation behavior of some metallic polycrystals, e.g., 6061-T6 aluminum alloy [1,2], AISI 4140T steel [1,2], 304 austenitic stainless steel [3], and commercially pure Al [4,5]. However, the microstructural complexities in those polycrystalline materials will undoubtedly preclude a distinct elucidation of the micro-mechanism for the pre-fatigue effect on the static mechanical behavior. In this sense, single crystals of high-purity metals would actually be more suitable materials for such investigations.

Quite recently, we investigated the effect of fatigue pre-deformation at different plastic strain amplitudes, γ_{pl}, on the uniaxial tensile behavior of the [017] critical double-slip-oriented Cu single crystal, and found that a low-cycle fatigue training treatment on this crystal at an appropriate plastic strain amplitude ($\gamma_{pl} = 7.0 \times 10^{-4}$) could effectively improve its static tensile properties; this phenomenon is called the strengthening effect by low-cycle fatigue training [6]. Actually, the dislocation structures induced by fatigue deformation are quite different, strongly depending upon the crystallographic orientation (especially for double-slip orientations) [7–9]. As differently-oriented Cu single crystals are adopted for fatigue pre-deformation treatments, the induced dislocation structures are different; in this case, it is important to know how the corresponding static mechanical properties will change. In light of this, the $[\bar{2}33]$ coplanar double-slip-oriented Cu single crystal was selected for the present study, focusing on the effect of fatigue pre-deformation at different γ_{pl} on its uniaxial tensile and compressive properties. Our previous work has demonstrated that the cyclic stress-strain (CSS) curve of the $[\bar{2}33]$ coplanar double-slip-oriented Cu single crystal shows a stress quasi-plateau over the γ_{pl} range of 3.0×10^{-4}–2.0×10^{-3} [10], and the fatigue dislocation structures consist mainly of dislocation cells, the size of which decreases with increasing γ_{pl} [11,12]. In this way, the effect of dislocation cell structures induced by pre-fatigue on the static mechanical properties of Cu single crystals can be exclusively examined.

2. Experimental Section

Cu single crystal was grown from oxygen free high conductivity (OFHC) copper of 99.999% purity by the Bridgman method. The fatigue specimens with a gauge section of 7 mm × 5 mm × 16 mm and 70 mm in length were spark-machined from the as-grown crystal. All specimens are oriented towards $[\bar{2}33]$, and the Laue back-reflection technique was used to determine the orientation of the specimen. Before fatigue tests, all specimens were annealed at 800 °C for 2 h in vacuum, and then electro-polished to produce a strain-free and smooth surface. Symmetric tension-compression fatigue pre-deformation tests were performed at room temperature in air using a Shimadzu servo-hydraulic testing machine (Shimadzu, Kyoto, Japan). A triangular waveform signal with a frequency range of 0.05–0.4 Hz was used for the constant plastic strain control. The plastic resolved shear strain amplitude γ_{pl} and shear stress τ were calculated by $\gamma_{pl} = \Delta\varepsilon_p/2\Omega$ and $\tau = \sigma\Omega$, where $\Delta\varepsilon_p/2$ is the axial plastic strain amplitude, Ω the Schmid factor of primary slip system and σ is an average value of the peak stresses in tension and compression. The crystal specimens were firstly pre-cycled at different plastic strain amplitudes ranging from 1.3×10^{-4} to 5.3×10^{-3} up to the occurrence of saturation. The pre-fatigue testing conditions and data are listed in Table 1. After pre-fatigue tests, the fatigue specimens were spark-machined into the tensile specimens with a gauge section of 5 × 2 × 16 mm^3 and 70 mm in length. The specimens were then polished electrolytically and subjected to tensile deformation to a final fracture at a strain rate of $10^{-3} \cdot s^{-1}$. After tensile tests, the surface deformation characteristics of these specimens were observed by using a SSX-550 scanning electron microscope (SEM, Shimadzu, Kyoto, Japan, and the dislocation structures were observed by using a Tecnai G^2 20 transmission electron microscope (TEM, FEI, Hillsboro, OR, USA) operated at 200 kV. TEM thin foils were first spark-cut from the gauge part (homogeneous deformation region near the fracture) of the fractured specimens, then mechanically thinned down to dozens of micron thick and finally polished by a conventional twin-jet method.

Table 1. Pre-fatigue testing conditions and data for the $[\bar{2}33]$ Cu single crystal [10]. Note that γ_{pl} is the plastic strain amplitude, N is the cyclic number, $\gamma_{pl,cum}$ is the cumulative plastic strain (=$4N\gamma_{pl}$), and τ_s is the saturation stress.

γ_{pl}	N	$\gamma_{pl,cum}$	τ_s, MPa
1.3×10^{-4}	65,000	33.8	25.0
3.4×10^{-4}	32,000	43.5	30.1
9.2×10^{-4}	12,000	43.5	31.7
3.5×10^{-3}	4700	65.8	33.5
5.3×10^{-3}	10,200	216.2	35.4

3. Results and Discussion

3.1. Effect of Pre-Fatigue on the Tensile and Compressive Properties

Figure 1 shows the tensile and compressive stress-strain curves of [$\bar{2}33$] Cu single-crystal specimens unfatigued (only annealed) and pre-fatigued at different γ_{pl}, and the CSS curve of this oriented crystal [10] is also reproduced here to clearly show the location of the applied γ_{pl} in the curve. It is apparent that there is an effect of pre-cycling, to a different extent, on the tensile or compressive yield strength and the ultimate tensile strength of the crystal, depending on the γ_{pl} imposed in pre-cycling.

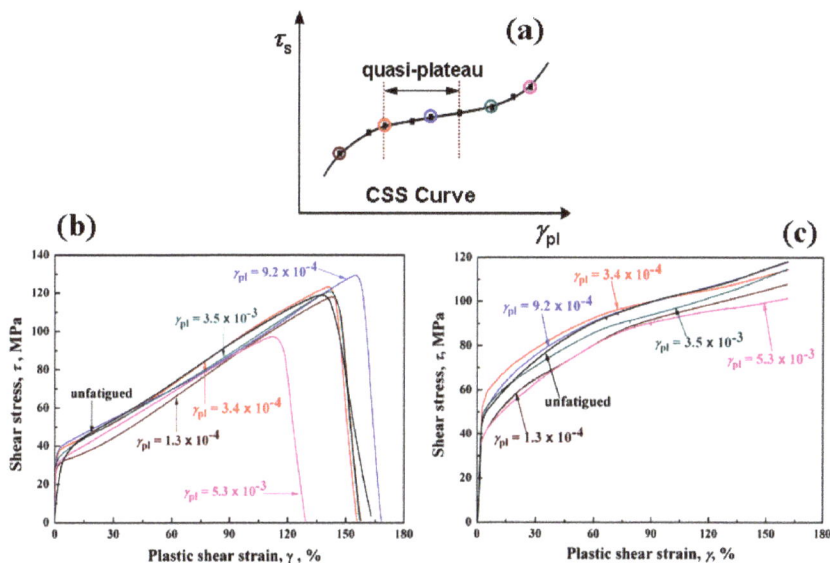

Figure 1. (a) CSS curve of [$\bar{2}33$] Cu single crystals [10]; (b) the tensile and (c) compressive stress-strain curves of [$\bar{2}33$] Cu single crystals unfatigued and pre-fatigued at different plastic strain amplitudes.

From Figure 1b it can be seen that, compared to the unfatigued specimen, the yield strength slightly increases but the tensile strength and elongation basically remain unchanged for the specimens pre-fatigued at two low γ_{pl} of 1.3×10^{-4} and 3.4×10^{-4}, which are located below the quasi-plateau region in the CSS curve (Figure 1a). As the pre-fatigue strain amplitude, e.g., 9.2×10^{-4}, is exactly located in the quasi-plateau region in the CSS curve, both strength and elongation are coincidently improved. As γ_{pl} increases to 3.5×10^{-3}, which is beyond the quasi-plateau region in the CSS curve, the yield and tensile strengths and elongation become comparable to those of the unfatigued specimen. With further increasing γ_{pl} to 5.3×10^{-3}, both strength and elongation decrease obviously, and even become much lower than those of the unfatigued specimen. Accordingly, only when the crystal is pre-fatigued at an intermediate γ_{pl} of 9.2×10^{-4}, the strengthening effect due to low-cycle fatigue training is the most notable.

From Figure 1c one can see that the general rule for the effect of pre-cycling on the compressive mechanical properties (e.g., compressive yield strength and flow stress) is basically similar to the tensile case (Figure 1b); however, the difference in the compressive properties of the specimens pre-cycled at different γ_{pl} is not so noticeable as compared to the difference in tensile properties.

3.2. Deformation Features on the Surface

Figures 2 and 3 present the essential surface deformation features of the unfatigued and pre-fatigued [$\bar{2}33$] Cu single-crystal specimens under tensile and compressive deformation, respectively.

Only a few slip bands appear in some regions and distribute inhomogenously on the specimen surface after pre-cycling at $\gamma_{pl} \leq 3.4 \times 10^{-4}$ [10], indicating that strain hardening induced by pre-cycling is not significant in this case, so that the tensile properties of the specimens pre-fatigued at these low γ_{pl} are nearly not improved as compared to those of the unfatigued specimen (Figure 1b). Correspondingly, the tensile deformation features indicated by Figure 1a–c are almost similar. It should be mentioned that [$\bar{2}33$] is a coplanar-slip orientation, i.e., the primary and secondary slip planes are the same plane (111); thus only single-oriented slip bands can be generally observed on the surface under pre-fatigue and tensile deformation (Figure 2a–c).

As γ_{pl} increases to 9.2×10^{-4}, a few deformation concentration regions, i.e., deformation bands (DBs), are formed, and the formation of such preliminary DBs may cause a notable cyclic hardening but does not bring about an obvious damage [10], and therefore, the working hardening capacity during subsequent tension is enhanced, as evidenced by the co-operation of the conjugate slip system $(\bar{1}11)[011]$ under tensile deformation (Figure 3d). Therefore, the tensile strength has been improved. It should be mentioned here that it is unnecessary for the operation of the conjugate slip system to accommodate plastic strain during pre-fatigue deformation [10]. Furthermore, the inducement of such preliminary DBs by pre-cycling should be beneficial to the subsequent tensile plastic deformation along these concentrated bands, so the tensile ductility (or elongation) is also somewhat increased (Figure 1b).

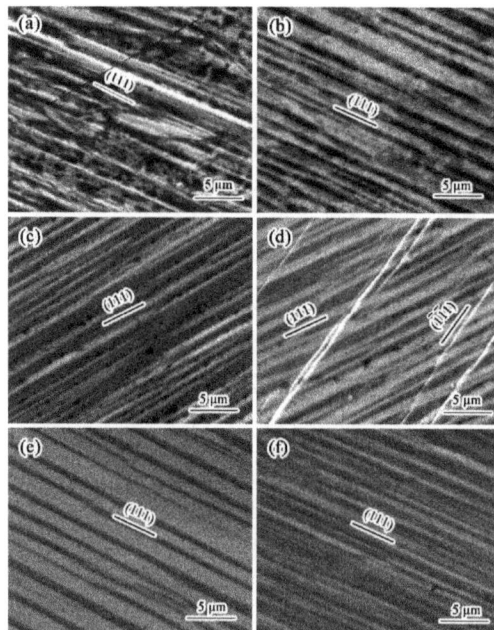

Figure 2. SEM images of the surface deformation features of unfatigued or pre-fatigued [$\bar{2}33$] Cu single crystals under uniaxial tension: (**a**) unfatigued; (**b–f**) pre-fatigued at $\gamma_{pl} = 1.3 \times 10^{-4}$, 3.4×10^{-4}, 9.2×10^{-4}, 3.5×10^{-3}, and 5.3×10^{-3}, respectively. (**a,b,e,f**) viewed from ($\bar{6}\,9\,\bar{1}3$); and (**c,d**) viewed from ($6\,\bar{9}\,13$). Note that the loading direction is horizontal.

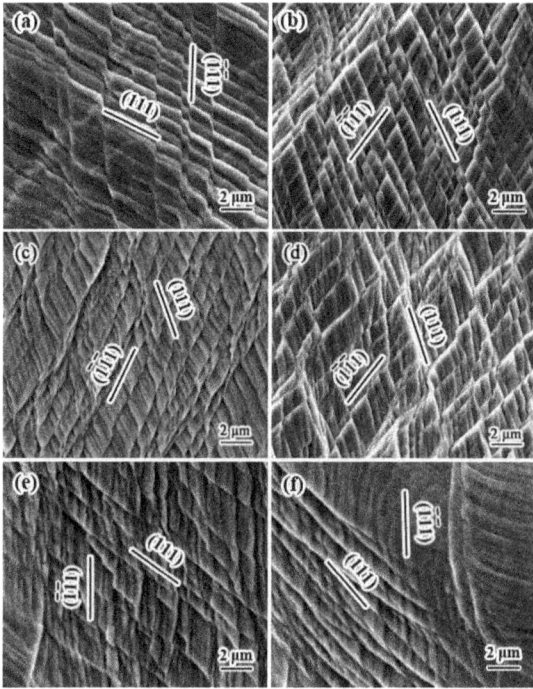

Figure 3. SEM images of the surface deformation features of unfatigued or pre-fatigued $[\overline{2}33]$ Cu single crystals under uniaxial compression: (**a**) unfatigued, (**b–f**) pre-fatigued at $\gamma_{pl} = 1.3 \times 10^{-4}$, 3.4×10^{-4}, 9.2×10^{-4}, 3.5×10^{-3}, and 5.3×10^{-3}, respectively. (**a,e,f**) viewed from $(\overline{6}\ 9\ \overline{13})$; and (**b–d**) viewed from (320). Note that the loading direction is horizontal.

However, at high strain amplitudes ($\gamma_{pl} \geq 3.5 \times 10^{-3}$), many highly concentrated DBs are formed and severe fatigue damage has been introduced [10], so that the subsequent tensile properties cannot be effectively enhanced. In particular, when γ_{pl} is as high as 5.3×10^{-3}, the induced permanent damage in the form of densely distributed DBs becomes extremely notable [10], thus causing a greatly detrimental effect on the subsequent tensile properties, i.e., the strength and ductility decrease obviously to much lower values than those of other specimens. Correspondingly, the density of slip bands is relatively low and, also, the conjugate slip system $(\overline{1}\overline{1}1)[011]$ is not activated (Figure 2e,f) after the tensile deformation of those specimens pre-fatigued at high γ_{pl}.

Compared to the case of tensile deformation, the stress state under compressive deformation is much more beneficial to the continuation of plastic flow deformation. Therefore, the secondary (conjugate) slip system $(\overline{1}\overline{1}1)[011]$ is commonly activated under compression (Figure 3). Furthermore, the subsequent compressive deformation behavior is not so sensitive to the damage induced by pre-cycling, so that the deformation features are basically similar for all the specimens unfatigued or pre-fatigued at different γ_{pl}, as shown in Figure 3, and no marked difference in the relevant compressive stress-strain curves is manifested (Figure 1c).

3.3. Dislocation Structures

Figure 4 shows the tensile dislocation structures of $[\overline{2}33]$ Cu single crystals unfatigued or pre-fatigued at different γ_{pl}, where **g** is the diffraction vector and **B** is the incident beam. It is found that the dislocation structures of those specimens unfatigued or pre-fatigued at low γ_{pl} of 1.3×10^{-4} and 3.4×10^{-4} are typically featured by dislocation cells (Figure 4a–c); however, for pre-fatigued specimens, the cell walls are relatively loose and the average cell diameter is ~0.5 μm (as representatively shown in Figure 4b,c), indicating that the plastic strain accommodated during the tensile process of specimens pre-fatigued at low γ_{pl} is not so high, and their strain-hardening abilities are thus basically equivalent to that of the unfatigued specimen (Figure 1b). As the pre-fatigue strain amplitude γ_{pl} increases to 9.2×10^{-4}, the dislocation density of the cell walls becomes obviously increased, and the cell diameter reduces to ~0.3 μm (Figure 4d), showing a stronger strain-hardening ability of the specimen pre-fatigued at such a γ_{pl} (Figure 1b). As γ_{pl} further increases to higher values of 3.5×10^{-3} or 5.3×10^{-3}, the dislocation cells become loose again, and the average cell diameter is increased to 0.6–0.7 μm (Figure 4e,f). Our previous work [11,12] has revealed that the fatigue dislocation structures of $[\bar{2}33]$ Cu single crystals were mainly composed of dislocation cells, the size of which decreased with increasing γ_{pl}. Moreover, as $\gamma_{pl} \geq 3.5 \times 10^{-3}$, the configuration of fatigue-induced dislocation cells exhibited a characteristic of a persistent slip band (PSB) ladder-like structure, and these dislocation cells were thus called PSB cells, showing a severe strain concentration and deformation damage [11,12]. In this way, during the subsequent tensile deformation, the serious strain concentration will easily occur at these PSB cells, thus leading to an earlier fracture failure at a lower tensile flow stress and strain levels (Figure 1b). In general, as shown in Figures 1b and 4, the strength values increase with the decreasing cell diameter; this phenomenon is consistent with the general relation of saturation stress vs. dislocation cell size for fatigued Cu single crystals [13].

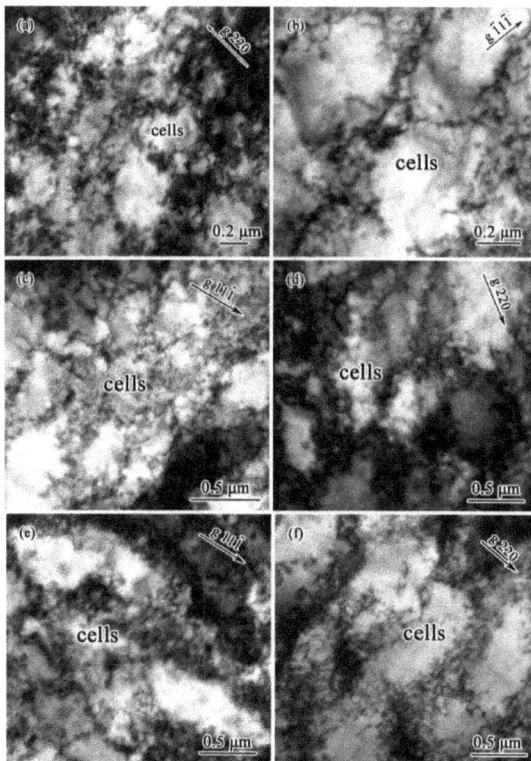

Figure 4. TEM images of dislocation configurations of unfatigued or pre-fatigued $[\bar{2}33]$ Cu single crystals under uniaxial tension: (**a**) unfatigued, **B** = $\bar{1}13$; pre-fatigued at (**b**) γ_{pl} = 1.3×10^{-4}, **B** = $\bar{1}12$; (**c**) γ_{pl} = 3.4×10^{-4}, **B** = 011; (**d**) γ_{pl} = 9.2×10^{-4}, **B** = $\bar{1}11$; (**e**) γ_{pl} = 3.5×10^{-3}, **B** = 011; and (**f**) γ_{pl} = 5.3×10^{-3}, **B** = $\bar{1}11$.

Figure 5 shows the compressive dislocation structures of $[\bar{2}33]$ Cu single crystals unfatigued or pre-fatigued at different γ_{pl}. The dislocation structure of the unfatigued specimen consists of equiaxed dislocation cells (Figure 5a). For the pre-fatigued specimen at $\gamma_{pl} = 1.3 \times 10^{-4}$, the dislocation cells become very loose (Figure 5b), and the relevant compressive flow stress is low (Figure 1c). As γ_{pl} increases to 3.4×10^{-4} and 9.2×10^{-4}, the dislocation density of cell walls increases greatly, and the compressive plastic flow undergoes relatively high stress levels (Figure 1c). It means that the strain-hardening ability can be somewhat improved after the $[\bar{2}33]$ Cu single crystal was pre-fatigued at these two γ_{pl}, although such an improvement is not as obvious as the case of tensile deformation. As γ_{pl} increases to 3.5×10^{-3}, the dislocation structure is featured by some loose dislocation cells (Figure 5e), corresponding to a decreased compressive flow stress (Figure 1c). As $\gamma_{pl} = 5.3 \times 10^{-3}$, the compressive dislocation structure becomes regularly arranged equiaxed dislocation cells (Figure 5f); however, the compressive flow stress level is not high, but at the lowest level (Figure 1c). Therefore, it is inferred that some of the equiaxed dislocation cells induced by pre-fatigue deformation [12] may be retained in the crystal subjected to subsequent compressive deformation, as shown in Figure 5f.

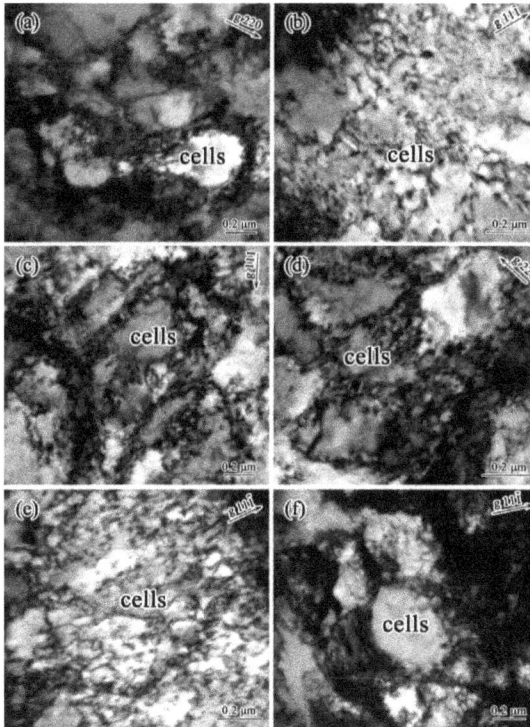

Figure 5. TEM images of dislocation configurations of unfatigued or pre-fatigued $[\bar{2}33]$ Cu single crystals under uniaxial compression: (a) unfatigued, $\mathbf{B} = \bar{1}11$; pre-fatigued at (b) $\gamma_{pl} = 1.3 \times 10^{-4}$, $\mathbf{B} = 011$; (c) $\gamma_{pl} = 3.4 \times 10^{-4}$, $\mathbf{B} = \bar{1}01$; (d) $\gamma_{pl} = 9.2 \times 10^{-4}$, $\mathbf{B} = \bar{1}11$; (e) $\gamma_{pl} = 3.5 \times 10^{-3}$, $\mathbf{B} = 011$; and (f) $\gamma_{pl} = 5.3 \times 10^{-3}$, $\mathbf{B} = 011$.

As is well known, the equal-channel angular pressing (ECAP) technique has recently become one of the most frequently used mechanical processing methods to produce ultrafine-grained (UFG) materials by introducing intensive plastic strain into materials after repetitive pressing [14,15]. It should be mentioned that the term "ultrafine-grained", widely used by the ECAP community, sometimes (e.g., UFG pure Cu) represents a microstructure comprising very fine, highly misoriented dislocation cells or subgrains [16–19]. Such ultrafine-celled (the cell diameter is around some hundred nanometers) microstructures produced by the ECAP severe plastic deformation technique normally exhibit an extraordinarily high strength but are followed by a great loss of ductility, although a subsequent, controlled short-term annealing treatment might cause a joint enhancement of strength and ductility [20]. In contrast, the cell structure developed under fatigue deformation is well recognized to be a typical type of heterogeneous dislocation distribution, and high-symmetry orientations, low friction stress, large deformation and the easy operation of cross slip favor the formation of dislocation cell structures [21]. For the present [$\bar{2}$33] orientation, dislocation reactions between the primary and coplanar slip systems produce the third Burgers vector contained in the same (111) plane, i.e., $(a/2)[\bar{1}01] + (a/2)[1\bar{1}0] \rightarrow (a/2)[0\bar{1}1]$. The produced dislocation $(a/2)[0\bar{1}1]$ is also glissile in the common primary slip plane (111). Such a dislocation reaction would promote the much easier formation of slightly misoriented dislocation cells during fatigue deformation [11,22], and the cell diameter is also around some hundred nanometers depending upon the applied strain amplitude [11,12]. Apparently, the dislocation cell structures directly produced by fatigue cycling are more uniformly distributed and the dislocation density is relatively lower compared with UFG microstructures. Accordingly, under the prerequisite that no obvious pre-damage is introduced, the inducement of such a uniformly distributed low-dislocation-density cell structure through an appropriate pre-fatigue deformation treatment may, more or less, improve the static strength and ductility together, e.g., pre-fatigue at $\gamma_{pl} = 9.2 \times 10^{-4}$ in the present work. This phenomenon is normally called strengthening by low-cycle fatigue (LCF) training.

4. Conclusions

The effect of pre-fatigue deformation at different plastic strain amplitudes on the monotonic tensile and compressive deformation behavior of the [$\bar{2}$33] coplanar double-slip-oriented Cu single crystal was experimentally investigated. The following conclusions can be drawn:

(1) As the [$\bar{2}$33] crystal specimen is pre-fatigued at a plastic strain amplitude γ_{pl} of 9.2×10^{-4}, which is located within the quasi-plateau of the CSS curve, its tensile strength and ductility can be coincidently improved, showing an obvious strengthening effect by LCF training, since uniformly distributed dislocation cell structures have been reasonably introduced by pre-fatigue cycling. In contrast, for the crystal specimens pre-fatigued at γ_{pl} lower or higher than the quasi-plateau region, due to a low pre-cyclic hardening or pre-induction of fatigue damage, there is no marked strengthening effect by LCF training, and a weakening effect by LCF damage occurs.

(2) The general rule for the effect of pre-fatigue deformation on the uniaxial compressive behavior follows a similar pattern as the case of uniaxial tension, but such an effect is not so significant, since the stress state under compressive deformation is much more beneficial to the continuation of plastic flow deformation and the compressive deformation behavior is insensitive to damage induced by pre-cycling.

Acknowledgments: This work was financially supported by the National Natural Science Foundation of China (NSFC) under Grant Nos. 51271054, 51231002 and 51571058.

Author Contributions: X.-W.L. designed the scope of the paper. X.-M.W., W.-W.G. and C.-J.Q. participated in the sample preparation and all experiments. X.-W.L., Y.Y. and X.-M.W. analyzed the experimental results. X.-W.L. wrote this paper.

Conflicts of Interest: The authors declare no conflict of interest.

References

1. Santana, U.S.; Gonzalez, C.R.; Mesmacque, G.; Amrouche, A.; Decoopman, X. Dynamic tensile behavior of materials with previous fatigue damage. *Mater. Sci. Eng. A* **2008**, *497*, 51–60. [CrossRef]
2. Santana, U.S.; Gonzalez, C.R.; Mesmacque, G.; Amrouche, A. Effect of fatigue damage on the dynamic tensile behavior of 6061-T6 aluminum alloy and AISI 4140T steel. *Int. J. Fatigue* **2009**, *31*, 1928–1937. [CrossRef]

3. Ye, D.Y.; Xu, Y.D.; Xiao, L.; Cha, H.B. Effects of low-cycle fatigue on static mechanical properties, microstructures and fracture behavior of 304 stainless steel. *Mater. Sci. Eng. A* **2010**, *527*, 4092–4102. [CrossRef]
4. Yan, Y.; Lu, M.; Li, X.W. Effects of pre-fatigue deformation on the uniaxial tensile behavior of coarse-grained pure Al. *Acta Metall. Sin.* **2013**, *49*, 658–666. (In Chinese) [CrossRef]
5. Yan, Y.; Lu, M.; Guo, W.W.; Li, X.W. Effect of pre-fatigue deformation on the thickness-dependent tensile behavior of coarse-grained pure aluminum sheets. *Mater. Sci. Eng. A* **2014**, *600*, 99–107. [CrossRef]
6. Li, X.W.; Wang, X.M.; Guo, W.W.; Qi, C.J.; Yan, Y. Effect of Cyclic pre-deformation on the uniaxial tensile deformation behavior of [017] copper single crystals oriented for critical double slip. *Metall. Mater. Trans. A* **2013**, *44*, 1631–1635. [CrossRef]
7. Li, X.W.; Zhang, Z.F.; Wang, Z.G.; Li, S.X.; Umakoshi, Y. SEM-ECC Investigation of dislocation arrangements in cyclically deformed copper single crystals with different crystallographic orientations. *Defect Diffus. Forum* **2001**, *188–190*, 153–170. [CrossRef]
8. Li, X.W.; Umakoshi, Y.; Gong, B.; Li, S.X.; Wang, Z.G. Dislocation structure in fatigued critical and conjugate double-slip-oriented copper single crystals. *Mater. Sci. Eng. A* **2002**, *333*, 51–59. [CrossRef]
9. Li, X.W. Monotonic and cyclic deformation in single crystal [A]. In *The Encyclopedia of Tribology*; Wang, Q.J., Chung, Y.-W., Eds.; Springer Science+Business Media: New York, NY, USA, 2013; Chapter 248; Volume 260, pp. 2313–2323.
10. Li, X.W.; Wang, Z.G.; Li, S.X. Cyclic deformation behavior of double-slip-oriented copper single crystals I. Coplanar double slip orientation on 011-$\bar{1}$11 side in the stereographic triangle. *Mater. Sci. Eng. A* **1999**, *260*, 132–138. [CrossRef]
11. Li, X.W.; Wang, Z.G.; Zhang, Y.W.; Li, S.X.; Umakoshi, Y. Dislocation structure in cyclically deformed coplanar double-slip-oriented copper single crystals. *Phys. Status Solidi* **2002**, *191*, 97–105. [CrossRef]
12. Zhou, Y.; Li, X.W.; Yang, Q.R. Study of Fatigue dislocation structures in [$\bar{2}$33] coplanar double-slip-oriented copper single crystals using SEM electronic channelling contrast. *Int. J. Mater. Res.* **2008**, *99*, 958–963. [CrossRef]
13. Suresh, S. *Fatigue of Materials*, 2nd ed.; Cambridge University Press: London, UK, 1998.
14. Valiev, R.Z.; Islamgaliev, R.K.; Alexandrov, I.V. Bulk nanostructured materials from severe plastic deformation. *Prog. Mater. Sci.* **2000**, *45*, 103–189. [CrossRef]
15. Valiev, R.Z. Materials science: Nanomaterial advantage. *Nature* **2002**, *419*, 887–889. [CrossRef] [PubMed]
16. Agnew, S.R.; Weertman, J.R. Cyclic softening of ultrafine grain copper. *Mater. Sci. Eng. A* **1998**, *244*, 145–153. [CrossRef]
17. Torre, F.D.; Lapovok, R.; Sandlin, J.; Thomson, P.F.; Davies, C.H.J.; Pereloma, E.V. Microstructures and properties of copper processed by equal channel angular extrusion for 1–16 passes. *Acta Mater.* **2004**, *52*, 4819–4832. [CrossRef]
18. Kong, M.K.; Kao, W.P.; Lui, J.T.; Chang, C.P.; Kao, P.W. Cyclic deformation of ultrafine-grained aluminium. *Acta Mater.* **2007**, *55*, 715–725.
19. Jiang, Q.W.; Li, X.W. Effect of pre-annealing treatment on the compressive deformation and damage behavior of ultrafine-grained copper. *Mater. Sci. Eng. A* **2012**, *546*, 59–67. [CrossRef]
20. Mughrabi, H.; Höppel, H.W.; Kautz, M.; Valiev, R.Z. Annealing treatments to enhance thermal and mechanical stability of ultrafine-grained metals produced by severe plastic deformation. *Z. Met.* **2003**, *94*, 1079–1083. [CrossRef]
21. Mughrabi, H. Dislocation wall and cell structures and long-range internal stresses in deformed metal crystals. *Acta Metall.* **1983**, *31*, 1367–1379. [CrossRef]
22. Jin, N.Y.; Winter, A.T. Cyclic deformation of copper single crystals oriented for double slip. *Acta Metall.* **1984**, *32*, 989–995. [CrossRef]

Article

Evaluation of Methods for Estimation of Cyclic Stress-Strain Parameters from Monotonic Properties of Steels

Tea Marohnić *, Robert Basan and Marina Franulović

Faculty of Engineering, University of Rijeka, Vukovarska 58, HR-51000 Rijeka, Croatia;
robert.basan@riteh.hr (R.B.); marina.franulovic@riteh.hr (M.F.)
* Correspondence: tmarohnic@riteh.hr; Tel.: +385-51-651-536

Academic Editor: Filippo Berto
Received: 11 November 2016; Accepted: 30 December 2016; Published: 7 January 2017

Abstract: Most existing methods for estimation of cyclic yield stress and cyclic Ramberg-Osgood stress-strain parameters of steels from their monotonic properties were developed on relatively modest number of material datasets and without considerations of the particularities of different steel subgroups formed according to their chemical composition (unalloyed, low-alloy, and high-alloy steels) or delivery, i.e., testing condition. Furthermore, some methods were evaluated using the same datasets that were used for their development. In this paper, a comprehensive statistical analysis and evaluation of existing estimation methods were performed using an independent set of experimental material data comprising 116 steels. Results of performed statistical analyses reveal that statistically significant differences exist among unalloyed, low-alloy, and high-alloy steels regarding their cyclic yield stress and cyclic Ramberg-Osgood stress-strain parameters. Therefore, estimation methods were evaluated separately for mentioned steel subgroups in order to more precisely determine their applicability for the estimation of cyclic behavior of steels belonging to individual subgroups. Evaluations revealed that considering all steels as a single group results in averaging and that subgroups should be treated independently. Based on results of performed statistical analysis, guidelines are provided for identification and selection of suitable methods to be applied for the estimation of cyclic stress-strain parameters of steels.

Keywords: estimation methods; monotonic properties; cyclic stress-strain parameters; Ramberg-Osgood; steel grouping; statistical analysis

---

1. Introduction

Development of computer technology and CAE software solutions have enabled performing complex simulations of material and product behavior under cyclic loading and fatigue life determination already during early stages of product development process. In recent years, rapid increases in computing power and availability of distributed and cloud-based resources have been seen. Within short time frame, complex simulations can be run for multiple materials. An example of these are strain-based, i.e., local strain-life fatigue, analyses which have been widely adopted in automotive, aeronautic, and power industry for fatigue life predictions of highly-loaded steel and aluminium components [1,2]. In order to perform these analyses, both cyclic stress-strain and strain-life fatigue curves and parameters that define them must be known. Well-accepted and widely used representation of stress-strain response of the majority of metallic materials is the cyclic Ramberg-Osgood (R-O) equation [3,4]:

$$\frac{\Delta\varepsilon}{2} = \frac{\Delta\varepsilon_e}{2} + \frac{\Delta\varepsilon_p}{2} = \frac{\Delta\sigma}{2E} + \left(\frac{\Delta\sigma}{2K'}\right)^{\frac{1}{n'}} \tag{1}$$

For determination of lifetime in both low-cycle and high-cycle regime, Coffin-Manson-Basquin (C-M-B) [4,5] approach is applied:

$$\frac{\Delta\varepsilon}{2} = \frac{\Delta\varepsilon_e}{2} + \frac{\Delta\varepsilon_p}{2} = \frac{\sigma_f'}{E}(2N_f)^b + \varepsilon_f'(2N_f)^c \tag{2}$$

In Equations (1) and (2) $\Delta\varepsilon$, $\Delta\varepsilon_e$ and $\Delta\varepsilon_p$ are true total, elastic and plastic strain ranges, respectively, $\Delta\sigma$ is true stress range, E is Young's modulus, K' is cyclic strength coefficient, and n' is cyclic strain hardening exponent. Furthermore, σ_f', ε_f', b and c are fatigue strength and ductility parameters obtained from fully reversed tension-compression fatigue tests.

Cyclic R-O and C-M-B parameters obtained through material testing are most accurate, but are very often unavailable due to long time and high costs associated with experimental characterization. Existing test results which are available in literature and materials databases often do not sufficiently correspond to the actual material under consideration. Hence, it has become common practice to estimate cyclic R-O and C-M-B parameters of the material from its monotonic properties early during product development. For estimation of C-M-B fatigue parameters from monotonic properties of materials, many methods have been developed [5–8] and evaluated in literature [9]. To the contrary, for estimation of cyclic R-O parameters of materials, only a limited number of methods are proposed [10–13]. For estimations of cyclic yield stress R_e' and R-O parameters (K' and n') various monotonic properties and their combinations are used, with ultimate strength R_m and yield stress R_e being the most common since they are readily available. Detailed overview of monotonic properties used for estimation of cyclic parameters of metallic materials and systematic study of their relevance for estimation purposes is provided in [14].

No independent and systematic evaluations of these methods can be found in the literature, the only ones available being those performed in respective papers where methods were proposed.

The main aim of this paper is to provide detailed analysis and evaluation of existing methods for estimation of cyclic yield stress R_e' and cyclic stress-strain parameters K' and n' from monotonic properties. For this purpose, a large and independent set of material data was collected from relevant sources. Since previous investigations [15–17] confirmed that dividing steels into different subgroups might improve estimation accuracy, this will also be taken into consideration. One-way Analysis of Variance (one-way ANOVA) and post hoc Tukey's test will be performed in order to check whether individual steel groups are statistically different regarding their cyclic parameters R_e', K' and n'. If such differences are confirmed to exist, in addition to evaluation of existing methods for all steels together, partial evaluations for each steel subgroup will be performed as well.

2. Overview of Existing Methods for Estimation of Cyclic Stress-Strain Parameters

2.1. Methods for Estimation of Cyclic Yield Stress R_e'

Li et al. [11] originally proposed estimation of cyclic yield stress R_e' of steels from ultimate strength R_m and reduction of area at fracture RA. Equation (3) was developed using monotonic and cyclic properties of 27, mostly unalloyed and low-alloy steels:

$$R_e' = (1 + RA)R_m \left(-\frac{0.002}{\ln(1 - RA)}\right)^{0.16} \tag{3}$$

Evaluation of proposed expression is performed on the same data used for developing the method, and is reported that estimated values of R_e' deviate at most 14% from their experimental counterparts.

Lopez and Fatemi [12] developed a number of relationships between Brinell hardness *(HB* or monotonic properties and cyclic yield stress R_e' of steels. These were developed and validated on a relatively large number of steels consisting mostly of unalloyed and low-alloy steels, covering a wide variation of chemical composition and mechanical properties, with ultimate stress R_m ranging from 279 to 2450 MPa and hardness from 80 to 595 HB. Materials were divided according to ultimate strength to yield stress ratio R_m/R_e, since it was shown that such division improves the accuracy of cyclic parameters estimation. Ratio R_m/R_e was originally proposed by Smith et al. [18] to be used for prediction of cyclic behavior (hardening, softening, stable behavior) of materials. Correspondingly, authors proposed a number of separate expressions for estimation of R_e' depending on value of R_m/R_e of which the most successful ones are:

$$R_e' = 0.75R_e + 82 \text{ for } R_m/R_e > 1.2 \tag{4a}$$

$$R_e' = 3.0 \times 10^{-4}R_e^2 - 0.15R_e + 526 \text{ for } R_m/R_e \leq 1.2 \tag{4b}$$

Additionally, single expression for all steels, regardless of the value of R_m/R_e is also proposed:

$$R_e' = 8.0 \times 10^{-5}R_m^2 + 0.54R_m. \tag{5}$$

Values of coefficient of determination R^2 for expressions (4a), (4b) and (5) were 0.88, 0.99, and 0.94 respectively. Evaluation was performed on a single dataset comprising data used for developing expressions and additional data (all together 121 materials, mostly unalloyed and low-alloy steels). It was established that 84% of estimated values of R_e' from yield stress R_e (Equations (4a) and (4b)) deviate up to $\pm20\%$ from experimental values while 79% of values of R_e' estimated from ultimate strength R_m (Equation (5)) deviated up to $\pm20\%$ from experimental values.

Motivated by findings from [12] that Equation (3) always underestimates cyclic yield stress R_e' when experimental value of R_e' exceeds 900 MPa, Li et al. [13] recently modified Equation (3) to:

$$R_e' = 0.089(1 + RA)^{1.35}R_m^{1.35} \times \left(-\frac{0.002}{\ln(1 - RA)}\right)^{0.216} + 120 \tag{6}$$

resulting in rather high coefficient of determination $R^2 = 0.961$. Analysis was performed on the majority of data used in [12]. For evaluation, data used for developing Equation (6) was complemented with additional data. Results showed that most values of R_e' estimated from Equation (6) deviate up to 20% from their experiment-based counterparts. It must be noted that [11] and [13] suggest that values of true fracture strength σ_f can be calculated using the expression:

$$\sigma_f = R_m(1 + RA) \tag{7}$$

which is recognizable as first part of Equations (3) and (6). However, a well-known approximation of the relationship between ultimate strength R_m and true fracture stress σ_f, recommended by Manson [5,9] is:

$$\sigma_f = R_m(1 + \varepsilon_f) \tag{8}$$

Therefore, caution is advised when applying expressions (3) and (6) for estimation of not only cyclic yield stress R_e', but also cyclic parameters K' and n' that will be discussed later in Section 2.2.

2.2. Methods for Estimation of Cyclic Parameters K' and n'

Zhang et al. [10] proposed several equations for estimation of K' and n' based on 22 steels, aluminium (Al), and titanium (Ti) alloys. For this purpose, materials were divided by value of so-called new fracture ductility parameter α:

$$\alpha_f = RA \times \varepsilon_f = -RA \ln(1 - RA) \tag{9}$$

proposed in [19]. Expressions were proposed for estimation of K' and n', Equation (10) through Equation (11c), when strength coefficient K and strain hardening exponent n are available:

$$K' = 57K^{0.545} - 1220 \tag{10}$$

$$n' = 1.06n\left(1 + \beta\left|1 - \frac{R_m}{R_{p0.2}}\right|\right) \text{ for } \alpha < 5\% \text{ or } 10\% \leq \alpha < 20\% \tag{11a}$$

$$n' = 1.06n\left(1 + \beta\left|1 - \frac{\sigma_f}{R_m}\right|\right) \text{ for } 5\% < \alpha < 10\% \tag{11b}$$

$$n' = \frac{R_{p0.2}}{\sigma_f - R_m}n \text{ for } \alpha > 20\% \tag{11c}$$

and Equation (12a) through Equation (13c) when K and n are not available:

$$K' = 57\left(\sigma_f \varepsilon_f^{-\frac{\log\left(\frac{R_m^2 \sigma_f^3}{R_{p0.2}^5}\right)}{3\log(500\varepsilon_f)}}\right)^{0.545} - 1220 \text{ for } \alpha < 5\% \text{ or } 10\% \leq \alpha < 20\% \tag{12a}$$

$$K' = 57\left(\frac{\sigma_f R_{p0.2}}{R_m}\varepsilon_f^{-\frac{\log\left(\frac{\sigma_f^2}{R_{p0.2}R_m}\right)}{2\log(500\varepsilon_f)}}\right)^{0.545} - 1220 \text{ for } 5\% < \alpha < 10\% \text{ or } \alpha > 20\% \tag{12b}$$

$$n' = 1.06\left(1 + \beta\left|1 - \frac{R_m}{R_{p0.2}}\right|\right)\frac{\log\left(\frac{R_m^2 \sigma_f^3}{R_{p0.2}^5}\right)}{3\log(500\varepsilon_f)} \text{ for } \alpha < 5\% \text{ or } 10\% \leq \alpha < 20\% \tag{13a}$$

$$n' = 1.06\left(1 + \beta\left|1 - \frac{\sigma_f}{R_m}\right|\right)\frac{\log\left(\frac{\sigma_f^2}{R_{p0.2}R_m}\right)}{2\log(500\varepsilon_f)} \text{ for } 5\% < \alpha < 10\% \tag{13b}$$

$$n' = \frac{R_{p0.2}}{\sigma_f - R_m}\frac{\log\left(\frac{\sigma_f^2}{R_{p0.2}R_m}\right)}{2\log(500\varepsilon_f)} \text{ for } \alpha > 20\% \tag{13c}$$

For both methods, parameter $\beta = 1$ for $\sigma_f/R_{p0.2} < 1.6$ and $\beta = -1$ for $\sigma_f/R_{p0.2} > 1.6$. As most successful expressions authors proposed estimation of K' based on strength coefficient K (Equation (10)) and estimation of n' based on ultimate strength R_m, yield stress R_e, true fracture stress σ_f and strain hardening exponent n (Equation (13a) through Equation (13c), depending on value of α). For steels, values of K' and n' estimated in such a way deviated up to 27% and 34%, respectively, from their experiment-based counterparts. Data tables with percentage deviation for aluminium and titanium alloys suggest even larger deviations of estimates of n' (up to 65%). They also suggested that, for stress amplitudes $\Delta\sigma/2$ calculated from estimated values of K' and n', besides percentage deviation of particular parameter, sign of deviation is also significant. If sign of deviations of K' and n' is the same, calculated and experimental cyclic stress-strain curves are in good agreement.

In [12], besides expressions for estimation of R_e', Lopez and Fatemi developed several relationships between Brinell hardness HB or monotonic properties and cyclic parameters K' and n' of steels. Steels are divided into two subgroups according to the value of the R_m/R_e ratio (as was the case for estimation of cyclic yield stress R_e') and different expressions are proposed accordingly. Equations (14a) and (14b) are denoted as most successful:

$$K' = 1.16R_m + 593 \text{ for } R_m/R_e > 1.2 \tag{14a}$$

$$K' = 3.0 \cdot 10^{-4}R_m^2 + 0.23R_m + 619 \text{ for } R_m/R_e \leq 1.2 \tag{14b}$$

$$n' = -0.37\log\left(\frac{0.75R_e + 82}{1.16R_m + 593}\right) \text{ for } R_m/R_e > 1.2 \tag{15a}$$

$$n' = -0.37\log\left(\frac{3.0 \times 10^{-4}R_e^2 - 0.15R_e + 526}{3.0 \times 10^{-4}R_m^2 + 0.23R_m + 619}\right) \text{ for } R_m/R_e \leq 1.2 \tag{15b}$$

Authors provided coefficients of determination R^2 only for expressions (14a) and (14b). It is worth noting that R^2 of expressions proposed for estimation of K' for steels with $R_m/R_e > 1.2$ is 0.75 which is significantly lower than 0.90 obtained for steels with $R_m/R_e \leq 1.2$. About 73% values of K' estimated using Equations (14a) and (14b) deviate less than $\pm 20\%$ from their experimental values. As for n', percentage of values estimated from Equations (15a) and (15b) that deviate less than $\pm 20\%$ from their experiment-based counterparts is around 60%.

Lopez and Fatemi [12] proposed additional expression for estimation of n' valid for all steels:

$$n' = -0.33(R_e/R_m) + 0.40 \tag{16}$$

for which R^2 obtained was 0.79. Percentage of values of n' estimated from Equation (16) that deviated up to $\pm 20\%$ from experimental values was 68%.

Both methods proposed in [12] for estimation of n' provide reasonably good results, so in further evaluations in this paper, both methods will be taken into account: first using Equation (14a) through Equation (15b), and second using Equations (14a), (14b) and (16).

Li et al. [13] proposed expressions for estimation of cyclic parameters K' and n':

$$K' = 500^{n'} R_e' \tag{17}$$

$$n' = \frac{\log(K') - \log(R_e')}{\log 500} \tag{18}$$

where R_e' is estimated using Equation (6). However, Equations (17) and (18) can be used only when either K' or n' are available, so in the same paper an alternative method for estimation of these parameters was proposed. Cyclic strength coefficient K' should be estimated using Equations (19a), (19b) or (19c) first, then cyclic strain hardening n' exponent is calculated from estimated values of K'.

$$K' = 2.16 \cdot 10^{-4}(R_m)^{2.1} + 738 \text{ for } R_m/R_e \leq 1.2 \tag{19a}$$

$$K' = 3.63 \cdot 10^{-4}(R_m)^2 + 0.68 R_m + 570 \text{ for } 1.2 < R_m/R_e < 1.4 \tag{19b}$$

$$K' = 1.21 R_m + 555 \text{ for } R_m/R_e \geq 1.4 \tag{19c}$$

Equation (19a) through Equation (19c), when used in combination with Equation (18), yielded reasonable results with most of estimated values of K' deviating up to $\pm 20\%$ from experimental values. Obtained coefficients of determination R^2 for Equation (19a) through Equation (19c) decrease with higher values of R_m/R_e, which is in accordance with findings from [12]. R^2 obtained for steels with $R_m/R_e \leq 1.2$ is 0.921, while for steels with $1.2 < R_m/R_e < 1.4$ and $R_m/R_e \geq 1.4$ coefficients of determination are $R^2 = 0.813$ and $R^2 = 0.712$, respectively. Again, caution is advised when using Equation (18) due to the suggested way of estimating R_e' that was already discussed at the end of Section 2.1.

2.3. Conclusions

A review of methods for estimation cyclic parameters shows that sets of material data on which most of them were developed and evaluated differ significantly regarding their size and material groups included. In this sense, they can still be considered adequate with the exception of expressions for estimation of K' and n' proposed in [10] which were developed using a quite small and heterogeneous set of material data (22 datasets for steels, aluminium and titanium alloys) and expression for estimation of R_e' proposed in [10] which was developed using only 27 steel datasets.

In an attempt to further improve estimation accuracy, most methods address steels separately from other kinds of metallic materials [11–13] and all methods divide materials into separate subgroups using different criteria [10,12,13]. For this purpose, Zhang et al. [10] used new fracture ductility parameter α that was originally developed to predict materials' cyclic behavior [10,19]. Lopez and

Fatemi [12] and Li et al. [13] divided steels into two, i.e., three subgroups according to the ratio of ultimate strength to yield stress R_m/R_e.

Lack of general consensus regarding the treatment of individual material subgroups as well as different methodologies for evaluation of estimation methods implemented in their respective papers makes comparison of their performance quite difficult. In order to determine which estimation method is most suitable for estimation of cyclic parameters of steels, systematic and consistent evaluation of presented methods will be performed on an independent set of material data.

Different delivery, i.e., testing conditions of material, can be obtained for example through different processing method or heat treatment and this can strongly impact both monotonic and cyclic/fatigue material properties and behavior [5,20]. This is an important aspect which none of the discussed methods takes into account directly. One of the possible reasons is a multitude of conditions of steel materials which were used for development of these methods, particularly in [12,13]. For certain materials, such information, even if available, was of a rather general nature (for example heat treated, modified, etc.).

In practice, steels are commonly divided according to the content of alloying elements into unalloyed, low-alloy and high-alloy steels. Already Baümel and Seeger [6] considered unalloyed and low-alloy steels separately from other metallic materials when they developed Uniform Material Law for estimation of C-M-B parameters. Hatscher et al. [8] also mentioned the prospect of such division for estimation of fatigue C-M-B parameters. Results of detailed analysis performed on a large number of material data done by Basan et al. [15] showed that there is statistically significant difference among individual C-M-B fatigue parameters as well as strain-life behavior ($\Delta\varepsilon$–$2N_f$ relationships) of unalloyed, low-alloy and high-alloy steels. Also, preliminary investigations on cyclic parameters in [16,17] showed that dividing steels by alloying content could result in more accurate estimations of cyclic parameters and hence, more accurate estimations of cyclic stress-strain curves of materials.

For that reason, statistical analysis of steel subgroups (unalloyed, low-alloy and high-alloy steels) will be performed in order to determine if their cyclic parameters differ significantly. If confirmed, individual steel subgroups will be taken into account during evaluation and comparison of estimation methods.

3. Methods and Data

3.1. Methods for Statistical Analysis

To test whether statistically significant differences exist among experimental data for cyclic yield stress R_e' and cyclic stress-strain parameters K' and n' of unalloyed, low-alloy and high-alloy steels, one-way Analysis of Variance (one-way ANOVA) is performed. One-way ANOVA is a technique that provides a statistical test of whether or not means of several (typically three or more) groups are all equal. If results obtained by one-way ANOVA show that statistically significant difference exists between cyclic parameters of analyzed groups, post hoc analysis by Tukey's multiple comparison method will be performed in order to determine pairwise differences between groups. Significance level α for one-way ANOVA is set to 0.05, while overall significance (family error rate) in Tukey's multiple comparison test is set to 0.05 to counter type I error for a series of comparisons. Procedure for both one-way ANOVA and Tukey's multiple comparison test are given in [21]. Statistical analyses were performed in statistical package MINITAB [22].

To evaluate predictive accuracy of estimation methods and to facilitate their comparison, deviations of estimated values from their experimentally obtained counterparts were used as relevant indicators. Deviations up to ± 10, ± 20 and $\pm 30\%$ were used as in [12,13] to facilitate comparison with results reported there. Instead of directly comparing estimated Ramberg-Osgood parameters K' and n' to their experiment-based counterparts as in [12,13], much more useful information regarding the predictive accuracy of these methods can be obtained by comparing values of stress amplitudes $\Delta\sigma/2$, i.e., points on cyclic stress-strain curves as in [10,16,17]. Therefore, values of stress amplitudes $\Delta\sigma/2$

were calculated using experimental values of K' and n' and their estimations for series of total strain amplitudes $\Delta\varepsilon/2$: 0.1, 0.2, 1 and 2%.

3.2. Data to Be Analyzed

Experimental data for three representative groups of steels: unalloyed (UA), low-alloy (LA) and high-alloy (HA) steels from [6] and [23] were obtained through the MATDAT Materials Properties Database [24]. Only results of strain-controlled, fully reversed ($R = -1$) axial cyclic tests performed in the air at room temperature were considered. Furthermore, only data for materials tested at more than four different strain amplitudes and at range of total strain amplitudes larger than 0.4% were used in the analysis.

Only datasets that contained all experimental values of monotonic properties needed for estimation of cyclic parameters by each method were used. An exception was made with the high-alloy steel group. Since most datasets did not contain values of true fracture stress σ_f which is necessary for calculation of parameters by Zhang et al. method [10], values were calculated by their relationship between ultimate strength R_m and true fracture strain ε_f, according to Equation (8). Also, if a dataset contained only reduction of area at fracture RA, true fracture strain ε_f was calculated by the relationship between these two properties:

$$\varepsilon_f = -\ln(1 - RA) \tag{20}$$

In total, 34 unalloyed steels, 47 low-alloy steels and 35 high-alloy steel datasets were available for analysis. Wide variety of conditions resulting from different processing and heat treatment were present in materials used for statistical analysis and evaluation of existing methods. This is consistent with datasets used for development of methods in their respective papers.

Detailed material data are given in Appendix A in Tables A1–A3. Data used for statistical analysis in [14] were complemented with values of true fracture strain ε_f. Additionally, data for high-alloy steels were complemented with values of true fracture stress σ_f, strength coefficient K and strain hardening exponent n since those are required so that evaluations of particular existing methods could be performed.

4. Analysis and Results

4.1. Results of One-Way ANOVA and Tukey's Multiple Comparison Test

Performing one-way ANOVA for three cyclic parameters (R_e', K' and n') of unalloyed, low- and high-alloy steel subgroups showed that statistically significant differences exist between steel subgroups regarding cyclic yield stress R_e' ($F(2, 113) = 32.25$; $p < 0.05$), cyclic strength coefficient K' ($F(2, 113) = 22.61$; $p < 0.05$), and cyclic strain hardening exponent n' ($F(2, 113) = 72.00$; $p < 0.05$).

Since steel subgroups were confirmed to be significantly different regarding their cyclic parameters R_e', K' and n', post hoc Tukey's test was performed to determine which subgroups are mutually different. Results showed that unalloyed and low-alloy steels as well as low-alloy and high-alloy steels differ significantly regarding the cyclic yield stress R_e'. No such difference was determined between unalloyed and high-alloy steels. Statistically significant difference was also found for cyclic strength coefficient K' of unalloyed and high-alloy steels, as well as low-alloy and high-alloy steels, while no such difference was found between unalloyed and low-alloy steels. Cyclic strain hardening exponent n' differs between pairs of all three groups.

4.2. Evaluation of Methods for Estimation of Cyclic Yield Stress R_e' and Ramberg-Osgood Parameters K' and n' of Steels

Since unalloyed, low-alloy and high-alloy steels subgroups were proved to be significantly different, steel subgroups will be considered separately for evaluation of estimation methods. Evaluation will also be made for all steels as a single group to check for differences between results of analyses performed on individual subgroups and to enable comparison and determination of potential discrepancies with results reported in original papers.

Methods for estimation of cyclic yield stress R_e' and R-O parameters K' and n' of steels whose predictive accuracy will be evaluated are listed in Table 1. For every material, expressions for estimation of cyclic yield stress R_e' and R-O parameters K' and n' will be used according to ranges defining the applicability of the models regarding criteria for grouping of materials in original papers (new fracture ductility coefficient α, ultimate strength to yield stress ratio R_m/R_e).

4.2.1. Evaluation of Methods for Estimation of Cyclic Yield Stress R_e'

Percentages of values of R_e' estimated according to selected methods (Table 1) that deviate up to 10%, 20% and up to 30% from experiment-based values were calculated and are given in diagrams on Figure 1.

Table 1. Methods for estimation of R_e' and K' and n' which will be evaluated.

Evaluated Value	Method	Estimated Parameters	Originally Proposed for	Equation Number
R_e'	Lopez and Fatemi 1 [12]	R_e'	steels divided by R_m/R_e	(4a,b)
	Lopez and Fatemi 2 [12]	R_e'	all steels	(5)
	Li et al. [13]	R_e'	all steels	(6)
$\Delta\sigma/2$	Zhang et al. 1 [10] (K and n available)	K'	steels, Al and Ti alloys	(10)
		n'	steels, Al and Ti alloys divided by α	(11a,b,c)
	Zhang et al. 2 [10] (K and n not available)	K'	steels, Al and Ti alloys divided by α	(12a,b)
		n'	steels, Al and Ti alloys divided by α	(13a,b,c)
	Lopez and Fatemi 1 [12]	K'	steels divided by R_m/R_e	(14a,b)
		n'	steels divided by R_m/R_e	(15a,b)
	Lopez and Fatemi 2 [12]	K'	steels divided by R_m/R_e	(14a,b)
		n'	all steels	(16)
	Li et al. [13]	K'	steels divided by R_m/R_e	(19a,b,c)
		n'	steels divided by R_m/R_e	(18)

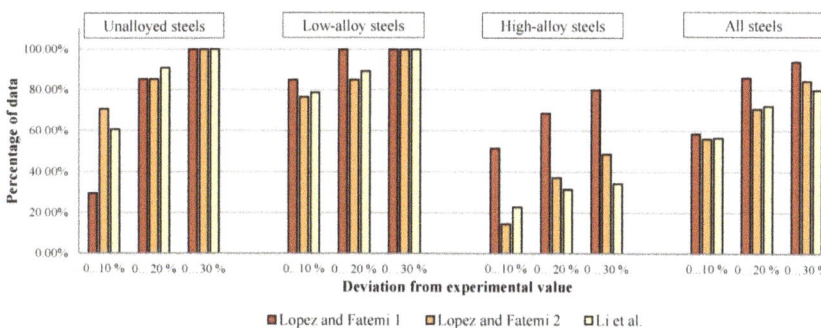

Figure 1. Percentage of R_e' values estimated by selected methods that deviate up to 10%, 20% and 30% from experiment-based values.

Reasonable results are obtained for All steels group with about 70%–80% of data deviating up to 20% from their experimental counterparts for each method.

For unalloyed steels, for each method about 80%–90% of estimates of cyclic yield stress R_e' deviate up to 20%, while all estimates fall within $\pm30\%$ deviation from corresponding experimental values. Highest percentage of data that deviate only up to 10% is obtained by Lopez and Fatemi 2 (about 70%).

For low-alloy steels, results obtained using any of the three selected methods were even more accurate than for unalloyed steels. Best results for estimation of R_e' of low-alloy steels are obtained using Lopez and Fatemi 1 method, for which all estimates deviate 20% or less from their experiment-based counterparts.

As for high-alloy steels, no method proved to be sufficiently accurate. Lopez and Fatemi 1 provides reasonable results with about 70% of estimated data deviating up to 20%, although 20% of data deviate more than 30% from experimental values. Estimations made by other two methods result with less than 50% of estimates deviating below 30% from experimental values (below 40% by Li et al.).

4.2.2. Evaluation of Methods for Estimation of Ramberg-Osgood Parameters K' and n' of Steels

As was explained in Section 3.1, accuracy of estimates of Ramberg-Osgood parameters K' and n' will be determined by evaluating values of stress amplitudes $\Delta\sigma/2$ calculated using estimated K' and n' opposed to those obtained using experimental values of those parameters.

Percentages of estimated values of $\Delta\sigma/2$ as calculated by selected methods (Table 1) that deviate up to 10%, 20% and up to 30% from experiment-based values are given in Figure 2.

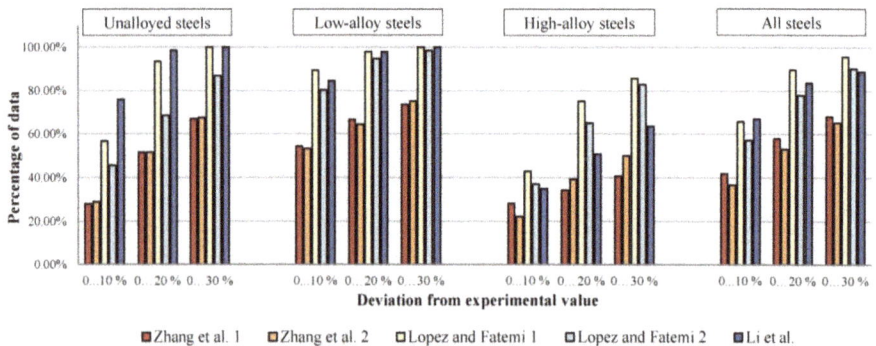

Figure 2. Percentage of $\Delta\sigma/2$ values estimated by selected methods that deviate up to 10%, 20% and 30% from experiment-based values.

Results obtained using Lopez and Fatemi 1 and Li et al. method for estimation of K' and n' both provide very good results for estimation of stress amplitudes $\Delta\sigma/2$ of unalloyed steels, with over 90% of data deviating 20% or less from experimental values. For both methods, all estimates of $\Delta\sigma/2$ for unalloyed steels fall within $\pm 30\%$ deviation from experimental values, with Li et al. method providing as much as 75% of data within $\pm 10\%$ deviation.

The same methods yield even better results for low-alloy steels, with more than 80% of data deviating up to 10% from experiment-based counterparts. Although somewhat less accurate than previous two, Lopez and Fatemi 2 method also provides very good results for estimation of $\Delta\sigma/2$ of low-alloy steels.

Both methods by Zhang et al. gave significantly inferior results for both unalloyed and low-alloy steel subgroups. For unalloyed steels, by either method, only about 50% of data deviate 20% or less from experimental values, while about one-third of data deviate more than 30% from corresponding experimental values. Results are somewhat better for low-alloy steels, with a higher percentage of data deviating 20% or less from experimental values (about 65%). Still, only 75% of estimates obtained using both methods by Zhang et al. deviate less than 30% from experimental values.

As for high-alloy steels, no method provided estimates on the level of accuracy observed for unalloyed and low-alloy steels. Results obtained using either method by Lopez and Fatemi are the most acceptable, with around 75% of values estimated using Lopez and Fatemi 1 deviating less than 20%. Estimations by Li et al. resulted in more than 35% of data deviating more than 30% from experimental values, while the most inaccurate results were obtained by either of Zhang et al. methods. More than half of the estimates obtained by these methods deviate at least 30% from corresponding experimental values.

Overall evaluation for all steels provided averaged results as was the case for estimates of R_e'. Lopez and Fatemi 1 method is the most accurate while both Zhang et al. methods are least successful, as was the case for individual steel subgroups.

5. Discussion

Estimation methods investigated in this paper were developed on datasets comprising all steels [11–13], and even some other groups of metallic materials (steels, aluminium and titanium alloys) [10]. However, in order to improve accuracy of estimations, most methods propose some criterion for grouping of materials. In [12,13] authors divided steels by rather easily available R_m/R_e ratio. In [10], grouping criteria used was the new fracture ductility parameter α, which is cumbersome to use since true fracture stress ε_f or reduction of area RA needed for its calculation are often unavailable.

Much more usable, and often encountered in practice, is division of steels by their alloying content into unalloyed, low-alloy and high-alloy steels. It was shown in [6,8,15–17] that dividing steels in this manner contributes to improvement of estimations of steel behavior from monotonic properties of steels. Results of analysis performed in Section 4.1 confirmed that statistically significant differences exist between cyclic yield stress R_e' and cyclic parameters K' and n' of mentioned group of steels. According to these findings, authors propose evaluation of existing methods for estimation of R_e', K' and n' to be performed for each group individually, in addition to all steels together.

For estimation of R_e', K' and n' of both unalloyed and low-alloy steels, methods by Lopez and Fatemi [12], and by Li et al. [13] provide very good results. However, estimations for high-alloy steels are notably worse, especially those obtained using the Li et al. method which is not surprising since both methods are developed on the same set of data, consisting mostly of unalloyed and low-alloy steels.

Values of K' and n' estimated with two methods developed by Zhang et al. [10], are generally unsatisfactory and provide poor results for all steel subgroups, especially high-alloy steels. This can be attributed to the fact that methods were developed on a modest number (22) of heterogeneous data (steels, aluminium and titanium alloys). Another drawback of methods proposed by Zhang et al. are intricate expressions requiring monotonic properties which are often unavailable, especially during the early stages of product development.

Evaluations of existing methods in original references are performed for all materials together. Also, in some cases, evaluation of existing methods is performed on the same sets of data that were used for development of the method—a practice which can be considered less than objective.

Results of evaluations strongly depend on structure and amount of data available. Proposed consideration of individual subgroups provides valuable additional information. Figure 3 shows that, although methods by Lopez and Fatemi and Li et al. are suitable for all steels, individual results for unalloyed and low-alloy steels are even better.

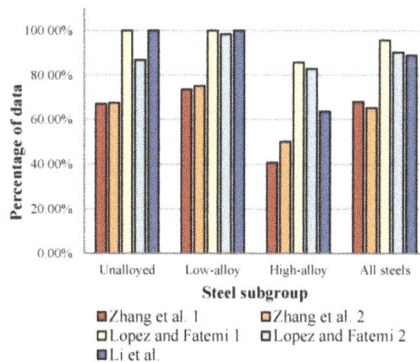

Figure 3. Percentage of estimates of $\Delta\sigma/2$ deviating up to 30% from experimental values.

As mentioned earlier, this was expected since the similar set of data, consisting of mostly unalloyed and low-alloy steels, was used in both papers. If only results for all steels group were observed, information about lower accuracy for high-alloy steels would go unnoticed, particularly for the method by Li et al. This example shows how evaluations of estimation methods performed for all steels together could be misinterpreted due to averaging.

Table 2 provides recommended methods for estimation of cyclic parameters of each group of steels, noting that great care is advised when estimating cyclic parameters of high-alloy steels. Ordinal number preceding the method indicates suggested selection priority. In cases where multiple methods are given without order simpler methods requiring a smaller number of monotonic properties (such as methods by Lopez and Fatemi) might be preferred.

Table 2. Recommended methods for estimation of cyclic yield stress R_e' and cyclic parameters K' and n' of steels.

Steel Subgroup	Estimation of R_e'	Estimation of $\Delta\sigma/2$ (K', n')
Unalloyed steels	Li et al. Lopez and Fatemi 2	1. Li et al. 2. Lopez and Fatemi 1
Low-alloy steels	1. Lopez and Fatemi 1 2. Li et al. 3. Lopez and Fatemi 2	Lopez and Fatemi 1 Lopez and Fatemi 2 Li et al.
High-alloy steels	Lopez and Fatemi 1	Lopez and Fatemi 1

6. Conclusions

Available methods for estimation of cyclic yield stress R_e' and cyclic stress-strain parameters K' and n' of steels and their applicability to individual steel subgroups and to steels as a general group were studied. A large, independent set of steel data was collected in order to perform the study as it was shown that number and type of materials used for development of estimation methods have significant influence on their performance and evaluation results.

Statistically significant differences were determined to exist among unalloyed, low-alloy and high-alloy steels regarding their cyclic stress-strain behavior and parameters. Such division of steels based on their content of alloying elements is also commonly encountered in practice, so it was used for performed evaluation of studied estimation methods.

Comparison of values of stress amplitudes $\Delta\sigma/2$ calculated using experimental and estimated cyclic parameters (K' and n') is proposed as a more suitable criterion for evaluation instead of direct comparison of corresponding individual cyclic parameters R_e', K' and n'.

Having all steels in a single group for evaluation purposes causes significant averaging of the results so unalloyed, low-alloy and high-alloy steels were treated separately. Considerable differences were determined in accuracy and applicability of different methods for different steel subgroups.

For estimations of R_e' of unalloyed and low-alloy steels methods proposed by Li et al. and Lopez and Fatemi were found to provide very good results, while for high-alloy steels, only the method dividing steels by R_e/R_m ratio proposed by Lopez and Fatemi provides reasonably accurate estimates. The method for estimations of K' and n' of unalloyed steels proposed by Li et al. gives the best estimates followed closely by the Lopez and Fatemi method which considers the R_e/R_m ratio. For low-alloy steels, both methods by Lopez and Fatemi and the method by Li et al. provide excellent results. Of all methods, only the method proposed by Lopez and Fatemi considering the R_e/R_m ratio can be considered for use with the high-alloy steels group.

Estimation accuracy of all studied methods for R_e', K' and n' was notably lower for high-alloy steels in comparison to other two subgroups, which can be attributed to the fact that high-alloy steels were found to be underrepresented in material datasets used for development of estimation methods.

Acknowledgments: This work has been supported in part by the University of Rijeka (projects number 13.09.1.2.09 and 13.09.2.2.18) and Croatian Science Foundation (scientific project number IP-2014-09-4982).

Author Contributions: R.B. and T.M. conceived and formulated the research; R.B., T.M. and M.F. acquired and systematized the material data, T.M. systematized and reviewed the estimation methods; T.M. designed and performed statistical analysis; T.M. and R.B., with the support of M.F., evaluated and interpreted estimation results; T.M., with the support of R.B. and M.F., prepared the manuscript. All co-authors contributed to the manuscript proof and submission.

Conflicts of Interest: The authors declare no conflict of interest. The founding sponsors had no role in the design of the study; in the collection, analyses, or interpretation of data; in the writing of the manuscript, and in the decision to publish the results.

Appendix A

Detailed material data used for analysis in this paper are given in Tables A1–A3.

Table A1. Monotonic and cyclic properties for unalloyed steels [6,23,24].

Material Designation	Monotonic Properties									Cyclic Parameters		
DIN or SAE/other	E (MPa)	R_e or $R_{p0.2}$ (MPa)	R_m (MPa)	R_m/R_e (-)	RA (%)	K (MPa)	n (-)	σ_f (MPa)	ε_f (-)	R_e' (MPa)	K' (MPa)	n' (-)
1038 (SAE)	207,000	347	610	1.758	55.5	511	0.071	956	0.590	332	1207	0.208
Armco (other)	210,000	207	359	1.734	64	675	0.285	653	1.030	280	858	0.18
C 20	190,000	224	414	1.848	70	330	0.061	953	1.190	239	1050	0.238
C 10	217,510	435	566	1.301	68	659	0.073	1205	1.130	463	1381	0.176
Ck 15	196,793	263	392	1.490	55	711	0.224	746	0.806	249	824	0.193
Ck 15	204,500	320	434	1.356	67.5	394	0.067	848.7	1.126	269	813	0.178
Ck 15	202,000	431.3	615.2	1.426	54	598	0.045	1011.7	0.776	492	1296	0.156
Ck 15	203,000	660	828	1.255	2.6	863	0.042	850.5	0.026	687	1165	0.085
Ck 25	210,000	346	507	1.465	63	926	0.264	1027	0.994	280	1345	0.252
Ck 25	210,000	307	464	1.511	65	924	0.276	982	1.050	278	1111	0.223
Ck 25	210,000	366	527	1.440	60	1033	0.264	997	0.916	303	1217	0.224
Ck 35	210,000	414	617	1.490	58	1216	0.258	1150	0.868	328	1355	0.229
Ck 35	210,000	394	593	1.505	62	1168	0.257	1169	0.968	333	1460	0.238
Ck 35	210,000	396	565	1.427	63	1134	0.264	1134	0.994	316	1534	0.254
Ck 35	210,000	587	780	1.329	67	1356	0.186	1514	1.109	463	1106	0.14
Ck 35	210,000	480	656	1.367	74	1196	0.207	1468	1.347	393	1033	0.156
Ck 35	210,000	596	733	1.230	71	1170	0.152	1541	1.238	447	1027	0.134
Ck 35	210,000	542	730	1.347	68	1311	0.2	1473	1.139	430	1087	0.149
Ck 35	210,000	513	669	1.304	70	1121	0.18	1417	1.204	387	1081	0.165
Ck 45	206,000	540	790	1.463	60	730	0.047	1400	0.916	481	980	0.115
Ck 45	210,500	531	790	1.488	60	1219	0.0151	1271	0.777	462	1078	0.133
Ck 45	199,700	622	915	1.471	59	1606	0.18	1784	0.900	591	2407	0.226
Ck 45	199,700	622	915	1.471	59	1606	0.18	1784	0.900	538	1762	0.191
Ck 45	201,500	380	684	1.800	36.8	735	0.092	987	0.460	336	1414	0.231
Ck 45	205,000	760	1018	1.339	0	1141	0.059	1018	0.000	722	2075	0.17
Ck 45	199,000	466	737	1.582	54	1469	0.248	1296	0.777	368	1486	0.225
Ck 45	207,000	462	672	1.455	61	1288	0.235	1298	0.942	354	1391	0.22
Ck 45	208,000	588	730	1.241	70	1154	0.148	1540	1.204	420	1194	0.168
Ck 45	207,000	551	774	1.405	68	1297	0.166	1559	1.139	464	1235	0.158
Ck 45	206,000	728	844	1.159	64	1208	0.108	1582	1.022	516	1217	0.138
Ck 45	210,000	652	787	1.207	68	1200	0.129	1568	1.139	472	1285	0.161
Ck 45	204,000	702	863	1.229	66	1268	0.118	1651	1.079	526	1243	0.138
St 37	210,000	295	435	1.475	64	829	0.275	835	1.020	273	988	0.207
St 52-3	210,000	400	597	1.493	63	1061	0.225	1083	0.980	389	1228	0.185

Note: Shaded values are calculated by Equation (20).

Table A2. Monotonic and cyclic properties for low-alloy steels [6,23,24].

Material Designation DIN	Monotonic Properties E (MPa)	R_e or $R_{p0.2}$ (MPa)	R_m (MPa)	R_m/R_e (-)	RA (%)	K (MPa)	n (-)	σ_f (MPa)	ε_f (-)	Cyclic Parameters R_e' (MPa)	K' (MPa)	n' (-)
100 Cr 6	207,000	1927	2016	1.046	12	2281	0.031	2230	0.120	1341	3328	0.146
11 NiMnCrMo 55	210,000	745	852	1.144	57	1277	0.124	1327	0.834	663	1145	0.088
14 Mn 5	206,000	580	697	1.202	68	858	0.067	1222	1.150	537	1436	0.158
16 NiCrMo 3 2	209,000	891	939	1.054	63	963	0.011	1491	0.994	617	1080	0.09
17 MnCrMo 33	214,000	833	929	1.115	58	1285	0.099	1446	0.867	663	1252	0.102
20 Mn 3	206,000	910	960	1.055	43	1190	0.06	1090	0.561	675	1313	0.107
22 MnCrNi 3	198,000	1200	1510	1.258	42	2447	0.114	2034	0.549	1046	2149	0.112
22 MnCrNi 3	195,000	1200	1586	1.322	3	2586	0.115	1669	0.026	1193	2759	0.135
23 Mn 4	207,000	1008	1091	1.082	61	1185	0.026	1616	0.950	656	1616	0.145
23 NiCr 4	208,531	725	808	1.114	66	762	0.007	1215	1.080	541	1221	0.131
25 Mn 3	200,000	351	540	1.538	67	992	0.236	1173	1.100	322	1219	0.214
25 Mn 5	207,000	904	1008	1.115	49	1138	0.033	1284	0.680	608	1900	0.183
28 MnCu 6	204,000	330	580	1.758	64	938	0.19	950	1.030	347	1151	0.193
30 CrMo 2	221,000	780	898	1.151	67	1117	0.063	1692	1.120	579	1366	0.138
30 CrMo 2	200,250	1360	1429	1.051	55	1661	0.033	2085	0.790	814	1758	0.124
30 CrMoNiV 5 11	212,000	605	773	1.278	62	717	0.027	1332	0.968	497	894	0.094
30 CrNiMo 8	206,000	700	910	1.300	66	1128	0.079	1168	0.708	573	972	0.085
30 CrNiMo 8	206,000	700	910	1.300	66	1128	0.079	1168	0.708	522	995	0.095
30 MnCr 5	206,000	820	950	1.159	64	1250	0.097	1445	1.068	576	1618	0.166
34 CrMo 4	193,000	1017	1088	1.070	65	1344	0.056	1903	1.050	692	1310	0.103
34 CrMo 4	188,000	847	939	1.109	69	1215	0.074	1795	1.171	624	1008	0.077
34 CrMo 4	190,000	893	978	1.095	67	1338	0.089	1787	1.109	650	987	0.067
34 CrMo 4	197,000	980	1078	1.100	61	1382	0.07	1818	0.942	711	1373	0.106
34 CrMo 4	194,000	780	881	1.129	71	1299	0.116	1740	1.238	556	1198	0.124
4 NiCrMn 4	206,000	454	623	1.372	76	753	0.081	1229	1.450	505	1111	0.127
40 CrMo 4	208,780	840	940	1.119	64	1300	0.094	1440	1.035	583	1307	0.13
40 NiCrMo 6	201,000	1084	1146	1.057	59	1549	0.083	1857	0.890	758	1550	0.115
40 NiCrMo 6	190,000	910	1015	1.115	62	1372	0.089	1808	0.970	660	1392	0.12
40 NiCrMo 6	202,000	953	1029	1.080	62	1448	0.1	1724	0.970	659	1628	0.145
40 NiCrMo 6	193,000	998	1067	1.069	62	1474	0.092	1761	0.970	716	1292	0.095
40 NiCrMo 6	205,000	810	884	1.091	67	1378	0.142	1680	1.110	586	1303	0.129
40 NiCrMo 7	193,500	1374	1471	1.071	38	1796	0.04	1920	0.480	905	1890	0.118
40 NiCrMo 7	193,500	635	829	1.306	43	1175	0.098	1201	0.570	474	1332	0.167
41 MnCr 3 4	207,280	800	930	1.163	62	1350	0.112	1390	0.960	551	1340	0.143
42 Cr 4	195,000	903	1006	1.114	62	1293	0.068	1716	0.968	679	1153	0.085
42 Cr 4	194,000	813	921	1.133	65	1249	0.086	1674	1.050	613	1147	0.101
42 Cr 4	194,000	845	952	1.127	62	1288	0.086	1689	0.968	619	1207	0.107
42 Cr 4	192,000	833	943	1.132	65	1289	0.09	1690	1.050	621	1192	0.105
42 Cr 4	193,000	717	840	1.172	69	1240	0.118	1617	1.171	543	1161	0.122
42 CrMo 4	211,400	998	1111	1.113	60	1469	0.069	1525	0.496	716	1367	0.104
49 MnVS 3	210,200	566	840	1.484	19	1428	0.194	1152	0.380	520	1396	0.159
50 CrMo 4	205,000	970	1086	1.120	48.6	1132	0.026	1609	0.665	700	1568	0.13
50 CrMo 4	205,000	947	983	1.038	14.6	1042	0.018	926	0.157	774	1754	0.132
8 Mn 6	198,000	862	965	1.119	57	1227	0.054	1579	0.850	580	1256	0.125
8 Mn 6	198,000	821	869	1.058	53	1085	0.046	1434	0.750	674	1258	0.101
80 Mn 4	187,500	502	931	1.855	16	1100	0.127	1060	0.174	459	1859	0.225
WStE 460	210,000	560	667	1.191	61	1096	0.153	1171	0.932	514	1194	0.128

Table A3. Monotonic and cyclic properties for high-alloy steels [6,23,24].

Material Designation	Monotonic Properties									Cyclic Parameters		
DIN	E (MPa)	R_e or $R_{p0.2}$ (MPa)	R_m (MPa)	R_m/R_e (-)	RA (%)	K (MPa)	n (-)	σ_f (MPa)	ε_f (-)	R_e' (MPa)	K' (MPa)	n' (-)
X 10 CrNi 18 8	204,000	245	635	2.592	79	1416	0.362	1908	1.563	307	2397	0.331
X 10 CrNiNb 18 9	210,000	237	615	2.595	72			1398	1.273	271	1967	0.319
X 10 CrNiNb 18 9	210,000	237	615	2.595	72			1398	1.273	276	1667	0.289
X 10 CrNiTi 18 9	210,000	211	677	3.209	67			1428	1.109	455	8384	0.469
X 10 CrNiTi 18 9	210,000	182	668	3.670	68			1429	1.139	414	6179	0.435
X 10 CrNiTi 18 9	210,000	211	677	3.209	69			1470	1.171	496	3647	0.321
X 10 CrNiTi 18 9	210,000	177	516	2.915	74			1211	1.347	220	2264	0.375
X 10 CrNiTi 18 9	210,000	177	516	2.915	74			1211	1.347	250	1535	0.292
X 10 CrNiTi 18 9	210,000	214	529	2.472	74			1242	1.347	228	2086	0.357
X 10 CrNiTi 18 9	210,000	214	529	2.472	74			1242	1.347	251	1682	0.306
X 10 CrNiTi 18 9	210,000	177	535	3.023	77			1321	1.470	220	3080	0.424
X 10 CrNiTi 18 9	210,000	177	535	3.023	77			1321	1.470	241	2097	0.348
X 15 Cr 13	210,000	598	736	1.231	70			1622	1.204	475	1056	0.128
X 15 Cr 13	210,000	598	736	1.231	70			1622	1.204	497	987	0.11
X 15 CrNiSi 25 20	210,000	271	630	2.325	69			1368	1.171	289	2302	0.334
X 15 CrNiSi 25 20	210,000	271	630	2.325	69			1368	1.171	284	2242	0.332
X 2 CrNi 18 9	192,000	280	601	2.146	46	455	0.097	971	0.616	207	2807	0.419
X 20 CrMo 12 1	210,000	795	1013	1.274	47			1656	0.635	716	1325	0.099
X 20 CrMo 12 1	210,000	795	1013	1.274	47			1656	0.635	730	1301	0.093
X 25 CrNiMn 25 20	193,340	220	642	2.918	63	754	0.228	1360	1.010	421	2267	0.271
X 3 CrNi 19 9	172,625	746	953	1.277	69	1114	0.063	2037	1.160	882	2313	0.155
X 3 CrNi 19 9	186,435	255	746	2.925	74	548	0.136	1920	1.370	678	4634	0.309
X 5 CrNi 18 9	210,000	207	611	2.952	75			1458	1.386	197	3331	0.455
X 5 CrNi 18 9	210,000	207	611	2.952	83			1694	1.772	203	3001	0.434
X 5 CrNiMo 18 10	210,000	230	587	2.552	78			1476	1.514	256	1644	0.299
X 5 CrNiMo 18 10	210,000	231	587	2.541	78			1476	1.514	247	2755	0.388
X 5 CrNiMo 18 10	210,000	257	606	2.358	79			1830	1.561	313	2000	0.298
X 5 CrNiMo 18 10	210,000	228	665	2.917	81			1769	1.661	259	2081	0.336
X 5 CrNiMo 18 10	210,000	228	665	2.917	81			1769	1.661	259	2674	0.376
X 5 NiCrTi 26 15	210,000	777	1158	1.490	52			2008	0.734	713	1617	0.132
X 5 NiCrTi 26 15	210,000	777	1158	1.490	52			2008	0.734	711	1543	0.125
X 6 CrNi 19 11	183,000	325	650	2.000	80	1210	0.193	1400	1.610	267	1628	0.291
X 8 CrNiTi 18 10	204,000	222	569	2.563	76	349	0.062	1381	1.427	383	5234	0.421
X2 CrNiMo 18 10	210,000	373	700	1.877	75			1670	1.386	295	1232	0.23
X5 CrNi 18 9	198,000	242	666	2.752	82	484	0.113	2407	1.715	275	2872	0.378

Note: Shaded values are calculated by Equations (8) and (20).

References

1. Williams, C.R.; Lee, Y.-L.; Rilly, J.T. A practical method for statistical analysis of strain-life fatigue data. *Int. J. Fatigue* **2003**, *25*, 427–436. [CrossRef]
2. Blackmore, P.A. A critical review of Baumel-Seeger method for estimating strain-life fatigue properties of metallic materials. *Eng. Integr.* **2009**, *27*, 6–11.
3. Ramberg, W.; Osgood, W.R. *Description of Stress-Strain Curves by Three Parameters*; Technical Note No. 902; National Advisory Committee for Aeronautics (NACA): Washington, DC, USA, 1943.
4. Manson, S.S.; Halford, G.R. *Fatigue and Durability of Structural Materials*, 1st ed.; ASM International: Materials Park, OH, USA, 2005.
5. Manson, S.S. Fatigue: A complex subject—Some simple approximations. *Exp. Mech. SESA* **1965**, *5*, 193–226. [CrossRef]
6. Bäumel, A., Jr.; Seeger, T. *Materials Data for Cyclic Loading—Supplement 1*, 1st ed.; Elsevier: Amsterdam, The Netherlands, 1990.
7. Roessle, M.L.; Fatemi, A. Strain-controlled fatigue properties of steels and some simple approximations. *Int. J. Fatigue* **2000**, *22*, 495–511. [CrossRef]
8. Hatscher, A.; Seeger, T.; Zenner, H. Abschätzung von zyklischen Werkstoffkennwerten—Erweiterung und Vergleich bisheriger Ansätze. *Materialprufung* **2007**, *49*, 2–14.

9. Park, J.H.; Song, J.H. Detailed evaluation of methods for estimation of fatigue properties. *Int. J. Fatigue* **1995**, *17*, 365–373. [CrossRef]
10. Zhang, Z.; Qiao, Y.; Sun, Q.; Li, C.; Li, J. Theoretical estimation to the cyclic strength coefficient and the cyclic strain-hardening exponent for metallic materials: Preliminary study. *J. Mater. Eng. Perform.* **2009**, *18*, 245–254. [CrossRef]
11. Li, J.; Sun, Q.; Zhang, Z.; Li, C.; Qiao, Y. Theoretical estimation to the cyclic yield strength and fatigue limit for alloy steels. *Mech. Res. Commun.* **2009**, *36*, 316–321. [CrossRef]
12. Lopez, Z.; Fatemi, A. A method of predicting cyclic stress-strain curve from tensile properties for steels. *Mater. Sci. Eng. A Struct.* **2012**, *556*, 540–550. [CrossRef]
13. Li, J.; Zhang, Z.; Li, C. An improved method for estimation of Ramberg-Osgood curves of steels from monotonic tensile properties. *Fatigue Fract. Eng. Mater. Struct.* **2016**, *39*, 412–426. [CrossRef]
14. Marohnić, T.; Basan, R. Study of Monotonic Properties' Relevance for Estimation of Cyclic Yield Stress and Ramberg–Osgood Parameters of Steels. *J. Mater. Eng. Perform.* **2016**, *25*, 4812–4823. [CrossRef]
15. Basan, R.; Franulović, M.; Prebil, I.; Črnjarić-Žic, N. Analysis of strain-life fatigue parameters and behavior of different groups of metallic materials. *Int. J. Fatigue* **2011**, *33*, 484–491. [CrossRef]
16. Basan, R.; Marohnić, T.; Prebil, I.; Franulović, M. Preliminary investigation of the existence of correlation between cyclic Ramberg-Osgood parameters and monotonic properties of low-alloy steels. In Proceedings of the 3rd International Conference Mechanical Technologies and Structural Materials MTSM 2013, Split, Croatia, 26–27 September 2013; Živković, D., Ed.; Croatian society for mechanical technologies: Split, Croatia, 2013.
17. Marohnić, T.; Basan, R.; Franulović, M. Evaluation of the Possibility of Estimating Cyclic Stress-strain Parameters and Curves from Monotonic Properties of Steels. *Procedia Eng.* **2015**, *101*, 277–284. [CrossRef]
18. Smith, R.W.; Hirschberg, M.H.; Manson, S.S. *Fatigue Behavior of Materials under Strain Cycling in Low and Intermediate Life Range*; Technical Note D-1574; National Aeronautics and Space Administration (NASA): Washington, DC, USA, 1963.
19. Zhang, Z.; Wu, W.; Chen, D.; Sun, Q.; Zhao, W. New Formula Relating the Yield Stress-Strain with the Strength Coefficient and the Strain-Hardening Exponent. *J. Mater. Eng. Perform.* **2004**, *13*, 509–512.
20. Landgraf, R.W.; Morrow, J.; Endo, T. Determination of cyclic stress-strain curve. *J. Mater.* **1969**, *4*, 176–188.
21. Devore, J. *Probability and Statistics for Engineering and the Sciences*; Brooks/Cole Cengage Learning: Boston, MA, USA, 2011.
22. *Minitab 17 Statistical Software*, Product version 17.3.1; Computer Software; Minitab, Inc.: State College, PA, USA, 2010.
23. Boller, C.; Seeger, T. *Materials Data for Cyclic Loading, Part A–D*, 1st ed.; Elsevier: Amsterdam, The Netherlands, 1987.
24. Basan, R. MATDAT Materials Properties Database, Version 1.1. 2011. Available online: http://www.matdat.com/ (accessed on 15 January 2016).

metals **MDPI**

Article

Role of Microstructure Heterogeneity on Fatigue Crack Propagation of Low-Alloyed PM Steels in the As-Sintered Condition

Saba Mousavinasab and Carl Blais *

Département de génie des mines, de la métallurgieet des matériaux, Université Laval, Québec, QC G1V 0A6, Canada; saba.mousavinasab.1@ulaval.ca
* Correspondence: carl.blais@gmn.ulaval.ca; Tel.: +1-418-656-2049

Academic Editor: Filippo Berto
Received: 17 November 2016; Accepted: 14 February 2017; Published: 17 February 2017

Abstract: Due to their lower production costs, powder metallurgy (PM) steels are increasingly being considered for replacing wrought counterparts. Nevertheless, the presence of a non-negligible volume fraction of porosity in typical PM steels makes their use difficult, especially in applications where cyclic loading is involved. On the other hand, PM offers the possibility of obtaining steel microstructures that cannot be found in wrought. Indeed, by adequately using alloying strategies based on admixing, pre-alloying, diffusion bonding or combinations of those, it is possible to tailor the final microstructure to obtain a distribution of phases that could possibly increase the fatigue resistance of PM steel components. Therefore, a detailed study of the effect of different microstructural phases on fatigue crack propagation in PM steels was performed using admixed nickel PM steels (FN0208) as well as pre-alloyed PM steels (FL5208). Specimens were pressed and sintered to a density of 7.3 g/cm^3 in order to specifically investigate the effect of matrix microstructure on fatigue properties. Fatigue crack growth rates were measured at four different R-ratios, 0.1, 0.3, 0.5 and 0.7 for both PM steels. The negative effect of increasing the R-ratio on fatigue properties was observed for both alloys. The crack propagation path was characterized using quantitative image analysis of fracture surfaces. Measurements of roughness profile and volume fractions of each phase along the crack path were made to determine the preferred crack path. Weak Ni-rich ferritic rings in the FN0208 series (heterogeneous microstructure) caused a larger crack deflection compared to the more homogeneous microstructure of the FL5208 series. It was determined that, contrary to results reported in literature, crack propagation does not pass through retained austenite areas even though fatigue cracks propagated predominantly along prior particle boundaries, i.e., intergranular fracture.

Keywords: PM steels; heterogeneous microstructure; retained austenite; fatigue crack growth; crack path tortuosity

---

1. Introduction

Utilization of powder metallurgy (PM) steel components has significantly increased in recent years due their ability to be processed to near-net shapes, their intrinsic sustainability as well as their lower production costs. Most of the applications targeted by powder metallurgy are found in the automotive industry. Moreover, these applications increasingly involve undergoing cyclic loading that require better or equivalent fatigue properties compared to their wrought counterparts [1,2]. Fatigue is a complex phenomenon in PM steels since it is controlled by several characteristics. Porosity and matrix microstructure are the two most important parameters that need to be considered when studying the mechanical properties of PM parts [3].

Porosity reduces the effective load bearing cross section, through which mechanical properties such as strength and ductility are negatively affected [4–6]. Pores can also change the stress distribution at micro-levels, leading to stress concentration at the pores. Hence, the volume fraction of porosity is an important parameter that affects mechanical properties, especially fatigue resistance [7]. In low density parts, total volume of porosity is the major parameter on mechanical properties, whereas in high density ones, pore characteristics including size, shape and distribution as well as matrix microstructure happen to be the dominant parameters [8].

Apart from porosity, the other specific feature of PM materials is their heterogeneous microstructure. Microstructure of a PM material depends on the production process and the alloying technique. Excluding the pre-alloying technique, the other ones namely admixed, diffusion alloyed and hybrid will typically result in a heterogeneous microstructure [9]. This heterogeneity occurs mainly because of the incomplete diffusion of alloying elements that occurs due to low diffusion rates of alloying elements, large particle size, high repulsion between elements due to their chemical potential and some processing variables such as insufficient sintering temperature and time [9,10]. For a given density, fatigue properties of a PM part are mainly controlled by its microstructure. The effect of microstructure gets more attraction when involving a PM steel with heterogeneous microstructure.

Fatigue crack propagation is controlled by local properties at the crack tip, through which the fatigue behavior of a material is determined. A fatigue crack will face different microstructural phases while propagating in a heterogeneous microstructure PM steel. Due to the interaction of a crack with other cracks and microstructural barriers, such as porosity and grain boundaries, crack growth rate can decrease or cracks can even be arrested and stopped [11]. The behavior of fatigue cracks confronting these phases has not been characterized with exactitude, thus leading to divergent points of view in the literature.

In a study by Andersson and Lindqvist [12], fatigue behavior of a heterogeneous PM steel (Fe-4Ni-1.5Cu-0.5Mo-0.8C) was studied through which austenite was characterized as a ductile phase that can delay or stop the crack propagation. In another study on a diffusion alloyed PM steel as base powder with a microstructure of ferrite, pearlite, martensite in Cu/Mo rich areas (sinter necks) and austenite in Ni-rich areas (around the pores), it was also observed that short cracks are stopped by austenitic areas, whereas the long ones circumvented them. Therefore, it would appear that both mechanisms lower the propagation rate of fatigue cracks in PM steel [3]. Bergmark and Alzati [13], investigated the short cracks in Fe-4Ni-2Cu-1.5Mo-0.7C PM steel, using successive grinding to obtain 3D crack path observations. They found that Ni-rich austenite could not retard crack propagation by itself and the reason for not seeing this propagation might be related to plastic deformation and smearing.

Although there exist several studies on the effect of microstructure on fatigue crack propagation in PM steels, unified conclusions have not yet been reached. Different approaches used for demonstrating the effect of microstructure on fatigue crack growth as well as unsupported interpretations based on insufficient results could be the reason for these contradictory deductions. In this study, in order to have a complete comprehension of the effect of different phases on fatigue crack propagation behavior of PM steels, fatigue crack growth rate of two common alloys with the same density but different microstructures (homogeneous and heterogeneous) were measured at four *R*-ratios. The crack path was thoroughly analyzed in optical microscopy to characterize the crack behavior in facing different microstructural phases. The two-parameter approach and crack closure were also considered to complete the investigation.

2. Materials and Experimental Procedure

2.1. Materials

Two common PM steels produced by different alloying strategies were chosen in order to provide two types of microstructures (heterogeneous and homogeneous). The powder mixture utilized for obtaining the heterogeneous microstructure contained nickel (Vale Canada type 123) that was admixed to an iron powder (Rio Tinto Metal Powders Atomet 1001), whereas the powder mixture that yielded the homogeneous microstructure had chromium and molybdenum as alloying elements and was obtained using the pre-alloying technique (North American Höganäs Astaloy CrL). These alloys were selected based on their inclusions in MPIF standard 35 and by the fact that they have a similar content in alloying elements. Graphite additions (Asbury natural graphite grade 1651) were adjusted to yield a combined carbon content of 0.6 wt. % for both mixtures. The exact chemical compositions of the materials studied are given in Table 1.

The specimens were made using a double-press/double-sinter approach. The powder mixtures were pressed and sintered to achieve the same density of 7.3 g/cm^3. Final sintering was done at 1120 °C for 30 min in a 90%N$_2$-10%H$_2$ atmosphere. Archimedes' principle (MPIF standard 42) was used to measure the density of the samples. Porosity of the samples was also characterized in terms of size and shape distribution by image analysis.

Table 1. Main alloying elements of the studied materials (wt. %).

Alloy	Fe	C	Ni	Mo	Cr
Admixed (FN-0208)	bal.	0.6	2	-	-
Pre-alloyed (FL-5208)	bal.	0.6	-	0.2	1.4

2.2. Fatigue Crack Growth Tests

The specimens preparation as well as the fatigue crack growth testing were perfomed in accordance with ASTM-E647-13 standard [14]. The sintered samples were machined into standard compact C(T) specimens using electrodischarge machining (EDM). Fatigue crack growth experiments were conducted at room temperature on an MTS servo-hydraulic testing machine (model 810 with a load cell of 10 KN) operating at a sinusoidal frequency of 10 Hz. Humidity was recorded at all times and tests were conducted in relative humidity ranging between 25% and 38%.

In order to provide a sharpened fatigue crack and also to remove the effect of the machining at the notch, fatigue pre-cracking was perfomerd with an a/w ratio of 0.2 using a constant ΔK procedure. The fatigue crack growth rates experiments were conducted at constant R-ratio following a constant-force-amplitude test procedure. This test procedure is suitable for crack growth rates above 10^{-8} m/cycle (Paris regime), which is the area of interest in this research. In order to study the effect of microstructure at different fatigue conditions, four different R-ratios (0.1, 0.3, 0.5 and 0.7) were also considered. Crack lengths were recorded at recommended intervals using the compliance method.

2.3. Characterization Techniques

Investigations were performed on the fatigue fracture surfaces as well as sections taken from the fractured specimens. The fractured specimens were cut vertically, i.e., along the crack path in order to follow and study the crack propagation path. An electroless Ni-plating process was used to coat the fracture surfaces before the cut was made. This was done to protect the details of the crack path. The transverse cross-sections were then prepared using the standard procedure of metallographic sample preparation for PM specimens.

Digital image analysis techniques were used on the as-polished and the etched specimens to characterize pore morphology and measure the volume fraction of each phase respectively. The etched transverse cross-sections were also used to follow the crack path in optical microscopy. In order to perform the quantitative analysis of the fracture surface, the volume fraction of each phase as well as the crack length along the crack path in the Paris regime were measured using an image analysis software (Clemex Vision, Clemex Technologies Inc., Longueuil, QC, Canada). Microhardness tests were performed using a test force of 25 gf applied for 13 s to help ascertain phase determination. Fatigue fracture surfaces were also characterized using scanning electron microscopy (SEM, JEOL Ltd., Tokyo, Japan).

3. Results

3.1. Microstructure

The typical microstructure of admixed and pre-alloyed PM steels considered in this study is shown in Figure 1. It can be seen that the admixed nickel-containing PM steel shows a heterogeneous microstructure of pearlite (P), Ni-rich ferrite (F) and retained austenite (RA) while the pre-alloyed one has a homogeneous pearlitic microstructure. The volume fractions of microstructural phases and the microhardness values of each phase is shown in Table 2.

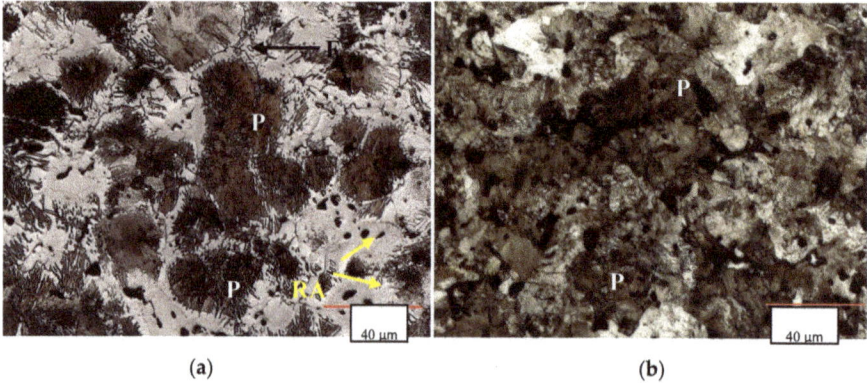

(a)   (b)

Figure 1. Microstructure of (**a**) admixed and (**b**) pre-alloyed powder metallurgy (PM) steels, etched by Nital 2%.

Table 2. Volume fraction and micro-hardness values of the phases present in the two series of specimens studied.

Alloy/Phase Hardness	Pearlite	Retained Austenite	Ni-Rich Ferrite
FN0208	69.5	9.7	20.8
Microhardness (HV)	280 ± 6	177 ± 10	193 ± 13
FL5208	98.5	1.5	0
Microhardness (HV)	290 ± 12	-	-

Low diffusion rate of nickel into an iron matrix at conventional sintering time and temperature as well as the repulsion between nickel and carbon atoms are responsible for the non-uniform distribution of nickel when this element is admixed. The Ni distribution gradient throughout the matrix leads to the formation of different microstructural phases upon cooling and consequently a heterogeneous microstructure is generated [15,16]. Ni-rich ferritic areas around pearlitic grains, bright areas in Figure 1, form due to the lack of carbon (Ni-rich/C-lean) for pearlite formation. Retained austenite regions are mainly present where prior Ni particles were located and nickel concentration is relatively high (more than 10 wt. %) [17].

Along with the microstructure, pore morphology (pore size and shape distribution) can also affect the fatigue crack propagation in high density PM steels [8]. Pore morphology was characterized on as-polished samples of both types of PM steels using optical microscopy. The filled area of each pore was measured and used as a criterion for its size. Pore shape factor was also calculated using the $F = 4\pi A/P^2$ expression, where A and P are the area and the perimeter of pores respectively. Figure 2 shows the size and shape distribution of the admixed and pre-alloyed PM steels. It can be seen that the difference between the pore morphology of the two alloys is negligible. Therefore, it can be concluded that any difference in their fatigue crack propagation behavior could be attributed to their microstructures. In other words, the effect of microstructure can be studied and compared between the two alloys without the interference of the pore morphology effect.

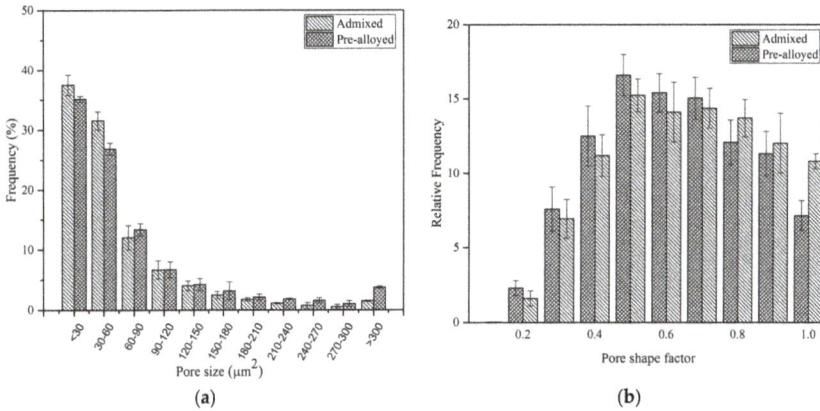

Figure 2. Distribution of the pore (**a**) size and (**b**) shape for both admixed and pre-alloyed sinter-hardened PM steels.

3.2. Fatigue Crack Growth Rate

Fatigue crack growth rate data acquired from the tests at four *R*-ratios are plotted against the stress intensity factor range on a logarithmic scale (Figure 3). Paris law ($da/dN = C\Delta k^m$) was then employed to the steady-state region of the plots to obtain m and C, the so-called Paris components, which are the empirical constants depending on material properties and testing conditions [18,19].

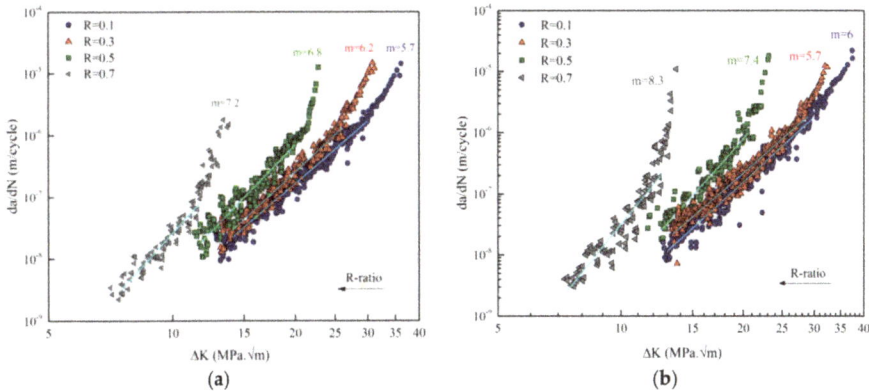

Figure 3. Fatigue crack growth rate vs. stress intensity factor range of the (**a**) admixed and (**b**) pre-alloyed PM steels at four tested *R*-ratios.

The variation of *m*—i.e., the slope of the linear section of the plot—versus *R*-ratio for both alloys is shown in Figure 4. According to Figures 3 and 4, it can be seen that the fatigue crack growth rate increases in both alloys by increasing the *R*-ratio due to the increase in minimum load. Nevertheless, this increase is more pronounced in the pre-alloyed PM steel, which is characterized by a homogeneous microstructure. In other words, the slope of the Paris regime, *m*, increases more rapidly by increasing the *R*-ratio. In pre-alloyed PM steels, *m* is almost the same for *R*-ratios of 0.1

and 0.3 but it increases intensively as the *R*-ratio increases from 0.3 to 0.7. This trend shows that the homogeneous microstructure is more sensitive to monotonic contribution since increasing the *R*-ratio indicates the higher dominant static effect [20].

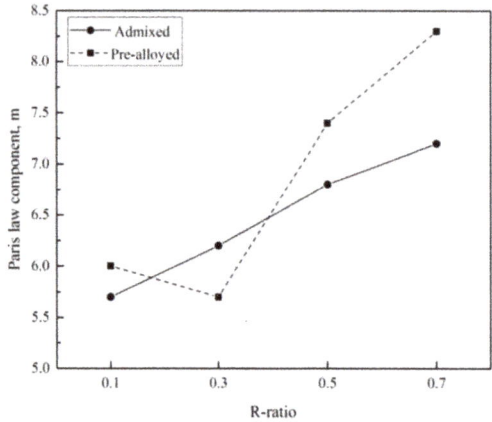

Figure 4. Variation of the slope (*m*) of the Paris regime versus the *R*-ratio.

4. Discussion

4.1. Fatigue Crack Growth (FCG) Behavior

The FCGs of the admixed and pre-alloyed PM steels should be compared quantitatively in order to understand the effect of microstructure homogeneity/heterogeneity on fatigue crack propagation. The FCGs for both types of microstructure were calculated using the Paris law and the derived Paris components, *m* and *C*. It is to be noted that *m* and *C* were derived from the fatigue crack growth rate (d*a*/d*N*) versus Δ*K* plots. The ratios of the calculated FCGs of the admixed PM steel to the FCG of the pre-alloyed PM steels were then obtained (for the purpose of comparison) and plotted against the stress intensity range (Δ*K*) at different *R*-ratios. Figure 5 shows the variation of the calculated ratios (d*a*/d*N* of the admixed to d*a*/d*N* of the pre-alloyed PM steels) against Δ*K* at four tested *R*-ratios.

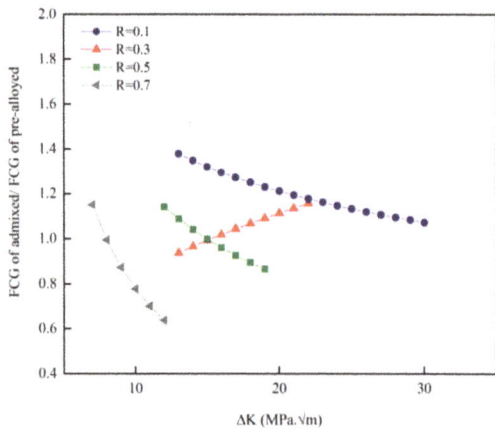

Figure 5. Variation of the fatigue crack growths' (FCGs') ratios (admixed/pre-alloyed) against Δ*K* in the Paris regime.

At $R = 0.1$, the FCG of the admixed PM steel is larger than the FCG of the pre-alloyed PM steel at low ΔKs (small crack lengths) but they became the same by increasing the ΔK. The calculated FCGs' ratio at $R = 0.3$ and 0.5 changes roughly between 0.8–1.2, throughout the applicable range of ΔK, which shows that the fatigue crack growth rates are almost the same for heterogeneous and homogeneous microstructures at these R-ratios. At $R = 0.7$, the fatigue crack growth rate in both types of alloys are similar at low ΔKs (small crack lengths) but it seems that by increasing the stress intensity factor range, the fatigue crack grew faster in the homogeneous microstructure since the calculated FCGs' ratio approaches the value of 0.6. The discussed trend of variation of fatigue crack growth rates with increasing R-ratio in both types of microstructures shows the higher sensitivity of the homogeneous microstructure to the localization of strain and plasticity at the crack tip.

In order to better understand the fatigue mechanisms involved in the tested PM steels, the two-parameter approach was also used. In this approach two parameters of K_{max} and ΔK are considered as the contributing parameters in fatigue crack growth. Therefore, there exists two crack tip driving forces and consequently two thresholds to be satisfied for crack growth to occur. ΔK vs. K_{max} plots that considered as fatigue maps are L-shaped plots that demonstrate the interplay of these two crack tip driving forces at any given crack growth rates [21]. These types of plots for both tested PM steels are shown in Figure 6. Since there were no experiments at negative R-ratios, the vertical parts of the plots are missing and the discussion will be made only using the horizontal part.

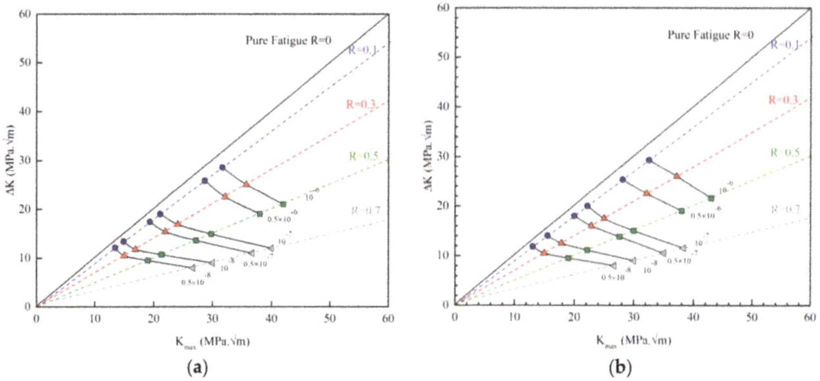

Figure 6. ΔK–K_{max} plots of (a) admixed (b) pre-alloyed PM steels, defining two limiting values at different crack growth rates.

By increasing the crack growth rate, the deviation (tilt) of the horizontal asymptote of the L-shape plots towards the K_{max} axis is slightly larger in the pre-alloyed PM steel compared to the one for the admixed PM steel. This shows that the pre-alloyed PM steel with a homogeneous microstructure is more affected by the K_{max}. In other words, this kind of microstructure is slightly more sensitive to monotonic contribution in fatigue damage. It can also be said that the crack tip is more sensitive to plasticity and strain localization in these kinds of microstructures [20–22].

Since there is a deviation (tilt) in horizontal asymptote of the L-shaped plots of both PM steels compared to a perfect L-shaped plot, it can be concluded that monotonic and/or environmental contribution were present for both types of microstructures [22]. In order to quantify this contribution, the crack growth trajectory maps can be used. By following the limiting values of the two driving forces (ΔK^* and K_{max}^*) at different crack growth rates in ΔK–K_{max} plots, the crack growth trajectory maps can be generated. A 45° line in these plots ($\Delta K^* = K_{max}^*$) is a pure fatigue line where only the cyclic damage is contributing in fatigue and any deviation from this line would determine the intrinsic mechanisms contributing in fatigue [21,22]. Figure 7 presents the trajectory maps of both types of alloys that have been derived from their ΔK–K_{max} plots.

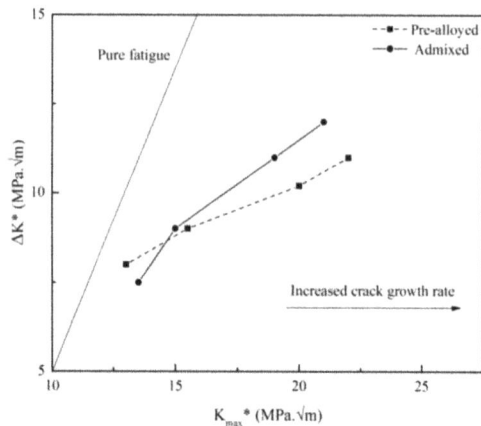

Figure 7. Crack growth trajectory maps of the admixed and pre-alloyed PM steels.

It can be seen that the trajectory path of the admixed PM steel is almost parallel to the pure fatigue line at low crack growth rates but deviates from it by increasing the crack growth rate. Being parallel to pure fatigue line at low crack growth rates indicates that fatigue is controlled only by cyclic plasticity whereas the divergence from this reference line at higher crack growth rates shows the effect of environment and monotonic contribution. The trajectory path of the pre-alloyed PM steel continuously diverges from the pure fatigue line with almost a constant angle.

Therefore, it can be concluded that the mechanisms are different in both types of microstructures, especially at low crack growth rates. The divergence from the pure fatigue line indicates that the environmental contribution increases with increasing crack growth rate. Besides, the deviation of the trajectory path towards the K_{max} axis, which is higher for the pre-alloyed PM steel, corresponds to superimposed K_{max} governing processes [20,22]. In other words, the amount of monotonic contribution is higher for the pre-alloyed steel with a homogeneous microstructure, which is in agreement with results of the variation of the slope of Paris regime, *m*, against the *R*-ratio.

4.2. Crack Closure Analysis

Fatigue crack closure should also be considered in studying the fatigue behavior of a material. Crack closure happens when two crack faces are in premature contact upon the unloading portion of a cycle. Elber [23] found out that due to the presence of the plastic zone at the wake of a fatigue crack, the crack could become closed far before the tensile stress reaches zero. A monotonic plastic zone, which is formed during loading to the maximum load, will lead to the permanent elongation in the loading direction. During unloading, this plastically elongated area will be under compression and consequently cause the premature crack closure even before reaching the minimum load. This type of crack closure is called plasticity-induced crack closure and it is reduced by increasing the *R*-ratio, i.e., increasing the minimum load [24,25].

When the crack is closed, the stress intensity at the crack tip decreases. Therefore, in order to analyze the degree of the crack closure, fatigue crack growth rates data should be plotted against ΔK_{eff}, which is equal to $K_{max}-K_{open}$. Fatigue crack closure were measured according to the ASTM E-647-13 guidelines using the load-displacement data obtained from the crack opening gauge during the test. Figure 8 shows the da/dN versus ΔK and ΔK_{eff} for both admixed and pre-alloyed PM steels at different *R*-ratios.

It can be seen that the amount of crack closure in pre-alloyed PM steels (Figure 8b) is the same at all *R*-ratios, i.e., it does not decrease by increasing the minimum load as expected [24,26]. This could be related to several causes such as different microstructure of PM steels compared to wrought ones and/or other crack closure controlling mechanisms such as the mismatch between upper and lower fracture surfaces and mixed mode sliding of the crack. The presence of porosity in the microstructure of PM steels most likely changes the expected trend of plasticity induced crack closure with increasing R-ratio. Nevertheless, according to previous studies [20,27], the decreasing trend of crack closure with R-ratio was observed in PM steels with either heterogeneous or homogeneous microstructure. Therefore, there should be another mechanism involved in the crack closure of these PM steels.

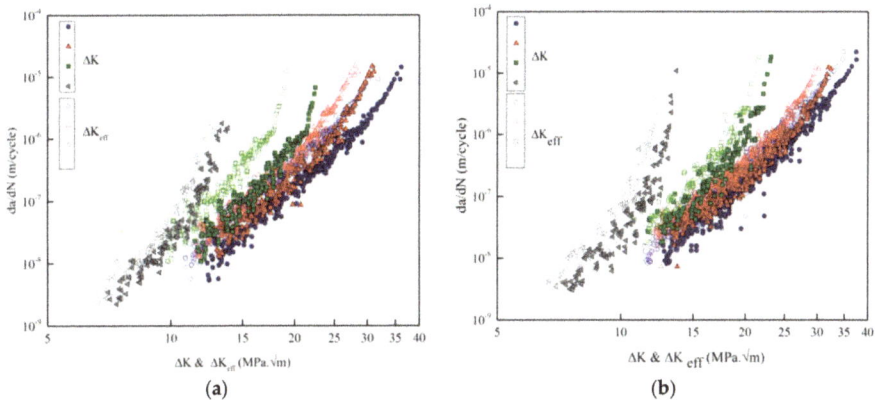

Figure 8. Fatigue crack growth rates of (**a**) admixed and (**b**) pre-alloyed PM steels versus ΔK and ΔK_{eff}.

Due to the fact that crack closure mechanisms could act synergistically, another mechanism or mechanisms are probably acting at the same time as the plasticity induced crack closure, causing the degree of the crack closure to be constant instead of decreasing (with increasing the R-ratio). The most probable mechanisms the mismatch between the upper and lower fracture surfaces that happens when two faces of the crack do not fit perfectly during unloading due to the crack deflection and/or the mixed mode sliding of the crack. This type of closure is called roughness-induced crack closure and is possible for both heterogeneous and homogeneous microstructures [24].

Figure 8a shows the degree of crack closure for the admixed PM steel at different R-ratios. It can be seen that the crack closure effect is almost constant for R-ratios of 0.1, 0.3 and 0.5, while it decreases when the R-ratio increases to 0.7. This diminution could be related to the formation of the reversed plastic zone that leads to the lower compression stress applied on the crack faces. The formation of this reversed plastic zone requires a sufficient increment in the local stress in the reversed direction and its size is around a quarter of the previous plastic zone [24]. The same as for the pre-alloyed PM steels, other closure mechanisms such as the mismatch between the upper and lower fracture surfaces and mixed mode sliding, might be contributing in the crack closure of the admixed PM steel that are causing the observed anomaly in the crack closure trend against the R-ratio. This phenomenon was observed in a recent article that studied similar PM steels [27].

4.3. Crack Path Analysis and Fractography

The effect of microstructure on fatigue crack growth behavior can be investigated using quantitative analysis of the fracture surface. This analysis can be done on a fracture surface as well as a fracture profile, which is the intersection of the fracture surface with a metallographic sectioning plane. A typical fracture profile is usually an irregular and complex line to which a profile roughness parameter (R_L) can be attributed. R_L is defined as follows:

$$R_L = \lambda_0 / L$$

where λ_0 is the actual measured length of fracture profile and L is equivalent straight path (projected length). R_L can vary between 1 and ∞ depending on the irregularity of the fracture profile [28,29]. This parameter can be used to quantify the crack deflection and could be an accurate criterion for comparing the fatigue crack paths in different microstructures. As mentioned above, metallographic sections along the crack paths (vertical sections) were made on each fracture surface. λ_0 was measured on micrographs obtained from the vertical sections and R_L was then calculated. Figure 9 shows the variation of profile roughness parameters for both alloys at different R-ratios.

It can be seen that fatigue crack has a tortuous path in both pre-alloyed and admixed PM steels with homogeneous and heterogeneous microstructures since R_L is larger than 1. Generally, a crack will change its path when confronting microstructural barriers or by passing through the grain and/or inter-particle boundaries. As previously observed by these authors [27] as well as being reported elsewhere in literature [20], a crack can be deflected by the Fe_3C lamellae of pearlite colonies. Since pearlite is the main constituent of the microstructure in both alloys (Table 1), it can be concluded that the reason for crack deviation in both types of microstructures is related to the presence of cementite lamellae. Moreover, R_L will be larger than one in the case of intergranular fracture in which the crack passes through boundaries, either grain boundaries or prior inter-particle boundaries for PM steels. Figure 10 shows the intergranular type of fracture at $R = 0.1$ and 0.7 for the admixed PM steel.

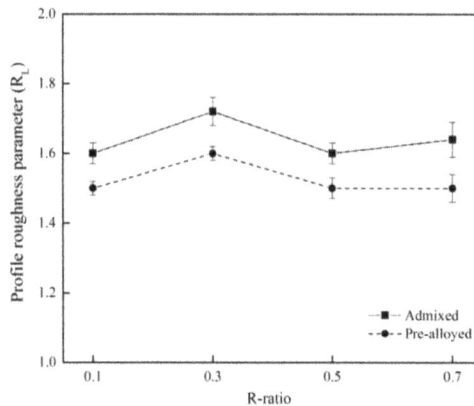

Figure 9. Profile roughness parameters for admixed and pre-alloyed at four R-ratios.

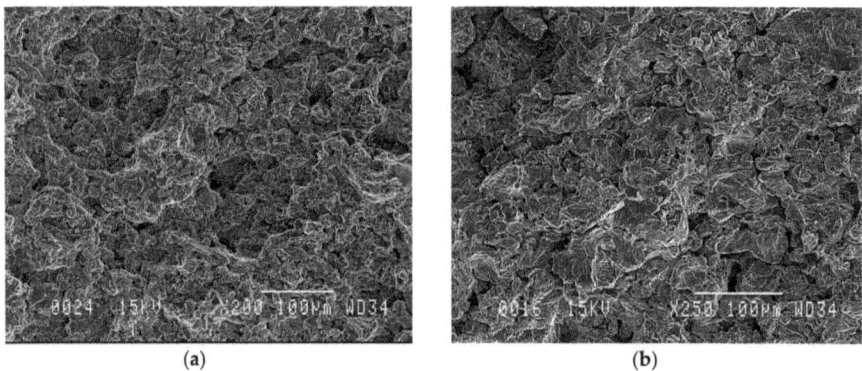

Figure 10. Fracture surfaces of the admixed PM steels tested at (**a**) $R = 0.1$; (**b**) $R = 0.7$. Intergranular fracture is the principal mode of crack propagation in both micrographs.

Profile roughness parameters at four tested R-ratios for the pre-alloyed and admixed PM steels show that the amount of R_L has a small increase at $R = 0.3$, which could be related to the interaction of the crack with the pulled-out cementite lamellae from their pearlitic matrix during fatigue. This was previously observed by these authors for the same alloys but in the sinter-hardened condition as well as being reported in the literature [20,27]. After an increase in roughness parameter at the mentioned R-ratio, there is a decrease for both types of microstructures that could be attributed to the increase in the percentage of transgranular fracture brought about by increasing the minimum load, i.e., increasing the R-ratio. Fracture type (intergranular/transgranular) can be better studied using scanning electron microscopy. Figure 11 shows the SEM micrograph of the admixed PM steel at $R = 0.7$ at higher magnification. It can be seen that by increasing the minimum load, the fracture surface shows more transgranular fracture proving that the crack has the tendency of propagating through the grains rather than at grain/inter-particle boundaries at high R-ratios. It should be mentioned that the principal mode of fracture is still the intergranular one and only the percentage of the transgranular fracture has been increased by increasing the R-ratio.

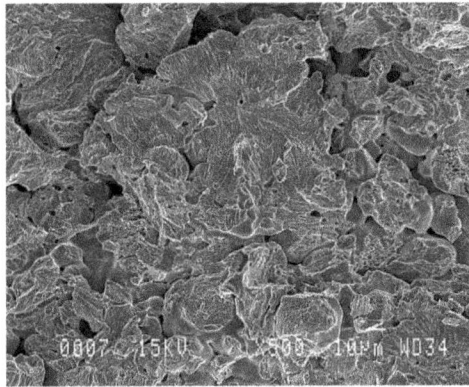

Figure 11. Fracture surface of the admixed PM steel at $R = 0.7$. The grains have been cut by the crack showing the transgranular fracture in some parts of the fracture surface.

The effect of microstructure heterogeneity/homogeneity can also be studied using the profile roughness parameter. As seen in Figure 9, there is a difference between the R_L of admixed and pre-alloyed PM steels at all R-ratios. The crack seems to be more deviated and tortuous in admixed PM steels with heterogeneous microstructure at all tested R-ratios although the difference is small. In order to investigate the reason for this difference, crack paths from the two types of microstructures were studied more precisely. Thus, the crack paths were followed in the area of interest (Paris regime) and volume fractions of the different phases found along the crack path were measured and compared to their amount in the bulk material. A ratio of the volume fraction of each phase in the crack path divided by its amount in bulk of the material was then calculated for all microstructural phases present to make the comparison easier. Figure 12 shows the calculated ratios for the admixed PM steels at four tested R-ratios. It should be noted that the mentioned ratio would be one in the case of the homogeneous microstructures since there was no difference between the volume fraction of microstructural phases along the crack path and in the bulk of the materials.

If there exists a difference between the volume fraction of a phase along the crack path compared to its amount in the material, the calculated ratio is then larger than one indicating that there is a preferred path for crack propagation in that specific phase [28]. As it can be seen in Figure 12, this ratio is three for Ni-rich ferrite in the admixed PM steels at all R-ratios. In other words, the amount of Ni-rich ferrite in the crack path is three times larger than its amount in the material indicating that this phase is a preferred path for a fatigue crack to propagate. These Ni-rich ferritic regions are formed around pearlitic grains due to the non-homogeneous distribution of nickel and carbon throughout the iron matrix. Due to the rapid diffusion of carbon into the iron matrix and the repulsion between carbon and nickel during sintering at 1120 °C, particle peripheries and sinter necks will be Ni-rich/C-lean austenite, whereas the particle interiors (cores) will be C-rich/Ni-lean austenite. The C-rich/Ni-lean austenitic regions will transform into pearlite during cooling and will be surrounded by the transformed Ni-rich ferritic rings from Ni-rich/C-lean austenitic regions [16]. The presence of these weak ferritic rings around pearlite caused the difference between the degree of crack deviation in heterogeneous and homogeneous microstructures.

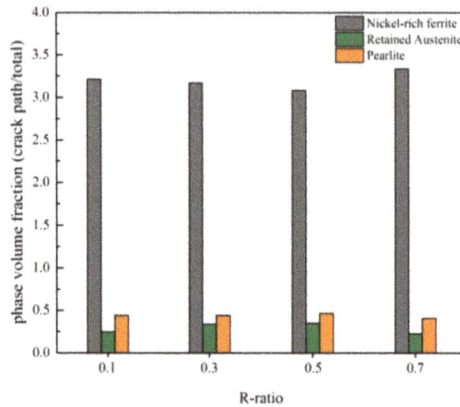

Figure 12. The ratios of the volume fraction of each phase in the crack path to its amounts in the material, for admixed PM steel at different *R*-ratios.

According to Figure 12, the crack has mostly passed through Ni-rich ferritic rings in the heterogeneous microstructure of admixed PM steels specimens. As stated above, Ni rich ferrite forms at the periphery of particles and/or sinter necks. Therefore, the results clearly indicate that the crack has propagated through prior particle boundaries in accordance with the intergranular nature of fracture characterized in SEM. Figure 13 shows a secondary crack in Ni-rich ferritic sinter necks in admixed PM steel. Retained austenite regions are formed mainly where the prior nickel particles were located. Therefore, these regions are also present at prior particle boundaries. Hence, it is expected that these regions were also being passed by the crack according to the intergranular nature of the fracture. On the other hand, the volume fraction of retained austenite along the crack path is less than half of its amount in the bulk material. This indicates that although these austenitic regions have relatively low strength, they are not preferred areas for fatigue crack propagation at least in comparison with Ni-rich ferritic regions. This can be related to the inherent ductility and toughness of retained austenite [8]. It should be mentioned that these conclusion were made based on the microstructure of one of the common admixed PM steels (FN0208) that usually have 10 vol. % of retained austenite and the results might be different when larger volume fractions of retained austenite are present.

These measures were only made on one half of the fractured surfaces, so it is important to understand and characterize the other half as well. In other words, to verify the above conclusions on fatigue cracks propagating preferably through specific microstructural phases, the other side of the crack path also needs to be investigated. On the other hand, aligning matching fracture surfaces is a daunting task especially when the latter where Ni-coated for metallography preparation. Thus, a strategy was devised using statistical techniques. Random micrographs were taken from each admixed PM steels' samples and the volume fractions of different prior particle boundaries were measured. Figure 14 shows a pie chart of the present prior particle boundaries in the admixed PM steels series. This chart shows the probability of the presence of different phases on the other side of the crack. It can be seen that the highest percentage of the prior particle boundaries (65.36%) is between Ni-rich ferritic regions, and 30% of the boundaries correspond to retained austenite and ferrite. Therefore, it can be said that the previous conclusions about the preferred crack path through the Ni-rich ferritic regions is valid due to the high possibility of the presence of the same phase on the other side of the crack.

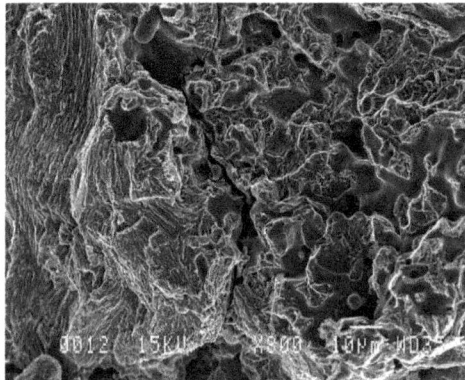

Figure 13. SEM micrograph of the fracture surface of the admixed PM steel at $R = 0.5$. A secondary crack has passed through a sinter neck.

Moreover, the amount of ferrite in the crack path will still be much larger than its amount in the bulk material even when considering the possibility of the presence of retained austenite in the other half. Figure 14 also shows that 4.5% out of 6% porosity of the admixed PM steels are located on ferritic boundaries. This amount should be subtracted from the calculated amount of ferrite along the crack path to obtain the exact volume fraction of this phase. This means that the crack has passed through pores as well as the microstructural phases but this was not distinguishable when measuring the volume fraction of each phase along the crack path.

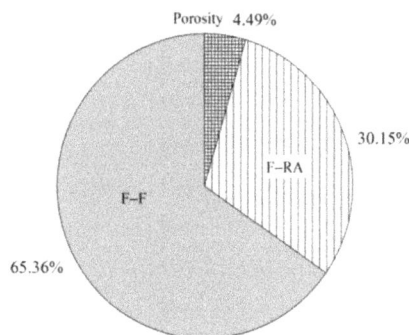

Figure 14. The pie chart of the present prior particle boundaries in admixed PM steels (F–F = Ferrite–Ferrite boundary; F–RA = Ferrite–Retained Austenite boundary).

According to Figures 5 and 9, it can be seen that at $R = 0.3$ and 0.5 fatigue crack growth rate is almost the same for both PM steels even though the crack path is more deviated in the admixed series with a heterogeneous microstructure. In the admixed PM steel with a heterogeneous microstructure made of pearlite, retained austenite and Ni-rich ferrite, the latter phase is responsible for crack deviation, whereas in the homogeneous pearlitic microstructure of the pre-alloyed PM steel, it is the cementite lamellae that are causing the tortuous crack path. Having the same FCGs while having different degrees of crack tortuosity could be related either to the small difference in their degree of crack tortuosity and/or the fact that the crack deviation is not the only parameter affecting the fatigue crack propagation rate. It was previously reported that crack deflection increases the fatigue resistance of a material but its contribution is small [30].

The reason for having the same FCGs in two microstructures with different crack path tortuosity is most likely related to the higher fatigue crack propagation rate in low strength Ni-rich ferrite compared to high strength pearlite. In other words, a fatigue crack propagated faster in Ni-rich ferrite with a more tortuous path and consequently the combination of these two caused the same fatigue crack growth rates as the one with a less tortuous crack path in pearlite with higher strength. Hence, inherent properties such as the strength of the microstructural phases in which the crack propagates should also be considered along with the crack path deviation. Higher fatigue crack growth rates in ferritic regions were also observed in a previous paper on the fatigue behavior of sinter-hardened PM steels [27] as well as reported by Deng et al. [20]. Therefore, care should be taken when selecting PM steels for applications where fatigue resistance is important. Chemistry as well as alloying strategy should be selected to minimize the formation of Ni-rich ferrite.

5. Conclusions

In the present study, the effect of microstructure heterogeneity on fatigue crack propagation was studied to better understand the fatigue crack growth behavior of two common PM steels with heterogeneous and homogeneous microstructures. The main findings of our work are summarized as follows:

Fatigue crack growth rates in both types of alloys increased by increasing the R-ratio. This was the result of the higher monotonic contribution and strain localization at the crack tip by increasing minimum loads. PM steel with a homogenous microstructure showed slightly higher sensitivity to the strain localization and crack tip plasticity.

Different crack closure mechanisms were involved in both PM steels, which have caused the observed anomaly in the variation trend of crack closure versus the R-ratio. Plasticity induced crack closure caused by the plastically deformed area in the wake of the crack as well as the roughness-induced crack closure caused by the crack deflection and/or the mixed mode sliding of the crack faces acting synergistically are the most probable mechanisms.

Profile roughness parameter (R_L) was larger than 1 for both PM steels at all tested R-ratios showing the tortuous and deflected crack path in both heterogeneous and homogeneous microstructures. The presence of a considerable amount of pearlite in both alloys was found to be the reason for the observed deflection, since cracks had their path significantly deviated when facing the hard cementite lamellae. The crack path was longer and more deviated in the heterogeneous microstructure of the admixed PM steel because of the Ni-rich ferritic rings around the pearlitic grains. The fatigue fracture surfaces showed that the prinicpal mode of fracture was the intergranular one, but by increasing the R-ratio the percentage of transgranular, fracture increased.

Although the presence of Ni-rich ferritic rings around the pearlitic grains in heterogeneous microstructure PM steel have caused the more deviated crack path, the fatigue crack growth rates in this alloy were almost identical to the pre-alloyed one with homogeneous microstructure, which indicates that the crack propagates faster in Ni-rich ferrite than in pearlite. This higher propagation rate is counterbalanced by the crack deviation and led to the same FCGs in both microstructures. The crack passed along the prior particle boundaries where the retained austenite was located, but it did not pass through this phase, indicating that austenitic regions do not constitute preferable paths compared to Ni-rich ferritic areas. The statistical analysis of the other half of the fracture surface verified the conclusions made on the effect of each phase on the crack propagation.

PM steels to be used in applications where fatigue resistance is a dominant factor should be engineered to minimize the formation of Ni-rich ferritic areas.

Acknowledgments: The authors would like to acknowledge the Natural Science and Engineering Research Council of Canada (NSERC (Grant No.: 210856817)) as well as the Network of Centers of Excellence—Auto21 (Grant No.: C502-CPM) for their financial support.

Author Contributions: Mousavinasab and Blais conceived and designed the experiments; Mousavinasab performed the experiments; Both authors analyzed the data; Mousavinasab wrote the paper and Blais reviewed it before submission.

Conflicts of Interest: The authors declare no conflict of interest.

References

1. Cimino, T.M.; Rutz, H.G.; Graham, A.H.; Murphy, T.F. The Effect of Microstructure on Fatigue Properties of Ferrous P/M Materials. *Adv. Powder Metall. Part. Mater.* **1997**, *2*, 13.
2. Rutz, H.; Murphy, T.; Cimino, T. The effect of microstructure on fatigue properties of high density ferrous P/M materials. *Adv. Powder Metall. Part. Mater.* **1996**, *4*, 13.
3. Abdoos, H.; Khorsand, H.; Shahani, A.R. Fatigue behavior of diffusion bonded powder metallurgy steel with heterogeneous microstructure. *Mater. Des.* **2009**, *30*, 1026–1031. [CrossRef]
4. Danninger, H.; Spoljaric, D.; Weiss, B. Microstructural features limiting the performance of P/M steels. *Int. J. Powder Metall.* **1997**, *33*, 43–53.
5. Holmes, J.; Queeney, R.A. Fatigue crack initiation in a porous steel. *Powder Metall.* **1985**, *28*, 231–235. [CrossRef]
6. Gerard, D.A.; Koss, D.A. Low cycle fatigue crack initiation: modeling the effect of porosity. *Int. J. Powder Metall.* **1990**, *26*, 337–343.
7. Polasik, S.; Williams, J.J.; Chawla, N.; Narasimhan, K.S. Fatigue crack initiation and propagation in ferrous powder metallurgy alloys. *Adv. Powder Metall. Part. Mater.* **2001**, 10–172.
8. Carabajar, S.; Verdu, C.; Hamel, A.; Fougeres, R. Fatigue behaviour of a nickel alloyed sintered steel. *Mater. Sci. Eng. A* **1998**, *257*, 225–234. [CrossRef]
9. Saritas, S.; Causton, R.; James, B.; Lawley, A. Effect of Microstructural Inhomogeneities on the Fatigue Crack Growth Response of a Prealloyed and Two Hybrid P/M Steels. *Adv. Powder Metall. Part. Mater.* **2002**, *5*, 5–136.
10. Murphy, T.F.; Lindsley, B.A.; Schade, C.T. A metallographic examination into fatigue-crack initiation and growth in ferrous PM materials. *Int. J. Powder Metall.* **2013**, *49*, 23–34.
11. Polasik, S.J.; Williams, J.J.; Chawla, N. Fatigue crack initiation and propagation of binder-treated powder metallurgy steels. *Metall. Mater. Trans. A* **2002**, *33*, 73–81. [CrossRef]
12. Andersson, O.; Lindqvist, B. Benefits of heterogeneous structures for the fatigue behaviour of PM steels. *Met. Powder Rep.* **1990**, *45*, 765–768. [CrossRef]
13. Bergmark, A.; Alzati, L. Fatigue crack path in Cu-Ni-Mo alloyed PM steel. *Fatigue Fract. Eng. Mater. Struct.* **2005**, *28*, 229–235. [CrossRef]
14. ASTM International. *Standard Test Method for Measurement of Fatigue Crack Growth Rates*; ASTM International: West Conshohocken, PA, USA, 2011.
15. Nabeel, M. Diffusion of Elemental Additives during Sintering. Master's Thesis, Royal Institute of Technology, Stockholm, Sweden, 2012.
16. Wu, M.; Hwang, K. Formation mechanism of weak ferrite areas in Ni-containing powder metal steels and methods of strengthening them. *Mater. Sci. Eng. A* **2010**, *527*, 5421–5429. [CrossRef]
17. Tougas, B.; Blais, C.; Larouche, M.; Chagnon, F.; Powders, R.T.M.; Pelletier, S. Characterization of the Formation of Nickel Rich Areas in PM Nickel Steels and Their Effect on Mechanical Properties. *Adv. Powder Metall. Part. Mater.* **2012**, *5*, 19–33.
18. Meyers, M.A.; Chawla, K.K. *Mechanical Behavior of Materials*; Cambridge University Press: Cambridge, UK, 2009; Volume 547.
19. Krupp, U. *Fatigue Crack Propagation in Metals and Alloys*; Wiley: New York, NY, USA, 2007.
20. Deng, X.; Piotrowski, G.; Chawla, N.; Narasimhan, K. Fatigue crack growth behavior of hybrid and prealloyed sintered steels: Part II. Fatigue behavior. *Mater. Sci. Eng. A* **2008**, *491*, 28–38. [CrossRef]

21. Sadananda, K.; Vasudevan, A.K. Crack tip driving forces and crack growth representation under fatigue. *Int. J. Fatigue* **2004**, *26*, 39–47. [CrossRef]
22. Sadananda, K.; Vasudevan, A.K. Fatigue crack growth mechanisms in steels. *Int. J. Fatigue* **2003**, *25*, 899–914. [CrossRef]
23. Elber, W. The significance of fatigue crack closure.   In *Damage Tolerance in Aircraft Structures*; ASTM International: West Conshohocken, PA, USA, 1971.
24. Schijve, J. *Fatigue of Structures and Materials*; Springer: New York, NY, USA, 2001.
25. Schijve, J. Fatigue of structures and materials in the 20th century and the state of the art. *Int. J. Fatigue* **2003**, *25*, 679–702. [CrossRef]
26. Bathias, C. *Fatigue of Materials and Structures*; Wiley: New York, NY, USA, 2013; Volume 53.
27. Mousavinasab, S.; Blais, C. Study of the effect of microstructure heterogeneity on fatigue crack propagation of low-alloyed PM steels. *Mater. Sci. Eng. A* **2016**, *667*, 444–453. [CrossRef]
28. Murphy, T.F. Evaluation of PM fracture surfaces using quantitative fractography. *Int. J. Powder Metall.* **2009**, *45*, 49–61.
29. Gokhale, A.; Underwood, E. A general method for estimation of fracture surface roughness: Part I. Theoretical aspects. *Metall. Trans. A* **1990**, *21*, 1193–1199. [CrossRef]
30. Piotrowski, G.B.; Deng, X.; Chawla, N.; Narasimhan, K.S.; Marucci, M.L. Fatigue-Crack Growth of Fe-0.85Mo-2Ni-0.6C Steels with a Heterogeneous Microstructure. *Int. J. Powder Metall.* **2005**, *41*, 31–41.

metals

MDPI

Article

Surface Characteristics and Fatigue Behavior of Gradient Nano-Structured Magnesium Alloy

Xiaohui Zhao [1], Yanjun Zhang [1] and Yu Liu [2],*

[1]   Key Laboratory of Automobile Materials, School of Materials Science and Engineering, Jilin University, Changchun 130025, China; zhaoxiaohui@jlu.edu.cn (X.Z.); yanjunzhang2016@163.com (Y.Z.)
[2]   School of Mechanical Science and Engineering, Jilin University, Changchun 130025, China
*   Correspondence: liuyuu@jlu.edu.cn; Tel.: +86-431-85095316; Fax: +86-431-8509-5316

Academic Editor: Filippo Berto
Received: 22 December 2016; Accepted: 14 February 2017; Published: 20 February 2017

Abstract: High-frequency impacting and rolling was applied on AZ31B magnesium alloy to obtain a gradient nano-structured surface. Surface characteristics were experimentally investigated, and the nanocrystallization mechanism is discussed in detail. Results showed that the gradient nano-structure with the characteristics of work hardening, compressive residual stress and a smooth surface was induced on the treated surface. Grains on the top surface were generally refined to around 20 nm. Twins, dislocations and dynamic recrystallization dominated the grain refinement process. Fatigue strength of the treated specimens corresponding to 10^7 cycles was increased by 28.6% compared to that of the as-received specimens. The work hardened layer induced by high-frequency impacting and rolling is the major reason to improve fatigue life.

Keywords: severe plastic deformation; hardening; twins; fatigue; compressive residual stress; nanocrystallization

1. Introduction

Magnesium alloys are extensively used in aerospace, electronics, military and other industry fields for weight reduction effectiveness [1–3]. However, the application of magnesium alloy has been limited by its low absolute strength and poor fatigue performance. The above problems will hopefully be resolved with the rapid development of surface strengthening technology [4–6]. Surface nanocrystallization induced by severe plastic deformation (SPD) has been drawing more and more attention in recent years [7–9]. Current surface strengthening methods generally include shot peening (SP), ultrasonic shot peening (USP), surface mechanical attrition treatment (SMAT), cryogenic burnishing, etc. The past studies showed that SPD on the material surface can effectively improve the fatigue performance of structural components. For example, Gao [10,11] investigated the influence of SP on the tension-tension fatigue properties of two kinds of high strength Ti alloys and found that fatigue strength for 1×10^7 cycles can be increased by 27%–29%. Arakawa et al. [12] studied the effect of USP on the fatigue characteristics of high strength structural materials of hydroelectric facilities, and the results showed that the fatigue limit of the treated material was approximately increased by 60%. For the high strength materials, the higher compressive residual stress can be formed and maintained after surface strengthening treatment, and the fatigue properties can be better improved [13,14]. As to low strength materials (magnesium or aluminum alloy), the lower compressive residual stress induced by surface strengthening treatment is easily relaxed at the moment a random or occasional high load appears, which will result in a limited improvement of fatigue properties. Therefore, it is unusual to improve the fatigue properties of low strength materials by surface strengthening treatment. For example, Lu et al. only studied the surface nanocrystallization mechanism of magnesium alloy based on SMAT technology [6]. Tsai et al. [15] also only analyzed the relationship between the microstructure and properties for different processing parameters of SMAT. Jordan Moering et al. studied the microstructural and textural development along the depth under the effect of SMAT by electron backscattered diffraction (EBSD) [16]. Pu et al. [17] proposed a new method called cryogenic burnishing. For the Mg-Al-Zn alloy treated by cryogenic burnishing, a large increase in microhardness from 0.86–1.35 GPa was obtained, and grains were refined from 12 μm down to 263 nm. However, the fatigue performance of Mg-Al-Zn alloy based on cryogenic burnishing treatment still has not been studied.

Generally, SPD methods are easy to result in the increase of surface roughness, cause a large stress concentration and deteriorate the improvement of fatigue life. High-frequency impacting and rolling (HFIR) as a kind of newly-developed surface nanocrystallization technology has been attracting more and more attention in recent years. HFIR treatment can generate a gradient nano-structured surface, as well as significantly reduce surface roughness. In this study, HFIR was used to treat AZ31B magnesium alloy to obtain a gradient nano-structured surface. Then, the nanocrystallization mechanism was discussed in detail, and the effect of surface nanocrystallization on the fatigue fracture mechanism of magnesium alloy was analyzed.

2. Material and Experiments

2.1. Material

The material used was AZ31B magnesium alloy, and the microstructure is shown in Figure 1, which consists of unevenly-equiaxed grains and a small amount of second β-$Mg_{17}Al_{12}$ phase precipitating along grain boundaries. The yield strength and the ultimate tensile strength through experimental measurement are about 174 MPa and 246 MPa, respectively.

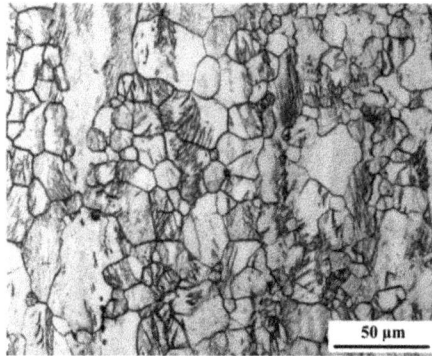

Figure 1. The microstructure of AZ31B magnesium alloy.

2.2. Fatigue Specimen and HFIR Treatment

The basic principle of HFIR treatment is similar to the ultrasonic surface rolling process proposed by Wang [18–20]. However, HFIR has a higher frequency, which can ensure the uniformity of grain refinement. Before fatigue tests, the hour glass specimen shown in Figure 2 was treated by HFIR.

The detailed processing parameters include impacting frequency of ~27 kHz, an amplitude of ~7.0 μm, a rotating speed of 110 r/min, a feed quantity of 0.1 mm/r and a suitable static force. The tension-tension axial fatigue tests of as-received and HFIR specimens were performed on a 20 kN high-frequency fatigue testing machine (CIMACH, Changchun, China) under constant amplitude load with the stress ratio $R = 0.5$ at room temperature to evaluate the effect of the applied treatments on the fatigue strength.

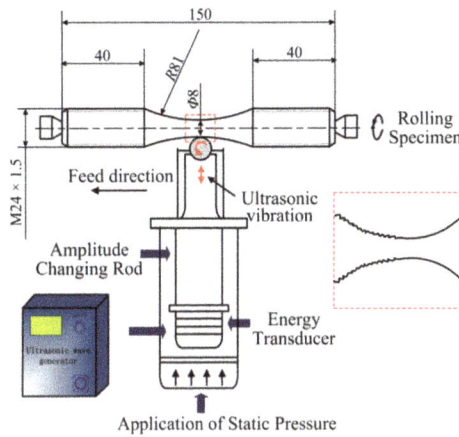

Figure 2. The hour glass fatigue specimen and high-frequency impacting and rolling (HFIR) schematic diagram.

2.3. Microstructure Observation

The overall appearance of the deformed layer was observed by an Axio Imager A1m type of optical microscope (OM, ZEISS, Oberkochen, Germany). The microstructures of the top surface, 80 µm and 160 µm from the treated surface were examined by a JEM-2100F type of transmission electron microscopy (TEM, JEOL, Tokyo, Japan). For TEM examination, flat specimen instead of round bar specimen was used to get the thin slices of different depths. HFIR parameters for the flat specimen are the same as those for the round bar specimen. Only rotating feed of the round bar specimen was replaced by the linear feed of the flat specimen. Fatigue fractures were investigated through a VEGA3 type of scanning electron microscope (SEM, TESCAN, Brno, Czech Republic).

2.4. Surface Roughness, Microhardness and Residual Stress Measurements

Surface roughness was detected using a 2201 type of surface roughness tester (Harbin Measuring and Cutting Tool Group Co., LTD, Harbin, China). Surface roughness (R_a) after HFIR was decreased from 1.38 down to 0.121. Such low surface roughness is hardly realized by other SPD methods. Microhardness variation from the top surface to the interior was determined by an MH-3 Vickers microhardness tester (Shanghai Hengqi Precision Machinery Plant, Shanghai, China) with a load of 100 g and loading time of 10 s. The X-ray scattering technique was used to test residual stress profiles [19].

3. Results and Discussion

3.1. Microstructure and Microhardness Distribution of the Work-Hardened Layer

Figure 3 presents the cross-sectional microstructure of HFIR specimen. Clearly, due to the effect of high-frequency impacting, the microstructure of material surface is distinctly smaller and denser compared with that of matrix. Meanwhile, the constant rolling pressure also caused the plastic slip of the surface material, which is marked by the yellow arrows.

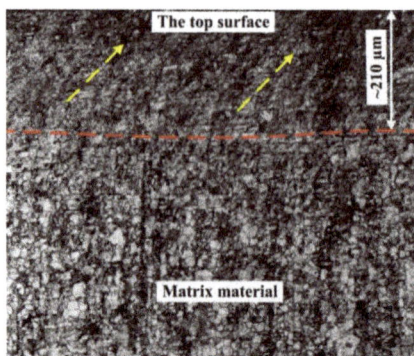

Figure 3. The cross-sectional microstructure of the HFIR specimen.

The thickness of work hardened layer is around 210 μm, which can be clearly distinguished from the matrix (see the red line in Figure 3). The thickness of the work-hardened layer was also estimated through the variation of microhardness along depth direction, as shown in Figure 4.

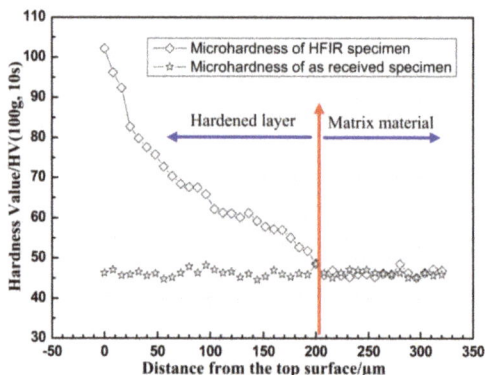

Figure 4. The cross-sectional microhardness of as-received and HFIR specimen.

The measurement results indicated that the highest microhardness was obtained on the top surface and then decreases gradually as going into the matrix. This microhardness variation was caused by the gradient nanostructure from the top surface to the matrix.

3.2. Discussion of Surface Strengthening Mechanism

With the increase of plastic deformation, twins, dislocations and dynamic recrystallization (DRX) gradually dominate the surface nanocrystallization process, which has been studied by Lu et al. [6] based on AZ91D alloy treated by SMAT. Here, the surface strengthening process of AZ31B alloy treated by HFIR was further investigated through TEM.

3.2.1. The 160-μm Deformed Layer from the Top Surface

The TEM bright field image at about 160 μm below the top surface of the treated AZ31B alloy is shown in Figure 5. As can be observed, the deformed twins or lath-like substructures with a width of 300–500 nm have been produced in some original grains. A large amount of dislocations and dislocation tangles can be observed in these twins or lath-like substructures in Figure 5a. In addition, dislocation cells were also found in this deformed layer (see Figure 5b). There is still a regional deformed feature under the condition of relatively homogeneous plastic deformation in the same deformed layer.

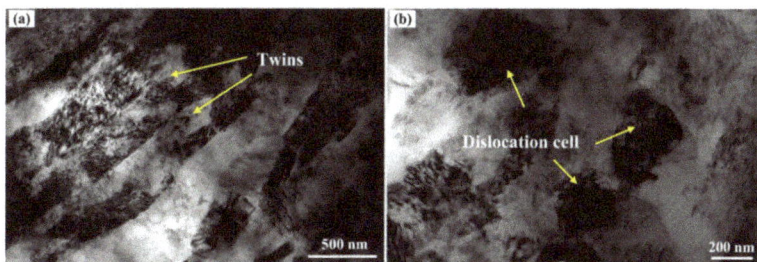

Figure 5. TEM image of the 160-µm deformed layer from the top surface. (**a**) Twins and (**b**) dislocation cells.

3.2.2. The 80-µm Deformed Layer from the Top Surface

The TEM bright field image at about 80 µm below the top surface is shown in Figure 6. Some clear boundaries of sub-grains have been formed in this layer, and the size of sub-grains has reached around 200 nm. Meanwhile, there is a high density of dislocations and dislocation tangles around black precipitated phases in part of the sub-grains. These sub-grains will be further separated by the above high density of dislocations at the next stage. For example, the smaller Sub-grains A and B in Figure 6 have been nearly formed under the effect of dislocations activities.

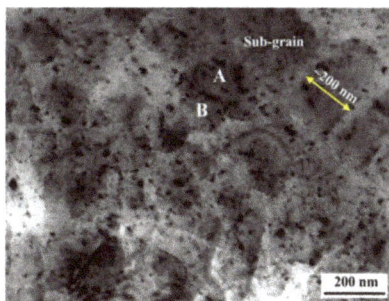

Figure 6. TEM image of the 80-µm deformed layer from the top surface.

3.2.3. The Top Surface after HFIR Treatment

Figure 7 showed the TEM bright field image and corresponding selected area diffraction (SAD) pattern obtained at the top surface of HFIR specimen. The bright field image indicates that the vast majority of sub-grains have been turned into equiaxed nanocrystals with clear boundaries, and the average size is measured to be around 20 nm. As shown in Figure 7b, the SAD pattern is composed of partially continuous diffraction rings, which further confirms that as-received large crystalline grains have been broken down to nanograins at this layer.

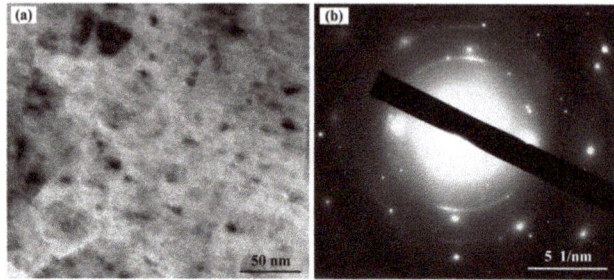

Figure 7. The TEM image of the top surface after HFIR treatment. (**a**) Nanocrystals and (**b**) SAD pattern corresponding to (**a**).

From the above TEM images, it can be concluded that the plastic deformation was governed by twinning at the early stage, and then, the non-basal dislocation slip systems were activated. With the increase of dislocation movement, sub-grains were gradually formed, and finally, nanocrystals were generated. In addition, Tan et al. [21], Myshlyaev et al. [22], Eddahbi et al. [23] and Galiyev et al. [24] reported that the DRX as an important grain refinement mechanism was observed in Mg alloys when grain size was refined to the micrometer range. Lu et al. [6] reported the effect of DRX on grain refinement of Mg alloy in the nanometer scale. HFIR has a higher impacting frequency, which can generate a higher strain rate and decrease the DRX temperature. Therefore, DRX is also an important grain refinement mechanism for the AZ31B magnesium alloy treated by HFIR technology. Figure 8 is a typical example of the newly-formed grains (A and B) in the severely deformed twin platelet near the twin interface.

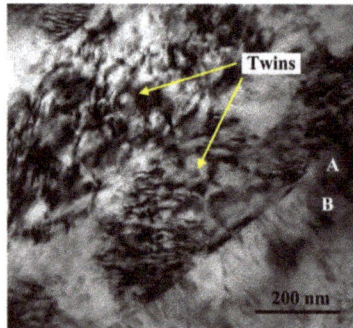

Figure 8. The TEM bright field image of the newly-generated grains through DRX.

3.3. S-N Curves

Fatigue data (nominal stress vs. cycles) of as-received and HFIR specimens were done in a linear fit in the double logarithm coordinate (confidence level of 95%). Fatigue *S-N* curves of as-received and HFIR specimens are shown in Figure 9. The run-out tests are marked by arrows. The fatigue strength of HFIR and as-received specimens corresponding to 10^7 cycles is about 180 MPa and 140 MPa, respectively. The former shows an improvement of almost 28.6% with respect to that of the latter. Although the improvement of 28.6% is not very high compared to the results of high strength steel treated by shot peening [10,12], such improvement is still considerable because of the low absolute strength of magnesium alloy.

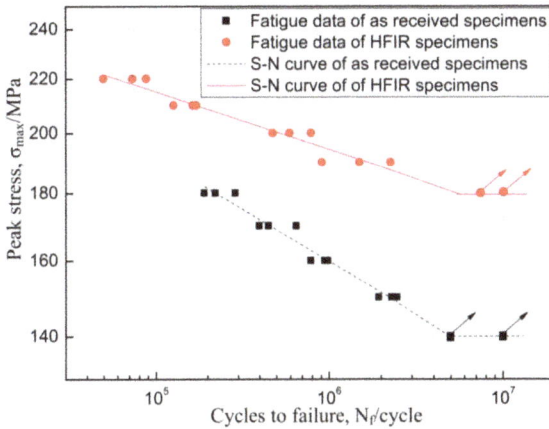

Figure 9. *S-N* curves of as-received and HFIR specimens.

3.4. Discussion of Fatigue Life Improvement

It is known that the fatigue fracture contains three stages, which are crack initiation, crack propagation and instantaneous fracture. The previous study had shown that the crack initiation stage could account for ~90% of the total fatigue life [25], and if the crack initiation point starts from the specimen interior, the fatigue properties will be significantly improved. Therefore, it is necessary to investigate the positions of fatigue crack sources of as-received and HFIR specimens. By comparing fracture surfaces of as-received and HFIR specimens, it is found that the positions of crack sources were different. All crack sources of as-received specimens initiated from the top surface (Figure 10a), while part of the crack sources of HFIR specimens initiated at the subsurface. For example, Figure 10b showed that one crack source of the HFIR specimen was found in the subsurface.

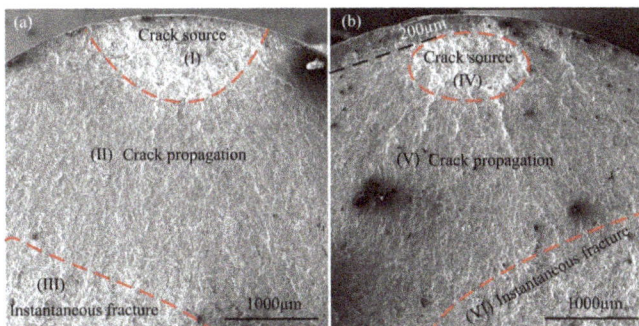

Figure 10. Fracture surfaces of as-received and HFIR specimen. (**a**) as-received specimens and (**b**) HFIR specimens.

First, the relatively higher surface roughness or worse surface smoothness of the as-received specimen generated by machining marks can easily cause stress concentration, and the crack source was prone to form at the top surface. After HFIR, the surface roughness of the specimen was decreased to R_a = 0.121. Generally, the decrease of surface roughness will effectively reduce surface stress concentration and delay crack initiation.

Second, except for the smooth surface, which can prolong the time of crack initiation, the in-depth residual stress also plays an important role. Figure 11 shows the residual stress distribution of the HFIR specimen. The results indicate that the top surface and subsurface of HFIR specimen are covered by the compressive residual stress, which can effectively hinder crack initiation and promote crack stopping; this viewpoint has been widely demonstrated [26,27]. Therefore, the compressive residual stress in the surface layer is a factor to change the position of the crack source and improve fatigue life.

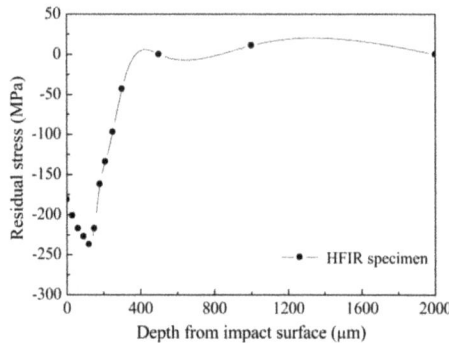

Figure 11. Residual stress distribution of the HFIR specimen.

In addition, McDowell Tanaka [28] illustrated that a high volume fraction of grain boundaries can stop dislocation sliding and restrain crack initiation. From Figures 6 and 7, it can be seen that grain refinement is quite obvious. Therefore, the grain refinement is also a factor to improve fatigue life by stopping dislocation sliding and restraining crack initiation. As the grain refinement process continues, the work hardened layer is formed (see Figure 4). The characteristics of high dense dislocation walls, dislocation tangles and sub-grains in the hardened layer can result in a lower crack propagation rate, which has been proven by Suresh and Li [25,26]. Thus, the work hardening is also a factor to influence fatigue life.

As we know, there are many methods related to fatigue life prediction and assessment [29,30]. However, in order to predict and evaluate the main factors that improve fatigue life after HFIR treatment, the correction formula of Baghrifard and Guagliano [31,32] based on the Eichlseder approach [30] by considering the ratio of the C_s coefficient and the ratio of *FWHM* for shot-peened specimen was referenced and described in Equation (1) [31].

$$\sigma_f = \sigma_{tf} \left[1 + \left(\frac{\sigma_{bf}}{\sigma_{tf}} - 1 \right) \left(\frac{\chi'}{2/b} \right)^{K_D} \right] \left(\frac{FWHM_P}{FWHM_{NP}} \right) \left(\frac{C_{sP}}{C_{sNP}} \right) \tag{1}$$

where σ_f is the fatigue limit; σ_{tf} is the fatigue limit in tension; σ_{bf} is the fatigue limit in bending; χ' is the relative stress gradient (RSG); K_D is the material parameter; b is the size parameter of the specimen; $FWHM_P$ is the full width at half maximum of the peening specimen; $FWHM_{NP}$ is the full width at half maximum of the non-peening specimen; C_{sP} is the surface roughness of the peening specimen; C_{NP} is the surface roughness of the non-peening specimen.

High-frequency impacting and rolling, the same as shot peening, also belongs to the surface strengthening method. Equation (1) is the fatigue criteria corresponding to the notched specimen. The ratio of the C_s coefficient and the ratio of *FWHM* are irrelevant to the stress gradient. Therefore, the effect rule of the ratio of C_s coefficient and the ratio of *FWHM* on fatigue life are also fit for the non-notched specimen.

The surface strain hardening index of *FWHM* was obtained from XRD analysis. In this study, the *FWHM* value of the HFIR specimen was improved by 20.1% compared to that of the as-received specimen. From Figure 9, we can know that HFIR specimens showed a fatigue strength improvement of almost 28.6% with respect to that of as-received specimens. Therefore, the work hardening is the most important factor to improve the fatigue properties of the HFIR specimen in this study.

As for the C_s coefficient, the surface of the as-received specimen before fatigue tests was polished by fine sandpaper along the axial direction of the specimen. Although the value of surface roughness is still larger, the micro-marks are parallel to the axial direction. For the tensile-tensile axial fatigue test, micro-marks in the axial direction will not produce a greater effect on the fatigue crack initiation.

For the residual stress, the value of compressive residual stress on the top surface of the magnesium alloy specimen treated by HFIR is lower than that of the high strength steel specimen treated by HFIR, and the compressive residual stress is easily relaxed when a random or occasional high load appears during fatigue tests.

To be sure, the smooth surface and compressive residual stress generated by HFIR played a positive role to improve fatigue life, but their effects are not the most important. The work hardening caused by HFIR is the key factor to influence the fatigue properties of magnesium alloy in this study.

3.5. Fatigue Fracture Process

The effect of the fatigue crack initiation stage on fatigue life has been systematically discussed in the previous section. As for the crack propagation and instantaneous fracture stages, the images corresponding to Figure 10 are shown in Figure 12. For the as-received specimen, fatigue cracks propagated in terms of the transcrystalline and crack propagation region presented fine striations. The orientation of fatigue striations is consistent and perpendicular to the direction of the crack propagation (Figure 12II). However, for the HFIR specimen, the orientation of the fatigue striations is relatively random due to the complex stress state in the subsurface (Figure 12V). The instantaneous fracture stages of the as-received and the HFIR specimen are characterized by transcrystalline ductile fracture with an equiaxed (Figure 12III) and tearing (Figure 12VI) type of dimple morphology, respectively.

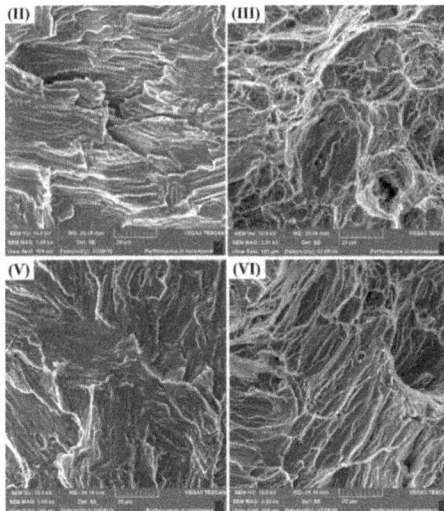

Figure 12. The high magnification images of as-received and HFIR specimens corresponding to Figure 10. (**II**) Crack propagation region of as-received specimen and (**III**) Instantaneous fracture region of as-received specimen. (**V**) Crack propagation region of HFIR specimen and (**VI**) Instantaneous fracture region of HFIR specimens.

4. Conclusions

The present work investigates the surface nano-enhanced mechanism and fatigue properties of AZ31B magnesium alloy subjected to HFIR treatment. After HFIR, grain refinement, work hardening, compressive residual stress and a smooth surface were induced on the surface of specimen. Grains were refined to equiaxed nanocrystals with an average size of 20 nm. The fatigue strength of HFIR specimens corresponding to 10^7 cycles was increased by 28.6% compared to that of as-received specimens. All crack sources of the as-received specimen initiated from the top surface, while part of the crack sources of the HFIR specimen initiated at the subsurface. The work hardening caused by HFIR is the most important factor to influence fatigue properties.

This study proved that the surface nano-enhanced method based on HFIR technology will have a good industrial application value in the prolonging life field.

Acknowledgments: This work was supported by the National Natural Science Foundation of China under Grant No. 51405182.

Author Contributions: Yanjun Zhang conceived and designed the experiments; Xiaohui Zhao, Yanjun Zhang, and Yu Liu performed the experiments, analyzed the data, and wrote the paper.

Conflicts of Interest: The authors declare no conflict of interest.

References

1. Asgari, H.; Szpunar, J.A.; Odeshi, A.G.; Zeng, L.J.; Olsson, E. Experimental and simulation analysis of texture formation and deformation mechanism of rolled AZ31B magnesium alloy under dynamic loading. *Mater. Sci. Eng. A* **2014**, *618*, 310–322. [CrossRef]
2. Li, Z.M.; Wang, Q.G.; Luo, A.A.; Fu, P.H.; Peng, L.M. Fatigue strength dependence on the ultimate tensile strength and microhardness in magnesium alloys. *Int J. Fatigue* **2015**, *80*, 468–476. [CrossRef]
3. Dogan, E.; Vaughan, M.W.; Wang, S.J.; Karaman, I.; Proust, G. Role of starting texture and deformation modes on low-temperature shear formability and shear localization of Mg-3Al-1Zn alloy. *Acta Mater.* **2015**, *89*, 408–422. [CrossRef]
4. Karimi, A.; Amini, S. Steel 7225 surface ultrafine structure and improvement of its mechanical properties using surface nanocrystallization technology by ultrasonic impact. *Int. J. Adv. Manuf. Technol.* **2016**, *83*, 1127–1134. [CrossRef]
5. Manafi, B.; Saeidi, M. Development of a novel severe plastic deformation method: Friction stir equal channel angular pressing. *Int. J. Adv. Manuf. Technol.* **2016**, *86*, 1367–1374. [CrossRef]
6. Sun, H.Q.; Shi, Y.N.; Zhang, M.X.; Lu, K. Plastic strain-induced grain refinement in the nanometer scale in a Mg alloy. *Acta Mater.* **2007**, *55*, 975–982. [CrossRef]
7. Trško, L.; Bokůvka, O.; Novýa, F.; Guagliano, M. Effect of severe shot peening on ultra-high-cycle fatigue of a low-alloy steel. *Mater. Des.* **2014**, *57*, 103–113. [CrossRef]
8. Fernández-Pariente, I.; Bagherifard, S.; Guagliano, M.; Ghelichi, R. Fatigue behavior of nitrided and shot peened steel with artificial small surface defects. *Eng. Fract. Mech.* **2013**, *103*, 2–9. [CrossRef]
9. Bagheri, S.; Guagliano, M. Review of shot peening processes to obtain nanocrystalline surfaces in metal alloys. *Surf. Eng.* **2009**, *25*, 3–14. [CrossRef]
10. Gao, Y.K. Influence of shot peening on tension-tension fatigue property of two high strength Ti alloys. *Surf. Eng.* **2006**, *22*, 299–303. [CrossRef]
11. Gao, Y.; Lu, F.; Yao, M. Influence of mechanical surface treatments on fatigue property of 30CrMnSiNi2A steel. *Surf. Eng.* **2005**, *21*, 325–328. [CrossRef]
12. Arakawa, J.; Kakuta, M.; Hayashi, Y.; Tanegashima, R.; Akebono, H.; Kato, M.; Sugeta, A. Fatigue strength of USP treated ASTM CA6NM for hydraulic turbine runner. *Surf. Eng.* **2014**, *30*, 662–669. [CrossRef]
13. Bagherifard, S.; Fernandez-Pariente, I.; Ghelichi, R.; Guagliano, M. Effect of severe shot peening on microstructure and fatigue strength of cast iron. *Int. J. Fatigue* **2014**, *65*, 64–70. [CrossRef]
14. Malaki, M.; Ding, H.T. A review of ultrasonic peening treatment. *Mater. Des.* **2015**, *87*, 1072–1086. [CrossRef]
15. Tsai, W.Y.; Huang, J.C.; Gao, Y.J.; Chung, Y.L.; Huang, G.R. Relationship between microstructure and properties for ultrasonic surface mechanical attrition treatment. *Scr. Mater.* **2015**, *103*, 45–48. [CrossRef]

16. Moering, J.; Ma, X.L.; Chen, G.Z.; Miao, P.F.; Li, G.Z.; Qian, G.; Mathaudhu, S.; Zhu, Y.T. The role of shear strain on texture and microstructural gradients in low carbon steel processed by Surface Mechanical Attrition Treatment. *Scr. Mater.* **2015**, *108*, 100–103. [CrossRef]

17. Pu, Z.; Yang, S.; Song, G.L.; Dillon, O.W., Jr.; Puleo, D.A.; Jawahir, I.S. Ultrafine-grained surface layer on Mg-Al-Zn alloy produced by cryogenic burnishing for enhanced corrosion resistance. *Scr. Mater.* **2011**, *65*, 520–523. [CrossRef]

18. Liu, Y.; Zhao, X.H.; Wang, D.P. Determination of the plastic properties of materials treated by ultrasonic surface rolling process through instrumented indentation. *Mater. Sci. Eng. A* **2014**, *600*, 21–31. [CrossRef]

19. Wang, T.; Wang, D.P.; Liu, G. Investigations on the nanocrystallization of 40Cr using ultrasonic surface rolling processing. *Appl. Surf. Sci.* **2008**, *255*, 1824–1829.

20. Liu, Y.; Zhao, X.H.; Wang, D.P. Effective FE model to predict surface layer characteristics of ultrasonic surface rolling with experimental validation. *Mater. Sci. Technol.* **2014**, *30*, 627–636. [CrossRef]

21. Tan, J.C.; Tan, M.J. Dynamic continuous recrystallization characteristics in two stage deformation of Mg-3Al-1Zn alloy sheet. *Mater. Sci. Eng. A* **2003**, *339*, 124–132. [CrossRef]

22. Myshlyaev, M.M.; McQueen, H.J.; Mwembela, A.; Konopleva, E. Twinning, dynamic recovery and recrystallization in hot worked Mg-Al-Zn alloy. *Mater. Sci. Eng. A* **2002**, *337*, 121–133. [CrossRef]

23. Eddahbi, M.; del Valle, J.A.; Pérez-Prado, M.T.; Ruano, O.A. Comparison of the microstructure and thermal stability of an AZ31 alloy processed by ECAP and large strain hot rolling. *Mater. Sci, Eng. A* **2005**, *410*, 308–311. [CrossRef]

24. Galiyev, A.; Kaibyshev, R.; Gottstein, G. Correlation of plastic deformation and dynamic recrystallization in magnesium alloy ZK60. *Acta Mater.* **2001**, *49*, 1199–1207. [CrossRef]

25. Suresh, S. *Fatigue of Materials*; Cambridge University Press: Cambridge, UK, 1998.

26. Li, D.; Chen, H.N.; Xu, H. The effect of nanostructured surface layer on the fatigue behaviors of a carbon steel. *Appl. Surf. Sci.* **2009**, *255*, 3811–3816. [CrossRef]

27. De los Rios, E.R.; Trull, M.; Levers, A. Modelling fatigue crack growth in shot-peened components of Al 2024-T351. *Fatigue Fract. Eng. Mater. Struct.* **2000**, *23*, 709–716. [CrossRef]

28. McDowell, D.L.; Dunne, F.P.E. Microstructure-sensitive computational modeling of fatigue crack formation. *Int. J. Fatigue* **2010**, *32*, 1521–1542. [CrossRef]

29. Zhu, S.P.; Lei, Q.; Huang, H.Z.; Yang, Y.J.; Peng, W.W. Mean stress effect correction in strain energy-based fatigue life prediction of metals. *Int. J. Damage Mech.* **2016**. [CrossRef]

30. Eichlseder, W. Fatigue analysis by local stress concept based on finite elementsresults. *Comput. Struct.* **2002**, *80*, 2109–2113. [CrossRef]

31. Bagherifard, S.; Guagliano, M. Application of different fatigue strength criteria on shot peenednotched parts. Part 2: Nominal and local stress approaches. *Appl. Surf. Sci.* **2014**, *289*, 173–179. [CrossRef]

32. Bagherifard, S.; Colombo, C.; Guagliano, M. Application of different fatigue strength criteria to shot peenednotched components. Part 1: Fracture Mechanics based approaches. *Appl. Surf. Sci.* **2014**, *289*, 180–187. [CrossRef]

metals

Article

Fatigue and Fracture Resistance of Heavy-Section Ferritic Ductile Cast Iron

Matteo Benedetti [1,*], Elisa Torresani [1], Vigilio Fontanari [1] and Danilo Lusuardi [2]

[1]   Department of Industrial Engineering, University of Trento, 38122 Trento, Italy;
      elisa.torresani@unitn.it (E.T.); vigilio.fontanari@unitn.it (V.F.)
[2]   Fonderie Ariotti S.p.A., 25030 Adro BS, Italy; danilo.lusuardi@fondariotti.it
*   Correspondence: matteo.benedetti@unitn.it; Tel.: +39-0461-282-457; Fax: +39-0461-281-977

Academic Editor: Filippo Berto
Received: 15 February 2017; Accepted: 6 March 2017; Published: 10 March 2017

Abstract: In this paper, we explore the effect of a long solidification time (12 h) on the mechanical properties of an EN-GJS-400-type ferritic ductile cast iron (DCI). For this purpose, static tensile, rotating bending fatigue, fatigue crack growth and fracture toughness tests are carried out on specimens extracted from the same casting. The obtained results are compared with those of similar materials published in the technical literature. Moreover, the discussion is complemented with metallurgical and fractographic analyses. It has been found that the long solidification time, representative of conditions arising in heavy-section castings, leads to an overgrowth of the graphite nodules and a partial degeneration into chunky graphite. With respect to minimum values prescribed for thick-walled ($t > 60$ mm) EN-GJS-400-15, the reduction in tensile strength and total elongation is equal to 20% and 75%, respectively. The rotating bending fatigue limit is reduced by 30% with respect to the standard EN-1563, reporting the results of fatigue tests employing laboratory samples extracted from thin-walled castings. Conversely, the resistance to fatigue crack growth is even superior and the fracture toughness comparable to that of conventional DCI.

Keywords: heavy-section ductile cast iron; chunky graphite; fatigue limit; fatigue crack growth resistance; fracture toughness

1. Introduction

Low production cost and excellent castability make ductile cast iron (DCI) the preferred material choice when low-to-moderately stressed mechanical components of complicated shape and large dimensions must be manufactured. It is used in many applications where strength and toughness are required, such as automotive parts, windmill parts, pipes, etc. The microstructural control of DCI is of paramount importance as it greatly influences a wide range of mechanical properties. The typical microstructure consists of graphite nodules dispersed in a matrix that can be ferritic, ferritic and/or pearlitic depending on the alloy formulation, the casting control and the final heat treatment [1,2]. The ferritic matrix displays usually the lower tensile and fatigue strength but also a noticeable ductility, which have promoted the DCI in the production of thick walled parts undergoing in service low/medium levels of mechanical stresses.

Casting large volume parts of weight on the order of tens of tons is particularly critical as the control of the final microstructure is rather difficult [3–5]. Consequently, solidification defects cannot be completely avoided, their presence must be somehow tolerated and different strategies are pursued to limit their detrimental effects. Typical design approaches are either to force the localization of the heterogeneous microstructure in regions undergoing very low "in service" stress levels or in regions that are removed during finishing operations. The choice is strongly connected to the knowledge of the detrimental effects of the microstructural defects on the mechanical behavior, specifically ductility and fatigue behavior [3]. The fatigue performance of DCI with different microstructures is widely analyzed in the technical literature [2–9]. Notably, Meneghetti et al. [6] showed the results of an extensive experimental study comparing different microstructures in a large range of fatigue lives. It can be observed that quite often, in research papers, fatigue tests are carried out on specimens taken from the Y-block or from small laboratory casting batches, thus accounting only partially for the heterogeneous microstructure arising in real parts. In fact, the final microstructure can be efficiently controlled in small materials batches, whereas, in real part, especially thick-walled components [3,4], some intrinsic defects are unavoidable. The difficult control of the casting process introduces into the material intrinsic defects, such as solidification cavities or poor graphite nodularity, dross and degenerated (chunky and spiky) graphite, which can be tolerated up to a certain level. Many works [1,3–5,7,10–12] found that these defects are preferential fatigue crack initiation sites; therefore, their distribution and the morphology are of particular importance when designing against fatigue. The results of laboratory tests are very useful for defining the role of the graphite morphology in the propagation of fatigue cracks through the specimen cross section [2,9–13]. In addition, the size and uniformity of distribution of graphite nodules can affect the properties of DCI, but with a lower effectiveness than graphite shape [14]. Indeed, graphite spheroids act like "crack-arresters" promoting the mechanical properties of the matrix; on the contrary, the presence of irregular shapes or non-spheroidal graphite such as chunky, exploded, spiky, vermicular and intercellular flake graphite worsens the mechanical properties. All these degenerated graphite morphologies are usually present in thick-walled components. The main factors that influence the degeneration of the graphite are the chemical composition and the cooling rate, especially in the case of the chunky graphite, which is the most frequent defect in heavy sections and is mainly observed in the thermal core [15–19]. In the technical literature, various theories have been proposed to explain the metallurgical processes responsible for the formation of chunky graphite [20–24] but none have received unanimous agreement. Nevertheless, it is well known that the chunky graphite is favored by high concentration of Si, Ni, Ca and carbon equivalent elements [25–27]. Small amounts of rare earth elements such as Ce, Nd, Pr, Sm and Gd induce graphite spheroidization [28–30]; however, if their concentration becomes excessive, they provoke the formation of chunky graphite due to the segregation at austenite grain boundaries [28,29]. The graphite morphology is influenced by another important factor that is also difficult to control in heavy-section castings, viz. the cooling rate [31–33]. Fast cooling allows a quick enveloping of graphite nodules by a shell composed of several austenite grains [34]. This helps to maintain the spheroidal graphite shape. The formation of the austenitic shell is delayed at low cooling rates. In this situation, liquid channels separate the austenitic grains and allow a longer contact time of the spherical graphite with the melt. Through these channels, carbon atoms diffuse towards the spheroidal graphite, causing their growth and degeneration [35]. Low-melting elements such as Al, Sn, and Pb tend to segregate thus delaying the formation of the austenite shell [34]. A similar effect is shown by segregating elements such as Mn, Mo, Cr, W, V and Ti, which are concentrated in the liquid channels [34]. A proper inoculation is an important factor to improve the nucleation factor of the melt, increase the nodularity and nodule count and reduce the carbides formation [36]. On the other hand, melt inoculation induces an increasing risk of chunky graphite formation in thick-walled components [37].

The mechanical properties are influenced by microstructure and solidification defects, and specifically the increase in non-nodular graphite content decreases the mechanical strength; indeed, the ultimate tensile strength and mostly the elongation to fracture are severely lowered, despite hardness and yield strength remain nearly unaffected [38,39]. Regarding the effect of the chunky graphite on the fatigue performance, different works have been published [11,40–42]. Mourujärvi et al. [40] analyzed the performance of a normalized DCI EN-GJS-800 containing different quantities of chunky graphite through rotating bending fatigue tests. They showed that the increase of chunky content up to 20% and more induces a significant decrease in fatigue lifetime, with a drop of fatigue limit on the order of 30%–40%. In another study [41], uniaxial pulsating fatigue tests were performed on ferritic DCI containing 40% of chunky graphite. In this case, a quite limited decrease in fatigue resistance is attained, with a reduction of fatigue limit of only 14%, whose main cause has been attributed to the presence of microporosities. Indeed, different works [43–45] showed that the fatigue limit is much more sensitive to surface defects than internal defects, where it is observed that micro-shrinkage cavities strongly influence the fatigue behavior of DCI castings. Foglio et al. [42] conducted fatigue tests on two different EN-GJS-400 DCI, in one of which the chunky graphite formation was induced by adding Ce-containing inoculants. Using a statistic analysis and applying the Murakami equation [39,46] to correct for the contribution of microporosity on the fatigue strength reduction, they estimated a decrease in the fatigue limit due to chunky graphite of 12%, which reflects a comparable drop in tensile strength equal to 14%.

From the above discussion, it is clear that the effect of chunky graphite on static and cyclic properties has been largely investigated in the literature. However, the design of high-added value components of large dimensions, such as those employed in MW-series wind turbines [47], cannot rely on the knowledge of these mechanical properties only. In fact, infinite life design approaches usually take large safety margins thus resulting in very heavy constructions, while safe life approaches suffer from large uncertainty on the actual lifetime of the component prior to its withdrawal from service. Damage tolerant design and structural health monitoring approaches have been then proposed to overcome these limitations. They were originally developed in the aeronautic context and have now been extended to the wind energy branch, especially in offshore and hardly-accessible installations [48,49]. According to these approaches, the fatigue life is exclusively thought of as the propagation of cracks up to a critical size leading to structural collapse or functionality loss. Therefore, they require a detailed characterization of the fracture toughness and the fatigue crack growth resistance of the castings in order to assess the critical crack size and the time to propagation to this critical value. In the present paper, we address this latter issue by investigating the crack growth resistance of a DCI produced under very slow cooling conditions, thus representative of the microstructural conditions arising in thick walled castings. For this purpose, static tensile, fatigue and fracture mechanics tests are performed to quantify the resistance to fatigue crack initiation and propagation as well as the fracture toughness. The paper is organized as follows. Section 2 describes the casting procedure followed to obtain a microstructure with degenerated chunky graphite and the experimental methods adopted for determining the mechanical properties. Section 3 illustrates the experimental results and discusses them on the basis of comparisons with literature data. The conclusions close the paper.

2. Material and Experimental Procedures

The experimentation is performed on an EN-GJS-400-type ferritic DCI, whose chemical composition is listed in Table 1. Small-scale castings are obtained from an innovative production line described in [50] and designed to reproduce the solidification conditions of heavy-section castings (with masses of tens of tons). It consists of an electric furnace where the solidification of a cylinder with 240 mm diameter and 260 mm height (about 80 kg mass) takes place under controlled conditions. The mold is made up of a special alumina-based refractory crucible, suitably designed to contain the liquid cast iron and to be handled with cast iron inside. The crucible is preheated at 1200 °C in the electric furnace, and then it is removed from the furnace and filled with the cast iron at 1350–1370 °C. The crucible is covered with a kerphalite lid and placed back into the furnace, setting the power control of the furnace to the desired solidification time. The furnace is maintained under inert Ar atmosphere to avoid melt exposure to oxygen. The cooling phase is monitored using a thermocouple inserted into the casting through the lid, and the cooling curve, shown in Figure 1, is recorded by means of a data logger with 30 s sampling time. It can be noted that the solidification time is about 12 h and that the eutectic phase transformation takes approximately 10 h. A specific testing campaign conducted in [50] on coupons extracted from both heavy-section and small-scale castings demonstrated the capability of the present experimental method to produce very similar microstructural conditions.

Table 1. Chemical composition of the investigated ferritic ductile cast iron (wt %), balance Fe.

C	Si	Mn	Cu	Sn	S	Mg	Cr	P
3.59	2.51	0.23	0.11	0.0031	0.0109	0.0462	0.043	0.044

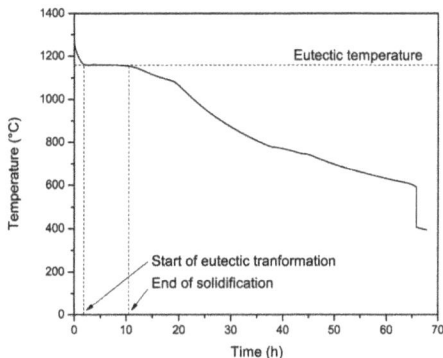

Figure 1. Record of the temperature measured by a thermocouple inserted into the casting during the solidification phase.

The microstructure is characterized by conventional metallographic analyses. Samples are mounted and ground from 220 to 4000 mesh SiC abrasive papers. The final polishing is done using a 3-micron diamond paste followed by a 0.04-micron alumina suspension. Nital etching is used to reveal the microstructure. Quantitative analysis is conducted using image analysis software ImageJ®.

Monotonic tensile tests are performed according to the standard UNI EN ISO 6892-1 on dog-bone coupons (14 mm gauge diameter, 84 mm parallel length), shown in Figure 2a, using a servo-hydraulic universal testing machine, equipped with hydraulic grips, a load cell of 200 kN. The yield strength is determined as the 0.2% offset yield stress.

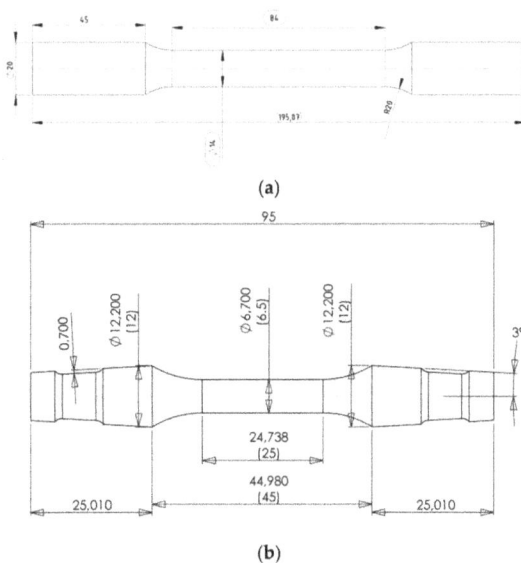

(a)

(b)

Figure 2. *Cont.*

101

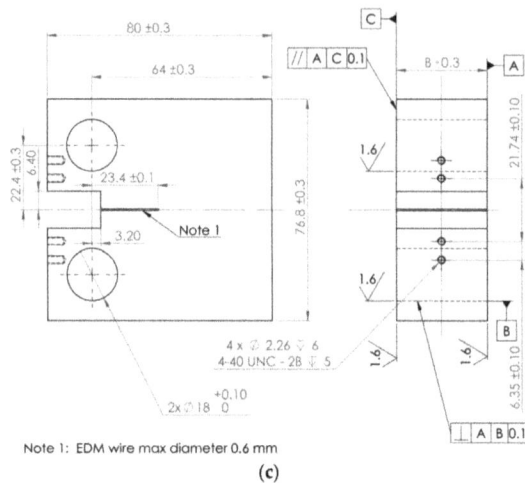

Note 1: EDM wire max diameter 0.6 mm

(c)

Figure 2. Geometry of the specimens used in the present experimentation. (**a**) Dog-bone samples used for monotonic tensile tests; (**b**) Uniform-gage test section samples used in rotating bending fatigue tests; (**c**) Compact tension C(T) specimens used in fracture mechanics tests. Dimensions are given in mm. The thickness *B* of the C(T) specimens is equal to 10 mm and to 32 mm for fatigue crack growth rate and fracture toughness tests, respectively.

Brinell hardness is measured by using a B3000 J hardness tester (Mechatronic Control System, Ichalkaranji, India) equipped with a 10 mm diameter sphere and applying a load of 29.4 kN. For each sample, at least three measurements are performed and the average value is calculated.

Fatigue tests are performed on hourglass specimens depicted in Figure 2b according to the standard ASTM E466. Rotating bending fatigue tests (load ratio $R = -1$) are carried out under load-control at a nominal frequency of 100 Hz in laboratory environment (25 °C, 60% R.H.). The fatigue strength corresponding to a fatigue life of 5×10^6 cycles is obtained by a staircase procedure, employing 15 samples and 10 MPa stress increments. Fatigue fracture surfaces have been investigated using a JEOL (Japan Electron Optics Laboratory Company, Ltd., Akashima, Japan) JSM-IT300LV scanning electron microscope (SEM) (Japan Electron Optics Laboratory Company, Ltd., Akashima, Japan) equipped with an EDXS probe for quantitative chemical analysis.

The fracture mechanics tests are carried out on compact C(T) specimens whose geometry, compliant with the standard ASTM E1820-09, is illustrated in Figure 2c. Specifically, K_{Ic} fracture toughness and fatigue crack growth rate tests are performed on 32 and 10 mm thick specimens, respectively. The initial notch of all C(T) specimens is Electro Discharge Machined (EDM) with a wire of 0.5 mm diameter to facilitate the follow-up pre-cracking process.

The fatigue crack growth rate testing is performed according to the standard ASTM E647-08. The experiments are conducted in the laboratory environment on a resonant testing machine Rumul (Russenberger Prüfmaschinen, AG, Neuhausen am Rheinfall, Switzerland) Testronic 50 kN. A sinusoidal pulsating load waveform is applied at a frequency of ~100 Hz. Tests are performed at three load ratios *R*, namely 0.1, 0.5 and 0.75. A Fractomat® (Augustine, FL, USA) apparatus based on the indirect potential drop method is used to continuously measure the crack length. For this purpose, a crack length foil (Krak gage® (Augustine, FL, USA)), consisting of a conducting layer on an electrically insulating backing, is bonded on one side of the C(T) sample and then connected to a signal conditioning module, whose output can be used to feed-back control the testing machine. Periodically, the crack length on the back face of the sample is inspected using a travelling microscope to make sure that the two crack lengths do not differ by more than 0.25 times the sample thickness, as prescribed by the standard ASTM E647-08. The crack growth rates are calculated from discrete crack length increments of 0.3 mm, necessary for sampling the fatigue crack growth resistance of the fairly coarse material microstructure. The tests are performed in two distinct phases. In the initial force-shedding phase, the applied ΔK is exponentially reduced by setting the decay constant $C = -\frac{1}{K}\frac{dK}{da} = 0.08$ mm^{-1} until reaching near-threshold fatigue crack growth conditions. A specific testing campaign, further described in the following, has been carried out to identify the minimum crack size requested to fully develop the extrinsic crack growth resistance mechanisms of the material, which greatly influence the fatigue crack growth threshold. The final part of the experiments is performed under constant force amplitude and the crack growth rate data are used to build the da/dN vs ΔK curves. After the fatigue tests, the specimens are sectioned across the thickness and metallographic samples are extracted in order to investigate the crack front profile. Fracture surfaces are investigated using a JEOL JSM-IT300LV SEM.

Specimens for fracture toughness tests are fatigue precracked to a crack-length-to-specimen-width of about 0.5 using the same resonant testing machine in laboratory environment under K-controlled conditions (load ratio $R = 0.1$) suitable to maintain the crack growth rate below 10^{-8} m/cycle. After precracking, to enforce the plane strain condition, 20% V-notch side grooves are machined on the samples, resulting in a net thickness (B_N) of 25.6 mm. Subsequently, the samples are monotonically loaded up to fracture to determine the K_c fracture toughness ($dK/dt = 1$ MPa $\sqrt{m}\cdot s^{-1}$, room temperature air) using a 100 kN servo-hydraulic testing machine equipped with a clip-gauge mounted on knife edges attached to the specimen to measure the crack mouth opening displacement. All data of load versus clip gauge displacement are acquired with a sampling rate of 50 Hz. After fracture, the crack length prior to final monotonic loading is measured using an optical microscope following the ASTM E1820-09 standard procedure.

3. Results and Discussion

3.1. Microstructure

The optical micrograph shown in Figure 3 illustrates a representative example of the material microstructure. It can be noted that the microstructure is highly inhomogeneous, being composed of a predominantly ferritic matrix in which both spheroidal and chunky graphite domains of different size are unevenly distributed. In some regions, the matrix is even pearlitic (visible in the center of Figure 3). To better characterize type, shape and distribution of the graphite particles, ten cross-sections randomly extracted from the samples are metallographically prepared and statistically analyzed. Specifically, the deviation from the spherical shape is estimated by the shape factor expressed as [47]:

$$f_0 = \frac{4\pi A}{U^2} \tag{1}$$

where A and U are the area and perimeter of the particles, respectively. Clearly, f_0 becomes 1 for an ideal sphere. Table 2 summarizes the results of the statistical analysis of the graphite morphology. It can be noted that about 1/3 of the graphite has degenerated into chunky graphite and that the remaining fraction of spheroidal graphite has overgrown to very large nodules with a mean diameter of about 300 micron. The shape factor of the spheroidal nodules equal to about 0.5 is pretty low in comparison to data available in the literature for DCI obtained under normal solidification conditions [47].

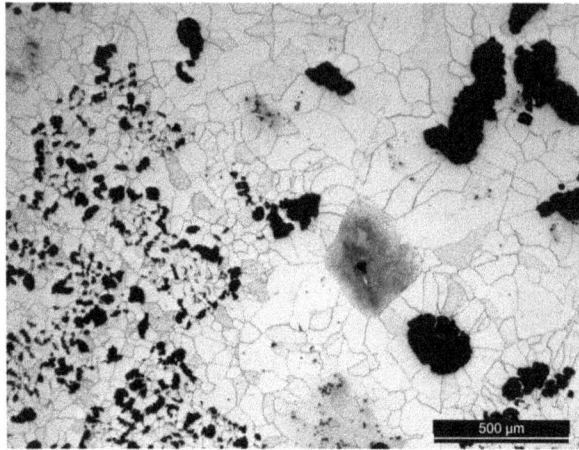

Figure 3. Optical micrograph of the microstructure (etching with 2% Nital reagent).

Table 2. Results of quantitative analysis done on 10 metallographic samples. Standard error corresponds to 1σ uncertainty band.

Total Graphite Content (%)	Fraction of Spheroidal Graphite (%)	Fraction of Chunky Graphite (%)	Diameter of Spheroidal Graphite (μm)	Shape Factor Spheroidal Graphite f_0
9.3 ± 2.1	71 ± 15	29 ± 15	310 ± 110	0.49 ± 0.19

3.2. Monotonic and High-Cycle Fatigue Properties

Mean values of materials parameters obtained from the analysis of tensile and hardness data are summarized in Table 3. It can be noted that the very long solidification time and the resulting microstructural alterations lead to low values of tensile strength and total elongation in comparison to conventional ferritic DCI; for instance, with respect to thick-walled ($t > 60$ mm) EN-GJS-400-15, the reduction in tensile strength and total elongation is about 20% and 75%, respectively. The yield-stress-to-tensile-strength ratio approaching the unity further denotes a very brittle material behavior, while, as expected, the yield stress is little affected [4], being the yield stress compliant with the minimum prescription of 240 MPa for EN-GJS-400-15.

Table 3. Monotonic tensile (based on four replicated tests) and rotating bending fatigue (from the staircase tests data listed in Table 4) properties of the present material. Standard error corresponds to 1σ uncertainty band.

σ_{YS} (MPa)	σ_U (MPa)	σ_{YS}/σ_U	T.E. (%)	HB	$\sigma_{lim,-1}$ (MPa)
275 ± 4	295 ± 8	0.93	2.9 ± 0.2	144 ± 2	138 ± 4

σ_{YS}: 0.2% yield stress; σ_U: ultimate tensile strength; T.E.: total elongation; HB: Brinell hardness. $\sigma_{lim,-1}$: rotating bending fatigue limit at five million cycles.

The results of the rotating bending fatigue tests are summarized in Table 4. The statistical elaboration of the fatigue data according to the staircase procedure leads to estimate the five million cycles fatigue limit (50% failure probability) equal to 138 MPa with 4 MPa standard deviation. In addition, in this case, it can be noted that the long solidification time leads to low fatigue properties, in fact the fatigue limit is 30% below the typical value of 195 MPa indicated in the European standard EN-1563 [51] for this DCI class on the base of rotating bending fatigue tests conducted on small-sized (φ = 10.6 mm) laboratory samples extracted from thin-walled castings ($t \leq 30$ mm), thus displaying ideal microstructural conditions. Furthermore, the investigations undertaken in [42] on a similar material showed that the presence of 33% of chunky graphite is responsible for a fatigue limit reduction of 14%. Therefore, it can be inferred that the remaining part of the fatigue limit decrement (about 16%) observed in the present paper can be imputed to solidification defects induced by the very slow cooling conditions.

Table 4. Results of the rotating bending fatigue tests.

Sample Number	Stress Amplitude, σ_a (MPa)	Number of Cycles to Failure, N_f	Remarks
5	130	8,396,603	Run-out
6	140	5,181,473	Run-out
7	150	3,870,378	
8	140	3,535,376	
9	130	8,110,087	Run-out
10	140	1,253,313	
11	130	24,732,626	Run-out
12	140	2,483,398	
13	130	8,143,060	Run-out
14	140	1,347,937	
15	130	6,989,257	Run-out
16	140	7,653,654	Run-out
17	150	2,851,237	
18	140	2,443,680	
20	130	8,891,424	Run-out

For this purpose, SEM analyses are carried out on the fracture surfaces of the fatigue specimens {6,7,8,14,17} to shed light on the damage mechanisms acting in the high-cycle fatigue regime. As shown in Figure 4a,b, shrinkage defects are present on the fracture surfaces, usually in regions of nearly complete absence of graphite. They can be either solidification cavities or pores located near the outer surface (Figure 4a) or in the interior (Figure 4b) of the sample, respectively. Kobayashi and Yamabe [5] showed that the presence and the size of these inherent defects control the fatigue limit of ductile cast irons. Specifically, the fatigue limit can be expressed as cyclic threshold stress at which the cracks emanating from the defects do not propagate. This type of analysis can be easily performed with the $\sqrt{area}$ parameter, defined as the square root of the area obtained by projecting a defect or a crack onto the plane perpendicular to the maximum tensile stress. Accordingly, Murakami and Endo [52] proposed the following equation to predict the fully-reversed fatigue limit ($R = -1$):

$$\sigma_{\text{lim},-1} = \frac{C(HV + 120)}{\left(\sqrt{area}\right)^{1/6}}; \quad C = \begin{cases} 1.43 & \text{surface defect} \\ 1.56 & \text{internal defect} \end{cases} \tag{2}$$

where HV is the Vickers hardness.

To check the applicability of this approach to the present material, the area of the largest pore is measured on the fracture surface of each of the above-mentioned samples. The mean and standard deviation of this parameter are estimated to be 164,000 and 78,000 μm^2, respectively. Assuming a Gaussian distribution of these critical defect sizes, the fatigue limit at different failure probabilities can be calculated using Equation (2). Specifically, the fatigue limits corresponding to 50%, 90%, and 10% failure probabilities are estimated to be equal to 149, 161, and 143 MPa, respectively, in very good agreement with the experimental values listed in Table 4.

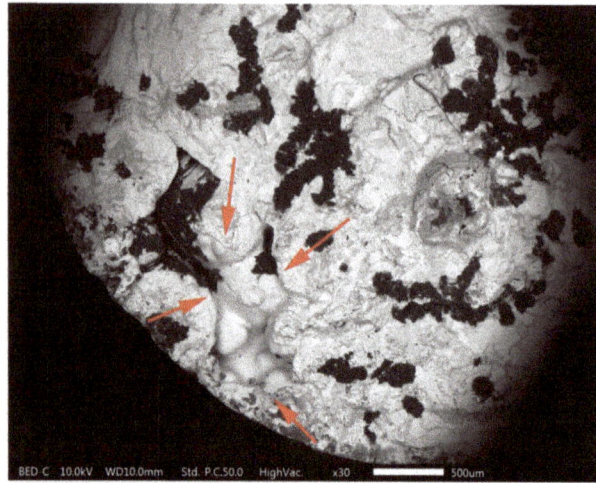

(a)

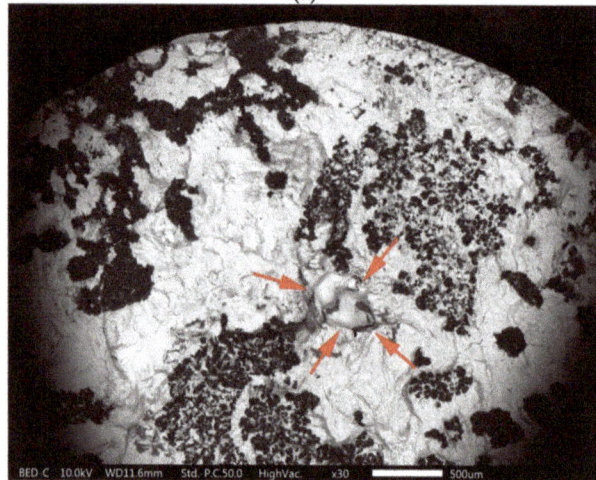

(b)

Figure 4. scanning electron microscope (SEM) micrographs of the fracture surfaces of rotating bending fatigue samples around the largest shrinkage pore marked by red arrows: (**a**) sample 6 (near-surface cavity); and (**b**) sample 8 (internal pore). Table 4 indicates stress amplitude and number of cycles to failure.

We must, however, outline that this powerful approach may oversimplify the actual high-cycle fatigue damage mechanism, as shrinkage porosity may not be the only crack nucleation site. To this regard, looking at Figure 4b, it can be noted that the pore marked by red arrows is located about 1.7 mm below the outer surface. It is therefore highly unlikely that this defect, experiencing a bending stress nearly 50% lower than the highest value achieved on the outer surface, acted as the first crack initiation site. Presumably, other microstructural defects, such as the network of chunky graphite in the vicinity of the pore, are involved in the fatigue damage mechanism and are responsible for the 14% reduction in fatigue limit estimated in [42].

3.3. Fatigue Crack Growth Curves

As anticipated in Section 2, fatigue crack growth da/dN-ΔK curves are measured considering experimental data referring to crack lengths sufficient for the material to fully deploy extrinsic crack growth resistance mechanisms [53,54]. In this way, a more realistic estimation of the actual crack growth resistance in heavy-section castings can be obtained. For this purpose, a specific testing campaign is carried out at $R = 0.1$ monitoring the crack growth rate at constant ΔK and increasing crack length. Representative results are shown in Figure 5. It can be noted that the material exhibits increasing crack growth resistance as the crack extends from the initial EDM crack starter notch of length a_0. The crack growth resistance tends to saturate when the crack extension is about 4 mm, viz. 40% of the specimen thickness. This value is considered in the following experimentation as the minimum crack extension for full development of extrinsic crack retardation mechanisms. Indeed, crack tip shielding mechanisms [53], such as crack closure and bridging, acting behind the crack tip, require the crack wake to be sufficiently long to be fully active. They are thus responsible for the development of rising resistance-curve (R-curve) behavior with increasing crack length. It is also known that eliminating by mechanical machining the crack wake behind the crack tip leads to the loss of such mechanisms and to a sudden crack acceleration as long as the crack wake is not fully restored [54]. Therefore, the second phase of crack growth testing performed under constant force amplitude is conducted only after the crack extension has overcome the minimum value of 4 mm. To this regard, it can be noted that Zambrano et al. [55] used C(T) samples with crack sizes less than 2 mm and reported very low fatigue crack growth resistance in comparison with other papers dealing with similar DCI [47,56,57].

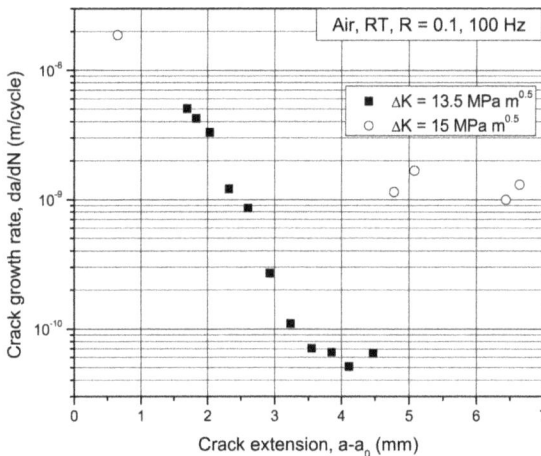

Figure 5. Results of fatigue crack growth tests conducted at constant ΔK and increasing crack extension from the initial Electro Discharge Machined (EDM) crack starter notch.

The resulting da/dN-ΔK curves obtained at $R = 0.1$, 0.5 and 0.75 are compared in Figure 6. The R-ratio effect is evident, indeed the curves move towards higher crack growth rates with increasing R, especially in the near-threshold regime. The experimental data are represented by the exponential law proposed by Klesnil and Lukáš [58]:

$$\frac{da}{dN} = C(\Delta K^m - \Delta K_{th}^m) \tag{3}$$

The best-fit parameters are given in Table 5 for the different load ratios used in the tests. It can be noted that Equation (3) well reproduces the material's crack growth behavior in Stages I (near-threshold regime) and II (*K*-controlled regime). When $K_{max} = \Delta K/(1 - R)$ exceeds ~45 MPa m$^{0.5}$, Stage III crack growth regime commences and *K*-dominance is lost.

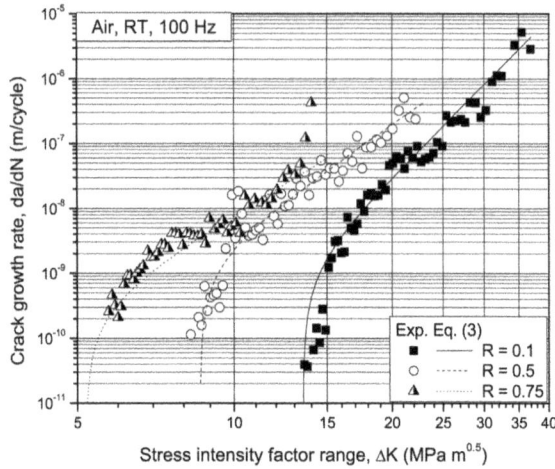

Figure 6. Fatigue crack growth curves plotted as a function of the nominal stress intensity factor range.

Table 5. Best-fit parameters of the crack propagation law expressed by Equation (3).

Load Ratio R	ΔK_{th} (MPa m$^{0.5}$)	C	m
0.1	13.5	7.72×10^{-19}	8.1
0.5	8.6	9.48×10^{-15}	5.6
0.75	5.2	9.59×10^{-14}	4.9

Units in Equation (3) are m/cycle and MPa m$^{0.5}$.

Figure 7a compares the curves obtained by fitting Equation (3) to the present experimental data and those measured by Mottitschka et al. [56] on an EN-GJS-400 DCI type with very similar chemical composition but obtained under fast solidification conditions leading to a mean graphite diameter of about 10 μm and therefore denoted as G10. Unexpectedly, it can be noted that the present material displays a noticeable higher resistance to fatigue crack growth, especially in the near-threshold regime. The *R*-ratio dependency of the threshold ΔK_{th} is well represented by the Walker equation [59]:

$$\Delta K_{th} = \overline{\Delta K}_{th}(1 - R)^{1-\gamma_{th}} \tag{4}$$

as shown in Figure 7b, where the least-square fit of Equation (4) is compared for both material types. Unfortunately, no information is provided in [54] about the size of the cracks monitored in the fatigue crack growth experiments. Therefore, the comparison of the results shown in Figure 7a,b could be impaired by different levels of deployment of the crack retardation mechanisms. Nevertheless, it seems that the presence of degenerated graphite has a beneficial, or at least non-detrimental, effect on the material fatigue crack growth resistance. This unexpected result can be explained looking at Figure 8a, which depicts a SEM overview of the fatigue facture surface in the vicinity of the near-threshold propagation region. The fracture surface is characterized by the presence of different graphite morphologies, ranging from nearly spherical nodules to fine dispersion of chunky graphite. Notably, the regions containing no graphite at all or surrounding graphite spheroids show large evidence of brittle cleavage (see the river pattern indicated by an arrow in Figure 8b) and intergranular fracture, as shown in more detail in Figure 8b,c, respectively. On the contrary, the presence of chunky graphite is associated to a less brittle fracture mode resulting in rougher fracture surface.

Similar indications are given by crack front profiles taken from regions under the same crack propagation regime and shown in Figure 9. Indeed, it can be noticed that where the crack front encounters regions characterized by degenerated graphite (Figure 9a) the crack profile is significantly rougher than where spherical graphite nodules only are present (Figure 9b). The former scenario will result in more effective crack retardation mechanisms, such as roughness-induced crack closure [53] and crack front geometry effects [54], with respect to the latter one representative of conditions occurring in conventional DCI.

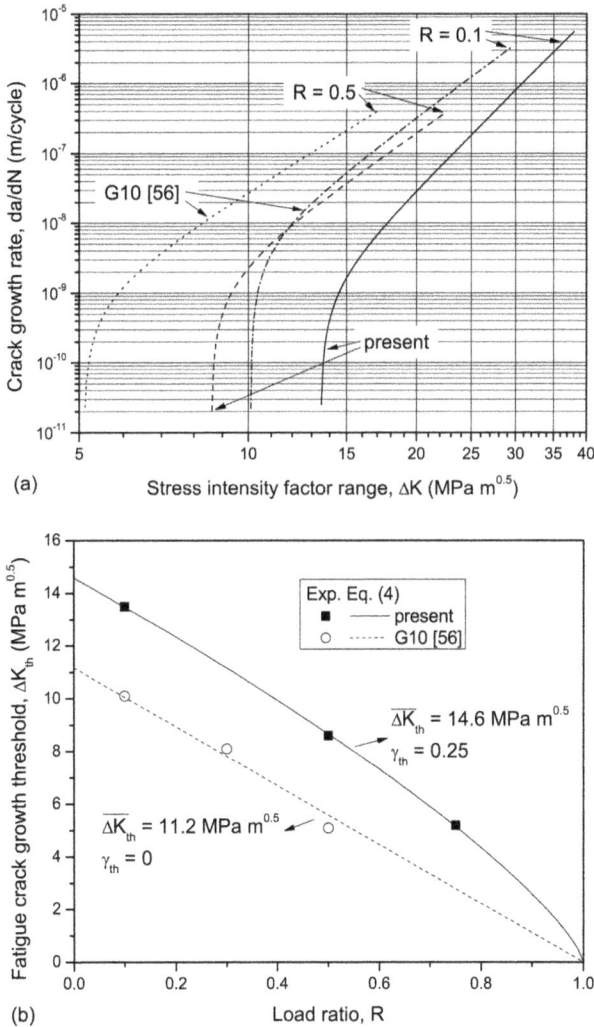

(a)

(b)

Figure 7. Comparison of fatigue crack growth resistance with data [54] referring to a similar ductile cast iron (DCI) subject to normal solidification conditions: (**a**) da/dN-ΔK curves; and (**b**) threshold dependency upon load ratio R.

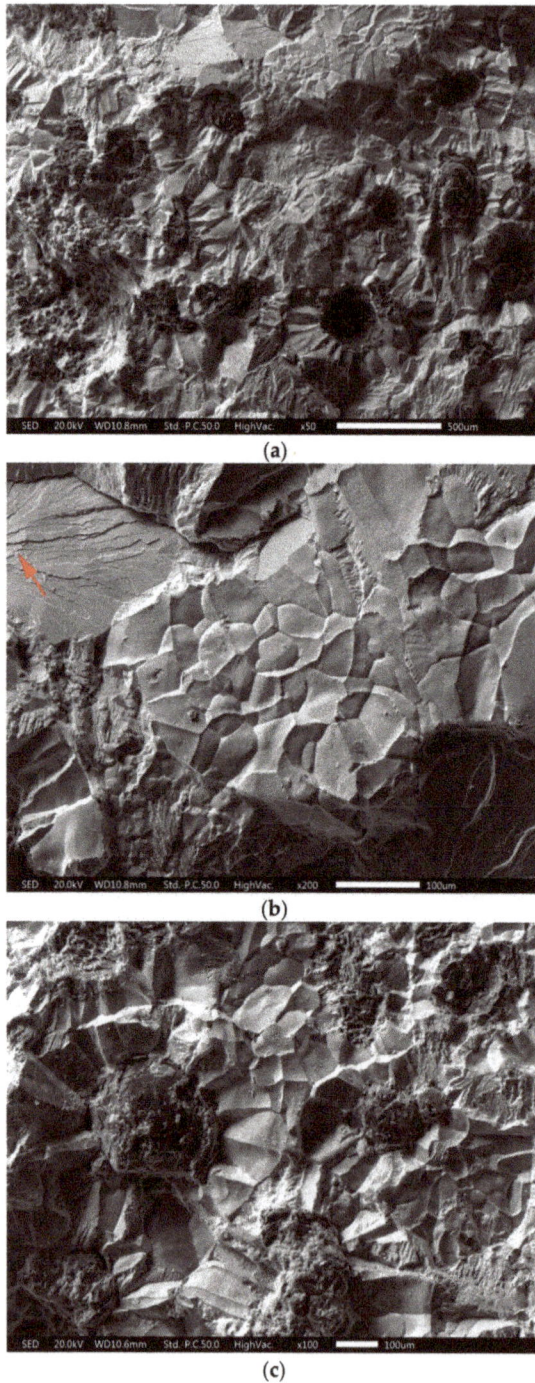

Figure 8. scanning electron microscope (SEM) micrographs of the fatigue fracture surfaces of the C(T) sample tested at $R = 0.1$: (**a**) overview of the near-threshold crack propagation region; (**b**) detail of cleavage (marked by an arrow) and intergranular fracture; and (**c**) detail of a graphite spheroidal nodule. Crack propagation from top to bottom.

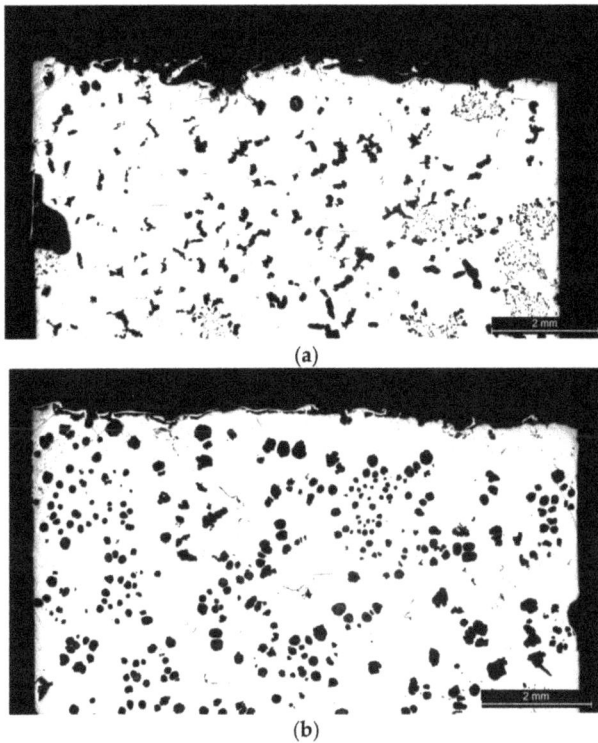

Figure 9. Light microscope images of crack front profiles taken in the near-threshold crack propagation region: (**a**) cross-section with significant presence of chunky graphite; and (**b**) cross-section with prevailing nodular graphite. Crack path direction is normal to paper plane.

3.4. Fracture Toughness

K_{Ic} fracture toughness tests are performed on C(T) specimens depicted in Figure 2c, whose geometry is taken as large as possible with the aim of satisfying the linear elastic fracture mechanics requirement expressed by [60]:

$$B, a, W - a > 2.5 \left(\frac{K_{Ic}}{\sigma_{YS}} \right)^2 \tag{5}$$

where a, B and W are the initial crack length, specimen thickness and width, respectively. Considering the value of the material yield strength σ_{YS} listed in Table 3, the maximum theoretical fracture toughness that can be estimated using this C(T) sample geometry is ≈ 32 MPa m$^{0.5}$, which is comparable with K_{Ic} values reported in the technical literature [56,57] for this DCI class.

Figure 10a,b shows the load versus crack mouth opening displacement recorded for specimen 1 and 2, respectively. In both cases, the record is of Type I according to ASTM-1820, in the sense that the load P_Q is determined as the intersection of the force versus crack opening displacement record with the 95% secant. Unexpectedly, despite the use of thick-walled and side-grooved specimens, the requirement on the ratio maximum load P_{max} to P_Q to be ≤ 1.10, which is necessary for K_Q to be qualified as linear elastic, plane-strain fracture toughness K_{Ic}, is not met. Therefore, the present tests result in thickness-dependent fracture toughness values, which can be used to safely design only components with thickness not larger than 32 mm. Apparently, the highly nonlinear stress–strain behavior displayed by cast irons makes very difficult to meet the ASTM-1820 requirement on P_Q, even though the conditions set by Equation (5) are fulfilled.

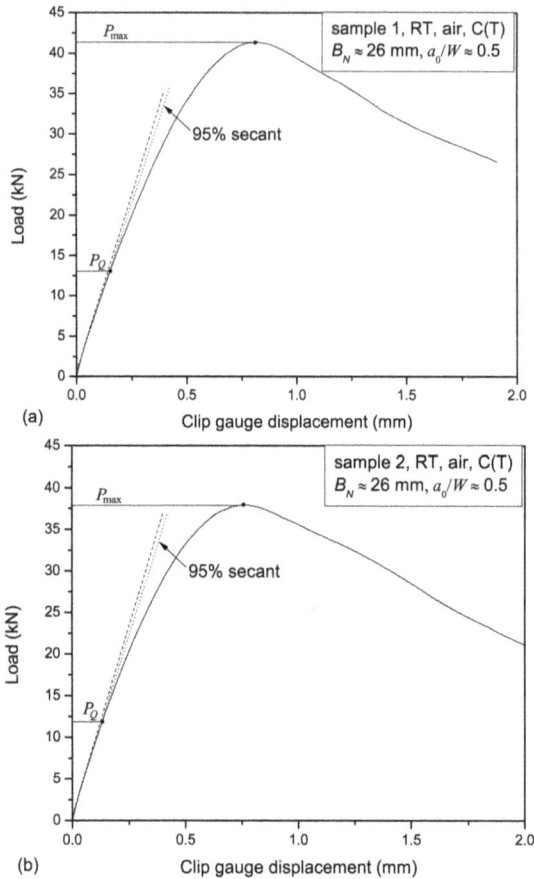

(a)

(b)

Figure 10. Load versus crack mouth opening displacement recorded for: specimen 1 (**a**); and specimen 2 (**b**) used in fracture toughness testing.

Table 6 summarizes the results of the two fracture toughness tests and compares them with data found in the technical literature regarding similar EN-GJS-400 DCI subjected to normal solidification conditions. The average fracture toughness K_Q is 17.3 MPa m$^{0.5}$, which is 14% lower than the value reported by Zambrano et al. [55]. It must be noted that this latter value was obtained using samples with thickness of only 10 mm, therefore at least part of the observed reduction in fracture toughness could be induced to a higher triaxiality of the crack tip stress field experienced by the samples tested in the present work. On the other hand, unfavorable microstructural conditions caused by the long solidification time could have a negative effect on the fracture toughness, even though to a lesser extent than that observed in Section 3.2 on tensile and fatigue strength.

Table 6. Results of the fracture toughness tests and comparison with literature data.

Specimen	Thickness, B (mm)	K_Q (MPa m$^{0.5}$)	K_{Pmax} (MPa m$^{0.5}$)	K_{Ic} * (MPa m$^{0.5}$)
1	32	18.0	57.3	-
2	32	16.6	50.2	-
C(T) [55]	10	20	39	-
WK1 [47]	10	-	-	32
G10 [56]	10	-	-	33.2

* Data inferred from fatigue crack growth rate vs. ΔK curves.

Hübner et al. [47] and Mottitschka et al. [56] did not perform true fracture mechanics tests but inferred the material fracture toughness K_{Ic} from fatigue crack growth tests and calculated it as $K_{Ic} = \Delta K_c / (1 - R)$, where ΔK_c is the asymptotic value of the stress intensity range at which the crack growth rate tends to infinite. Clearly, this is not a true stress intensity factor, as prescriptions and procedures set by ASTM-1820 are not met. Anyway, if the data illustrated in Figure 6 are elaborated in the same way, a value of $K_{Ic} \approx 45$ MPa m$^{0.5}$ is obtained, thus higher than the values reported by [47,56].

4. Conclusions

Solidification conditions typical of heavy-section castings were reproduced on a small-scale ductile cast iron (DCI) casting, from which laboratory coupons were extracted and used to carry out monotonic tensile, fatigue and fracture mechanics tests. The obtained results were discussed by comparison with data available in the technical literature. The following conclusions can be drawn:

(1) The microstructure consists of a prevailing ferritic matrix, in which 2/3 of graphite is interspersed in form of overgrown spheroidal nodules and the rest in form of chunky graphite.

(2) The tensile strength and the total elongation are reduced by 20% and 75% with respect to thick-walled ($t > 60$ mm) EN-GJS-400-15.

(3) The rotating bending fatigue limit at five million cycles is reduced by 30% with respect to the standard EN-1563 reporting the results of fatigue tests employing laboratory samples extracted from thin-walled castings. The fatigue limit is strongly correlated to the dimension of the largest shrinkage pore found on the fracture surface. Nevertheless, shrinkage pores are not the only defects involved in the fatigue damage, as also degenerated chunky graphite can be site of crack nucleation.

(4) The material displays a higher fatigue crack growth resistance, especially in the near-threshold regime, with respect to DCI subject to normal solidification conditions. This has been attributed, at least partially, to crack front geometry effects and crack closure phenomena induced by the high roughness and interlocking crack mating surfaces in the neighborhood of chunky graphite.

(5) The fracture toughness is 14% lower than that reported for DCI obtained under normal solidification conditions. On the other hand, the thickness of the samples used in the present experimentation is three times larger than that of specimens usually tested in the literature. The highly nonlinear stress–strain behavior displayed by cast irons makes very difficult to meet the ASTM prescription on the P_Q load.

Author Contributions: D.L. prepared the samples and performed static tensile and rotating bending fatigue tests. E.T. conducted metallographic and fractographic analyses. V.F. contributed to the analysis and discussion of the results. M.B. carried out the fracture mechanics tests, contributed to the analysis and discussion of the results and wrote the paper.

Conflicts of Interest: The authors declare no conflict of interest.

References

1. Luo, J.; Hardning, R.A.; Bowen, P. Evaluation of the fatigue behavior of ductile irons with various matrix microstructures. *Metall. Mater. Trans. A* **2002**, *33A*, 3719–3729. [CrossRef]

2. Di Cocco, V.; Iacoviello, F.; Rossi, A.; Cavallini, M.; Natali, S. Graphite nodules and fatigue crack propagation micromechanisms in a ferritic ductile cast iron. *Fatigue Fract. Eng. Mater. Struct.* **2013**, *36*, 893–902. [CrossRef]
3. Kaufmann, H.; Wolters, D.B. Zyklische Beanspruchbarkeit dickwandiger Bauteile aus ferritischem Gusseisen mit Kugelgraphit. *Konstr. Giess.* **2002**, *27*, 4–27.
4. Minnebo, P.; Nilsson, KF.; Blagoeva, D. Tensile, compression and fracture properties of thick walled ductile cast iron components. *J. Mater. Eng. Perform.* **2007**, *16*, 35–45. [CrossRef]
5. Kobayashi, M.; Yamabe, J. Influence of casting surfaces on fatigue strength of ductile cast iron. *Fatigue Fract. Eng. Mater. Struct.* **2006**, *29*, 403–415.
6. Meneghetti, G.; Ricotta, M.; Masaggia, S.; Atzori, B. Comparison of the los cycle and medium cycle fatigue behaviour of ferrtic, pearlitic, isothermed and austempered ductile irons. *Fatigue Fract. Eng. Mater. Struct.* **2013**, *36*, 913–929. [CrossRef]
7. Wang, Q.Y.; Bathias, C. Fatigue characterization of a spheroidal graphite cast iron under ultrasonic loading. *J. Mater. Sci.* **2004**, *39*, 687–689. [CrossRef]
8. Caldera, M.; Chapetti, M.; Massone, J.M.; Sikore, J.A. Influence of nodule count on fatigue properties of ferritic thin wall ductile iron. *Mater. Sci. Technol.* **2007**, *23*, 1000–1004. [CrossRef]
9. Andreiko, I.M.; Ostash, O.P.; Popovych, V.V. Influence of microstructure on the strength and cyclic crack resistance of cast irons. *Mater. Sci.* **2002**, *38*, 659–671. [CrossRef]
10. Greno, G.L.; Otegui, J.L.; Boeri, R.E. Mechanisms of fatigue crack growth in austempered ductile Iron. *Int. J. Fatigue* **1999**, *21*, 35–43. [CrossRef]
11. Borsato, T.; Ferro, P.; Berto, F.; Carollo, C. Mechanical and fatigue properties of heavy section solution strengthened ferritic ductile iron castings. *Adv. Eng. Mater.* **2016**, *18*, 2070–2075. [CrossRef]
12. Borsato, T.; Ferro, P.; Berto, F.; Carollo, C. Fatigue strength improvement of heavy-section pearlitic ductile iron castings by in-mould inoculation treatment. *Int. J. Fatigue* **2017**. [CrossRef]
13. Taylor, D.; Hughes, M.; Allen, D. Notch fatigue behavior in cast irons explained using a fracture mechanics approach. *Int. J. Fatigue* **1996**, *18*, 439–445. [CrossRef]
14. Davis, J.R. *Cast Irons*, 1st ed.; ASM International: Materials Park, OH, USA, 1996; p. 65.
15. Karsay, S.I. Control of Graphite Structure in Heavy Ductile Iron Castings. *AFS Trans.* **1970**, *78*, 85–92.
16. Strizik, P.; Jeglitsch, F. Contribution to the Mechanism of Formation of Chunky Graphite. *AFS Int. Cast Met. J.* **1976**, *1*, 23–30.
17. Liu, P.C.; Li, C.L.; Wu, D.H.; Loper, C.R. SEM Study of Chunky Graphite in Heavy Section Ductile Iron. *AFS Trans.* **1983**, *91*, 119–126.
18. Hoover, H.W., Jr. A literature survey on degenerate graphite in heavy section ductile iron. *Trans. Am. Foundrym. Soc.* **1986**, *94*, 601–608.
19. Stefanescu, D.M.; Alonso, G.; Larrañaga, P.; De la Fuente, E.; Suarez, R. On the crystallization of graphite from liquid iron-carbon-silicon melts. *Acta Mater.* **2016**, *107*, 102–126. [CrossRef]
20. Itofuji, H.; Uchikawa, H. Formation mechanism of chunky graphite in heavy-section ductile cast irons. *AFS Trans.* **1990**, *90*, 429–448.
21. Gagné, M.; Argo, D. Heavy section ductile cast iron castings part I and part II. In Proceedings of the an International Conference on Advanced Casting Technology, Kalamazoo, MI, USA, 12–14 November 1989; ASM International: Materials Park, OH, USA, 1989; pp. 231–256.
22. Nakae, H.; Junk, S.; Shin, H.-C. Formation mechanism of chunky graphite and its preventive measures. *J. Mater. Sci. Technol.* **2008**, *24*, 289–295.
23. Udroiu, A. Wedge theory: new approach to explain the formation of "chunky graphite" in ductile iron. In Proceeding of the 70th World Foundry Congress, Monterrey, Mexico, 25–27 April 2012; pp. 614–631.
24. Källbom, R.; Hamberg, K.; Wessén, M.; Björkegren, L.-E. On the solidification sequence of ductile iron castings containing chunky graphite. *Mater. Sci. Eng. A* **2005**, *413–414*, 346–351. [CrossRef]
25. Riposan, I.; Chisamera, M.; Stan, A. Control surface graphite degeneration ductile iron windmill applications. *Int. J. Metalcast.* **2013**, *7*, 9–20. [CrossRef]
26. Nakae, H.; Fukami, M.; Kitazawa, T.; Zou, Y. Influence of Si, Ce, Sb and Sn on chunky graphite formation. *China Foundry* **2011**, *8*, 96–100.
27. Lacaze, J.; Armendariz, S.; Larrañaga, P.; Asenjo, I.; Sertucha, J.; Suarez, R. Effect of carbon equivalent on graphite formation in heavy-section ductile iron parts. *Mater. Sci. Forum* **2010**, *636–637*, 523–530. [CrossRef]

28. Riposan, I.; Chisamera, M.; Uta, V.; Stan, S. The importance of rare earth contribution from nodulizing alloys and their subsequent effect on the inoculation of ductile iron. *Int. J. Metalcast.* **2014**, *8*, 65–80. [CrossRef]

29. Onsøien, M.I.; Grong, Ø.; Skaland, T.; Jørgensen, K. Mechanisms of graphite formation in ductile cast iron containing rare earth metals. *Mater. Sci. Technol.* **1999**, *15*, 253–259. [CrossRef]

30. Choi, J.O.; Kim, J.Y.; Choi, C.O.; Kim, J.K.; Rohatgi, P.K. Effect of rare earth element on microstructure formation and mechanical properties of thin wall ductile iron castings. *Mater. Sci. Eng. A* **2004**, *383*, 323–333. [CrossRef]

31. Zhe, L.; Weiping, C.; Yu, D. Influence of cooling rate and antimony addition content on graphite morphology and mechanical properties of a ductile iron. *China Foundry* **2012**, *9*, 114–118.

32. Diao, X.G.; Ning, Z.L.; Cao, F.Y.; Ren, S.Z.; Sun, J.F. Effects of antimony addition and section size on formation of chunky graphite in ductile iron. *Mater. Sci. Technol.* **2011**, *27*, 834–838. [CrossRef]

33. Gorny, M.; Tyrala, E. Effect of cooling rate on microstructure and mechanical properties of thin-walled ductile iron castings. *J. Mater. Eng. Perform.* **2013**, *22*, 300–305. [CrossRef]

34. Ferro, P.; Fabrizi, A.; Cervo, R.; Carollo, C. Effect of inoculant containing rare earth metals and bismuth on microstructure and mechanical properties of heavy-section near-eutectic ductile iron castings. *J. Mater. Process. Technol.* **2013**, *213*, 1601–1608. [CrossRef]

35. Zhang, Z.; Flower, H.M.; Niu, Y. Classification of degenerate graphite and its formation processes in heavy section ductile iron. *Mater. Sci. Technol.* **1989**, *5*, 657–664. [CrossRef]

36. Skaland, T.; Grong, Ø.; Grong, T. A model for the graphite formation in ductile cast iron. *Metall. Mater. Trans. A* **1993**, *24A*, 2321–2345. [CrossRef]

37. Asenjo, I.; Larrañaga, P.; Sertucha, J.; Suárez, R.; Gómez, J.-M.; Ferrer, I.; Lacaze, J. Effect of mould inoculation on formation of chunky graphite in heavy section spheroidal graphite cast iron parts. *Int. J. Cast Met. Res.* **2007**, *20*, 319–324. [CrossRef]

38. Kallbom, R.; Hamberg, K.; Björkegren, L.-E. Chunky graphite—Formation and influence on mechanical properties in ductile cast iron. In *Competent Design by Castings—Improvements in a Nordic Project*; Samuelson, J., Marquis, G., Solin, J., Eds.; VTT: Helsinki, Finland, 2005; pp. 63–86.

39. Endo, M.; Yanase, K. Effects of small defects, matrix structures and loading conditions on the fatigue strength of ductile cast irons. *Theor. Appl. Fract. Mech.* **2014**, *69*, 34–43. [CrossRef]

40. Mourujärvi, A.; Widell, K.; Saukkonen, T.; Hänninen, H. Influence of chunky graphite on mechanical and fatigue properties of heavy-section cast iron. *Fatigue Fract. Eng. Mater. Struct.* **2009**, *32*, 379–390. [CrossRef]

41. Ferro, P.; Lazzarin, P.; Berto, F. Fatigue properties of ductile cast iron containing chunky graphite. *Mater. Sci. Eng. A* **2012**, *554*, 122–128. [CrossRef]

42. Foglio, E.; Lusuardi, D.; Pola, A.; La Vecchia, G.M.; Gelfi, M. Fatigue design of heavy section ductile irons: Influence of chunky graphite. *Mater. Des.* **2016**, *111*, 353–361. [CrossRef]

43. Nadot, Y.; Mendez, J.; Ranganathan, N.; Beranger, A.S. Fatigue life assessment of nodular cast iron containing casting defects. *Fatigue Fract. Eng. Mater. Struct.* **1999**, *22*, 289–300. [CrossRef]

44. Collini, L.; Pirondi, A. Fatigue crack growth analysis in porous ductile cast iron microstructure. *Int. J. Fatigue* **2014**, *62*, 258–265. [CrossRef]

45. Collini, L.; Pirondi, A.; Bianchi, R.; Cova, M.; Milella, P.P. Influence of casting defects on fatigue crack initiation and fatigue limit of ductile cast iron. *Procedia Eng.* **2011**, *10*, 2898–2903. [CrossRef]

46. Murakami, Y. *Metal Fatigue: Effects of Small Defects and Nonmetallic Inclusions*, 1st ed.; Elsevier: Oxford, UK, 2002.

47. Hübner, P.; Schlosser, H.; Pusch, G.; Biermann, H. Load history effects in ductile cast iron for wind turbine components. *Int. J. Fatigue* **2007**, *29*, 1788–1796. [CrossRef]

48. Benedetti, M.; Fontanari, V.; Zonta, D. Structural health monitoring of wind towers: Remote damage detection using strain sensors. *Smart Mater. Struct.* **2011**, *20*, 055009. [CrossRef]

49. Benedetti, M.; Fontanari, V.; Battisti, L. Structural health monitoring of wind towers: Residual fatigue life estimation. *Smart Mater. Struct.* **2013**, *22*, 045017. [CrossRef]

50. Foglio, E.; Gelfi, M.; Pola, A.; Goffelli, S.; Lusuardi, D. Fatigue Characterization and Optimization of the Production Process of Heavy Section Ductile Iron Castings. *Int. J. Metalcast.* **2017**, *11*, 33–43. [CrossRef]

51. European Standard EN 1563:2011. *Founding—Spheroidal Graphite Cast Iron*; CEN: Brussels, Belgium, 2011.

52. Murakami, Y.; Endo, M. Effects of hardness and crack geometries on ΔK_{th} of small cracks emanating from small defects. In *The Behaviour of Short Fatigue Cracks*; Miller, K.J., de los Rios, E.R., Eds.; Institution of Mechanical Engineers: London, UK, 1986; pp. 275–293.

53. Ritchie, R.O. Mechanisms of fatigue-crack propagation in ductile and brittle solids. *Int. J. Fract.* **1999**, *100*, 55–83. [CrossRef]

54. Benedetti, M.; Heidemann, J.; Peters, J.O.; Lütjering, G. Influence of sharp microstructural gradients on the fatigue crack growth resistance of $\alpha + \beta$ and near-α titanium alloys. *Fatigue Fract. Eng. Mater. Struct.* **2005**, *28*, 909–922. [CrossRef]

55. Zambrano, H.R.; Härkegård, G.; Stärk, K.F. Fracture toughness and growth of short and long fatigue cracks in ductile cast iron EN-GJS-400-18-LT. *Fatigue Fract. Eng. Mater. Struct.* **2011**, *35*, 374–388. [CrossRef]

56. Mottitschka, T.; Pusch, G.; Biermann, H.; Zybell, L.; Kuna, M. Influence of overloads on the fatigue crack growth in nodular cast iron: Experiments and numerical simulation. *Procedia Eng.* **2010**, *2*, 1557–1567. [CrossRef]

57. Čanžar, P.; Tonković, Z.; Kodvanj, J. Microstructure influence on fatigue behaviour of nodular cast iron. *Mater. Sci. Eng. A* **2012**, *556*, 88–99. [CrossRef]

58. Klesnil, M.; Lukáš, P. Influence of strength and stress history on growth and stabilisation of fatigue cracks. *Eng. Fract. Mech.* **1972**, *4*, 77–92. [CrossRef]

59. Walker, E.K. The effect of stress ratio during crack propagation and fatigue for 2024-T3 and 7075-T6 aluminum. In *Effect of Environment and Complex Load History on Fatigue Life*; American Society for Testing and Materials: West Conshohocken, PA, USA, 1970; pp. 1–14.

60. Anderson, T.L. *Fracture Mechanics: Fundamentals and Applications*, 3rd ed.; CRC Press: Boca Raton, FL, USA, 2005.

![metals logo] *metals*

MDPI

Article

Very High Cycle Fatigue of Butt-Welded High-Strength Steel Plate

Hyunho Yeom, Byungjoo Choi, Taeho Seol, Moongu Lee and Yongho Jeon *

Department of Mechanical Engineering, Ajou University, 206 Worldcup-ro, Yeongtong-gu, Suwon, 16499, Korea; ysquall@ajou.ac.kr (H.Y.); dasom@ajou.ac.kr (B.C.); seol0817@ajou.ac.kr (T.S.); moongulee@ajou.ac.kr (M.L.)
* Correspondence: princaps@ajou.ac.kr; Tel.: +82-31-219-3652

Academic Editor: Filippo Berto
Received: 2 January 2017; Accepted: 6 March 2017; Published: 21 March 2017

Abstract: Welded parts fabricated from high-strength steel (HSS) require an almost infinite lifetime, i.e., a gigacycle (10^9). Therefore, it is necessary to test its high-cycle fatigue behavior. In this paper, an accelerated fatigue test method using ultrasonic resonance is proposed. This method reduces the experimental time required in comparison with a conventional fatigue test setup. The operating principle of the accelerated ultrasonic fatigue test involved the use of a 20-kHz resonant frequency. Therefore, it was necessary to design a specimen specifically for the test setup. In the study, ultrasonic fatigue testing equipment was used to test butt-welded 590- and 780-MPa ferrite–bainite steel plates. In order to design the specimen, a dynamic Young's modulus was measured using piezoelectric element, a laser Doppler vibrometer, and a digital signal analyzer. The S–N curves of fatigue behavior of the original and butt-welded specimens were compared. The fatigue test results showed that the infinite (i.e., gigacycle) fatigue strengths of the welded specimens were approximately 8% less than those of the original specimen.

Keywords: ultrasonic fatigue test; high-strength steel; plate specimen; butt welding; dynamic elastic modulus

---

1. Introduction

Industrial development has led to increasing demand for parts and elements with a longer service life. Given this trend, the standard value accepted for an infinite service life has increased from 10^7 to 10^9 cycles [1]. Previous studies investigated gigacycle fatigue regimes to assess fatigue behavior in applications in the aerospace, space, high-speed rail, automotive, and biomedical industries. These applications involve operation under severe environments and in situations where it is difficult or impossible to replace parts. Previous studies determined that differences exist between fatigue and fracture behaviors according to the test regime. Test regimes are divided into three segments: low cycle (~10^4), high cycle (~10^7), and very high cycle (~10^9) [2–5].

It is also necessary to increase the test speed to check gigacycle regime fatigue behaviors. Typically, a conventional hydraulic fatigue test machine runs in an approximate range of 2–15 Hz, while rotary bending and electro-force fatigue testers operate in an approximate range of 50–300 Hz. A gigacycle regime was reached in approximately 912 h during a test involving a frequency of 300 Hz. Hence, some researchers used an ultrasonic fatigue tester to expedite experiments [6–8].

Figure 1 shows the setup of an ultrasonic fatigue test using a frequency of 20 kHz. Through this setup, a gigacycle test can be completed within a period of only 14 h. Therefore, because of the time saved, this setup is suitable for very-high-cycle testing. In a test, the maximum load was applied at the center of the specimen, and the maximum displacement was observed at the free end. For this ultrasonic fatigue test, the design of the test specimen was critical because it was necessary for the specimen to resonate. Most previous studies were performed with hourglass-type solid cylindrical or plate-shaped test specimens that were manufactured from stainless steel or magnesium alloys [9,10].

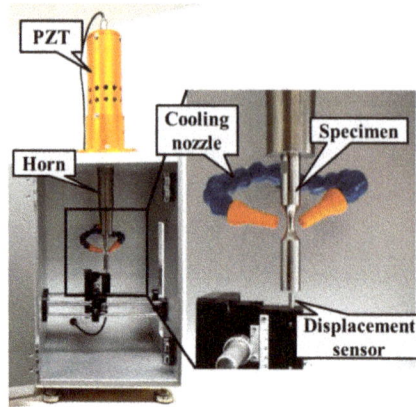

Figure 1. Ultrasonic fatigue tester.

In the present study, tests were performed on high-strength steels (590- and 780-MPa ferrite–bainite (FB)) that are used in several industries, and especially in the automotive and ship building industries. There has been previous research on fatigue tests of FB steel used in various fields. When comparing ferrite–martensite (FM) steel to FB steel, it was found that crack initiation was delayed rather than FM steel. However, fatigue test data is available only up to a 10^5 cycle test regime [11]. In the automotive industry, metal inert gas (MIG)/metal active gas (MAG) welding is mainly used for joining components, such as lower arms and subframes [12,13]. Parts such as lower arms, in which FB steel is mainly used, are subject to repetitive loads. Therefore, it is necessary to consider the fatigue life of the welded parts.

Previous studies showed that welded materials had a shorter fatigue life than raw materials. Additionally, a grounded weld toe sample had a longer fatigue life than a welding bead remain sample. The fatigue test results considered up to 10^7 cycles [14–16]. As indicated above, the currently required fatigue life is up to 10^9 cycles, and there are no previous studies on this requirement for welded FB steel plates.

While FB steels have only been studied up to high cycle fatigue (HCF), research on other welded steels has been performed up to the very high cycle fatigue (VHCF) regime. The VHCF performance of FV520B-I welded material was characterized by a longer fatigue life because the surface roughness value decreased and the inclusion size was smaller [17]. In addition, the results of the fatigue test of a Cr-Ni-Mo-V steel welded material showed that cracks were formed in the internal nonmetallic inclusions in the VHCF region [18]. EW36 steel welding results showed that the slope of the S–N curve decreased even after 10^7 cycles, which is considered the conventional fatigue limit. This means that it is necessary to study the VHCF behavior for developing safe designs. In addition, the crack propagation rate was rapid because of the larger internal defects in the welded region compared with the base metal [19]. Fatigue failure was observed up to 10^9 cycles for weld material of Q345, and the fatigue strength of the fusion zone and heat-affected zone decreased by 60% and 55%, respectively [20].

In this study, welded plate specimens were designed for ultrasonic fatigue tests, and finite element analysis on the free resonant mode was conducted to validate the test specimen design. Additionally, fatigue tests up to the gigacycle regime were performed both on the raw steel plates and on the welded plate specimens because the welded specimens were expected to exhibit different fatigue behaviors.

2. Materials and Methods

2.1. Materials

The selected materials included 2.8-mm-thick, 590-MPa and 780-MPa FB high-strength steel plates. Currently, these materials are commercially applied to the suspension, arms, and wheel disks of automobiles in the automotive industry because ferrite and bainite provide the required high elongation and high stretch-flangeability, respectively [21]. Table 1 summarizes the material compositions. The specimens were prepared using an electric discharge machining (EDM) process.

Table 1. Composition of ferrite–bainite (FB) steel.

Materials	C Max. (wt %)	Mn Max. (wt %)	Si Max. (wt %)	P Max. (wt %)	S Max. (wt %)	Fe
590 FB	0.090	1.550	0.150	0.030	0.003	balance
780 FB	0.050	1.800	1.250	0.015	0.003	balance

2.2. Dynamic Young's Modulus

Precise values for the dynamic Young's modulus of the specimens were required to ensure that the specimens resonated at 20 kHz during the ultrasonic fatigue test. The dynamic Young's modulus values were also used to calculate the appropriate loads that should be applied during the fatigue tests. Figure 2 shows the experimental set up for measurement. A piezoelectric element (Physik Instrumente(PI) GmbH & Co.KG, Karlsruhe, Germany), a laser Doppler vibrometer (LDV, OFV-352, Polytec GmbH, Waldbronn , Germany), and a digital signal analyzer (DSA, HP-35670A, Agilent Technologies, Inc., Santa Clara, CA, USA) were used during the experiments, following the steps below:

1. Suspend the bar-shaped original FB steel with a thread for free motion;
2. Place a piezoelectric element at the end of the bar with wax to provide the excitation frequency;
3. Measure the displacement using the LDV on the other side of the piezoelectric element; and
4. Measure the resonant frequency by analyzing the excitation frequency and displacement with the DSA.

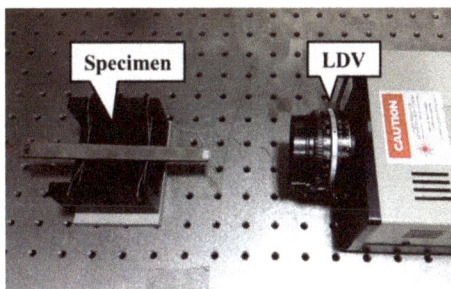

Figure 2. Measurement of dynamic Young's modulus.

The dynamic Young's modulus (E_d) was calculated using Equation (1). The density (ρ) was determined by mass measurement, and the acoustic velocity (V_c) was calculated using the resonant frequency (f_n) and the length of the wave (λ). Table 2 summarizes the calculated results. The calculation is as follows:

$$E_d = \rho V_c^2 \tag{1}$$

Where: $V_c = f_n \lambda_n$

Table 2. Physical properties of FB Steel.

Materials	Density (g/cm³)	Dynamic Young's Modulus (GPa)
590 FB	7.91	221.28
780 FB	8.01	221.98

2.3. Design of Plate Specimen

The waves applied to the specimen resulted in the exertion of stress and the occurrence of displacement. The elastic wave theory was used to calculate the displacement and stress [1]. If gravity is neglected, then the wave equation can be simplified to a single equation, as shown in Equation (2). In the equation, u, x, and t represent the rectangular components of displacement, Cartesian coordinates, and time, respectively.

$$E\frac{\partial^2 u}{\partial x^2} = \rho\frac{\partial^2 u}{\partial t^2} \tag{2}$$

$$l = \frac{1}{2f}\sqrt{\frac{E_d}{\rho}} \tag{3}$$

$$\rho S(x)\frac{\partial^2 u}{\partial t^2} = \frac{\partial f}{\partial x}, f = E_d S(x)\frac{\partial u}{\partial x} \tag{4}$$

$$y(x) = R_2, L_2 < |x| \le L \text{ where, } L = L_1 + L_2, l = 2L$$
$$y(x) = R_1 \exp(2\alpha x), |x| \le L_2 \tag{5}$$
$$\alpha = \frac{1}{2L_2}\ln\left(\frac{R_2}{R_1}\right), \beta = \sqrt{\alpha^2 - k^2} \text{ where, } k = \frac{\pi}{l}$$

The maximum displacements at both ends of the specimen were used as boundary conditions. The length of resonance (l) was calculated based on Equation (2) and is shown in Equation (3). The cross-sectional area $S(x)$ was reduced to ensure that the stress concentration occurred at the center of the specimen. Equation (4) shows the wave equation for the varying cross section. Figure 3a shows a schematic of the specimen. The curve of L_2 is set as exponential, and α and β are constants, as shown in Equation (5).

$$L_1 = \frac{1}{k}\arctan\frac{1}{k}\{\beta\cot h(\beta L_2) - \alpha\} \tag{6}$$

The plate specimen exhibited maximum stress in the middle, and the total length corresponded to the resonance length. Equation (6) shows L_1 with differentiable assumptions. Figure 3b shows the designed specimen for the FB steel plate.

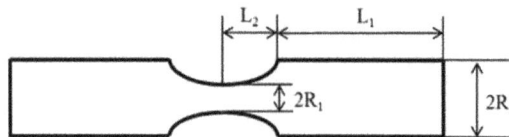

(a)

Figure 3. *Cont.*

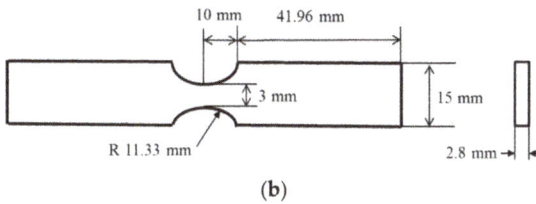

(b)

Figure 3. (**a**) Schematic of specimen; and (**b**) Specimen design of ferrite–bainite (FB) steel plate.

2.4. Welded Fatigue Specimen

As previously discussed, high-strength 590 FB and 780 FB steels were selected because they are widely used in the automotive industry. The pressed parts were welded and can be expected to exhibit lower strengths and fatigue lives.

A digital-waveform-controlled metal active gas (MAG) butt-welding process was used to prepare the welded specimens because it is capable of producing a superior weld quality with reduced spatter and heat input. A highly-accurate clamped welding jig and a semiautomatic welding system were used to ensure consistent weld quality. Figure 4 shows the weld setup. The welder used was the 350A DC CO_2/MAG (Acro Co., Sejong, Korea), and the welding torch was fixed on a 5-μm-accuracy linear stage (LP200-SC, LPK, Cincinnati, OH, USA) to attain stable movement. The linear stage was controlled with an AC servo-driver (MR-J3-70A, MISUBISHI, Tokyo, Japan) on a controller (PXI-1008B, National Instrument, Austin, TX, USA).

The welding wire used corresponded to the KC-28 (KISWELL) based on the American Welding Society standard AWS A5.18 ER70S-6. The welding conditions are summarized in Table 3. The welded specimens were prepared by a wire–EDM process, and half the designed plate specimens were precisely clamped on the jig and were butt-welded. The weld bead was milled to prevent any uncontrolled stress concentrations, as shown in Figure 5. Although multiple clamps and additional milling were applied, the samples still had some distortion or misalignment. However, the resonance frequency of the fabricated samples was under a 1.5% range difference. For this reason, it was assumed that the distortion or misalignment did not affect to the experimental results.

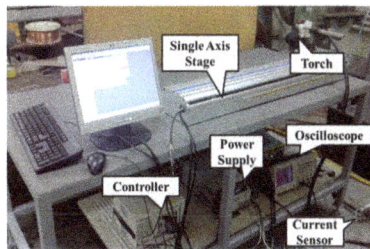

Figure 4. Semiautomatic welding machine.

| (a) | (b) |

Figure 5. (**a**) Welding plate; and (**b**) welding ultrasonic fatigue specimen.

Table 3. Conditions of welding experiment.

Current (A)	Voltage (V)	Feed (cm/min)	Shield Gas Ratio (%)	Shield Gas Injection Quantity (L/min)	Contact Tip to Work Distance (mm)	Torch Degree (°)
210	24.5	76	80(Ar):20(CO$_2$)	20	15	90

3. Results

3.1. Finite Element Method (FEM) Simulation for Specimen Resonance

It was necessary to resonate the ultrasonic fatigue plate specimen at a specified frequency (20 kHz). Hence, an FEM simulation was performed to confirm the proper design of the specimen. An ANSYS (V.17, ANSYS, Inc., Canonsburg, PA, USA) modal simulation was performed in a free vibration condition. Figure 6 shows a simulated model in which the number of nodes corresponded to 10,638. The material properties for the booster and horn corresponded to that of titanium Ti–6Al–4V, and the properties of general FB steel were used for the specimen (Table 4). The model was simulated from 15 kHz to 25 kHz, and the result indicated an axial directional mode at 19.927 kHz. The result revealed a difference of only 0.4% when compared with the target frequency (20 kHz), thus confirming that the specimen design was appropriate.

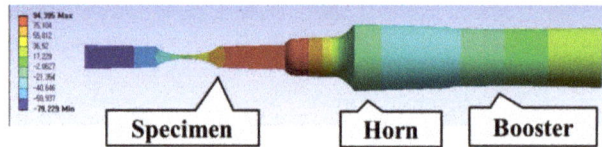

Figure 6. Modal simulation (19.927 kHz).

Table 4. Material properties for simulation.

Part	Material	Density (kg/m^3)	Young's Modulus (GPa)	Poisson's Ratio
Specimen	FB590	8170	221	0.3
Horn, Booster	Titanium Alloy	4620	96	0.36

3.2. Hardness Tests

Welding always generates a heat-affected zone that changes the specimen hardness. Therefore, hardness tests were conducted from the middle of the welded zone to the end of the base material. Figure 7 shows the results of the hardness tests. The specimens were cut using machining oil to prevent work hardening. A micro-Vickers tester (HM-122, Mitutoyo, Kawasaki, Japan) was used with a 500-g load and 5-s time condition.

In Figure 7a, the results indicated that the welded zones for both specimens had values of 225 HV because they were welded using the same welding wire. The 590 FB and 780 FB base materials had hardness values of 195 HV and 280 HV, respectively. The results confirmed that there was a linear correlation between the hardness and the yield strength of the base material. Figure 7b shows the results from butt-welding on 590 FB steel (on the left) and on 780 FB steel (on the right) for each side. The hardness of the 780 FB steel side could be attributed to an increased number of bainite structures.

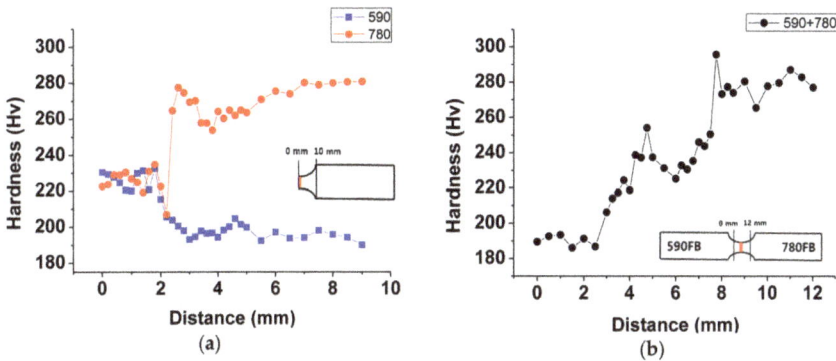

Figure 7. (a) 590 FB and 780 FB hardness; (b) 590 FB + 780 FB hardness.

3.3. Ultrasonic Fatigue Test

The ultrasonic fatigue test conditions of the specially designed FB steel plate specimens for resonance at 20 kHz are summarized in Table 5. The stress ratio and experimental resonant frequency corresponded to $R = -1$ and 19.914 kHz, respectively. The fatigue tests were performed at room temperature. However, heat was generated in the middle of the specimen owing to stress concentration and the internal friction of the steel material [22]. Compressed air was blown on the middle section of the specimen for cooling purposes to avoid experimental errors and any resulting changes in the resonant frequency from the generated heat. Additionally, the experiments were performed at intervals to prevent excessive heat generation.

Table 5. Conditions for ultrasonic fatigue test.

Stress Ratio (R)	Frequency (kHz)	Temperature (°C)	On/Off Time (s)
−1	19.914	25	5/2

Five cases were examined in the fatigue tests and included raw 590 FB steel, raw 780 FB steel, butt-welded 590 FB steel, butt-welded 780 FB steel, and 590 FB and 780 FB steels butt-welded to each other. The specimens failed in the middle section, as predicted (Figure 8).

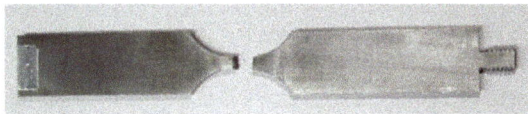

Figure 8. Specimen after the fatigue test.

Figure 9 shows the results of the fatigue tests. With respect to the raw 590 FB steel, the shortest and longest fatigue failures corresponded to 7.34×10^4 cycles with a load of 540 MPa and 1.41×10^9 cycles with a load of 460 MPa, respectively. The S–N curve followed a general trend. Thus, it could be used to determine that the fatigue tests were appropriately performed. A specimen with a load of 460 MPa was capable of reaching gigacycles, while another specimen reached 8.37×10^7 cycles under the same load. Based on these results, it was proposed that the raw 590 FB material is capable of an infinite fatigue life when a load less than 460 MPa is applied.

The butt-welded 590 FB steel had the shortest fatigue life, with failure occurring at 1.49×10^5 cycles at a load of 490 MPa. However, failure was not observed for a load of 420 MPa. A fatigue strength loss of approximately 7.97% was observed in the butt-welded 590 FB steel when compared with the raw 590 FB material. The S–N curves for the raw and butt-welded 590 FB materials exhibited similar trends. The raw 780 FB and butt-welded 780 FB materials exhibited fatigue strengths of 667 MPa and approximately 618 MPa, respectively, in the gigacycle test. In the case of the 780 FB steel, a fatigue strength loss of 7.38% was observed owing to welding, similar to the strength loss of the 590 FB steel.

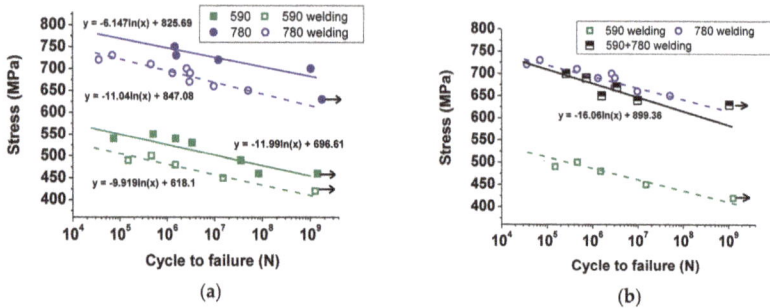

Figure 9. (**a**) S–N curves of 590 and 780 FB; (**b**) S–N curves of welding specimens (survival probability = 50%).

Figure 9b depicts the results of the 590 FB and 780 FB steels butt-welded to each other. The predicted fatigue strength was similar to that of the butt-welded 590 FB steel. However, the test results indicated that the fatigue life of the 590 FB and 780 FB steels butt-welded to each other was similar to that of the butt-welded 780 FB. This could occur because an increased stress was absorbed by the 780 FB steel region. Nevertheless, further studies should be conducted to examine this issue.

3.4. Fracture Surface

Figure 10 shows cross sections of the welded 590 FB and 780 FB specimens. The specimens were polished and etched using a solution containing 95 mL of methanol and 5 mL of HNO_3 [23]. The figure shows the base material, heat-affected zone, and welding zone separately. Larger grains were observed in the welding zone as a result of grain growth.

A field emission scanning electron microscope (FE-SEM, JSM-6700F, JEOL, Tokyo, Japan) was used to examine the fractured surface. Figure 11 shows the results. Each of the 590 FB steel and 780 FB steel sheet material specimens exhibited a clean and sharp fracture surface. By contrast, the welded specimens exhibited bumpy and wavy surfaces. This could have occurred because of the welding wire and brittle fracture caused by the grain growth [24,25]. Figure 12 shows the crack initiations with a magnified view of the arrow point of Figure 11. All fractures occurred at periods exceeding 10^7 cycles. A previous study proposed that cracks were initiated at a single point exceeding 10^7 cycles. This was verified by the results of the present study [26]. All cracks were initiated at the surface.

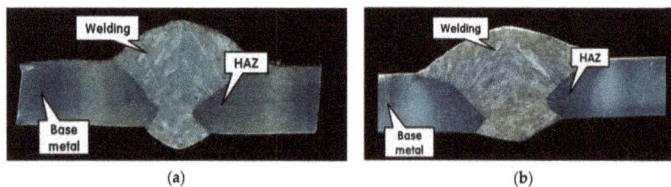

Figure 10. Cross-sections of (**a**) 590 FB steel and (**b**) 780 FB steel.

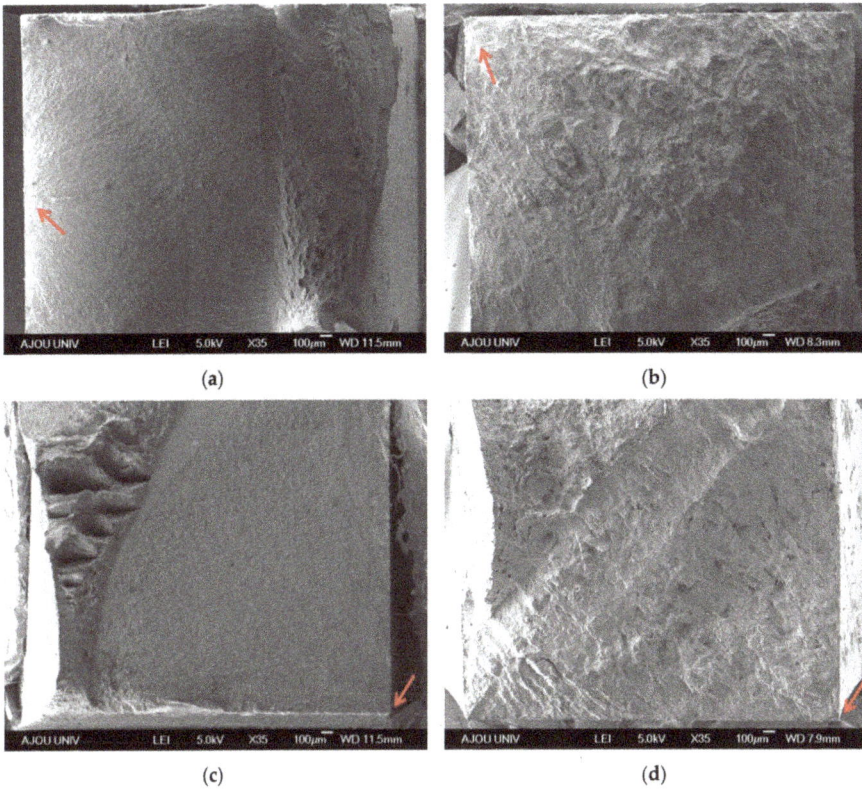

Figure 11. Fracture images ($\times$35): (**a**) 590 FB (490 MPa, 3.53×10^7 cycles); (**b**) welded 590 FB steel (450 MPa, 1.50×10^7 cycles); (**c**) 780 FB steel (700 MPa, 1.03×10^9 cycles); and (**d**) welded 780 FB steel (650 MPa, 5.08×10^7 cycles).

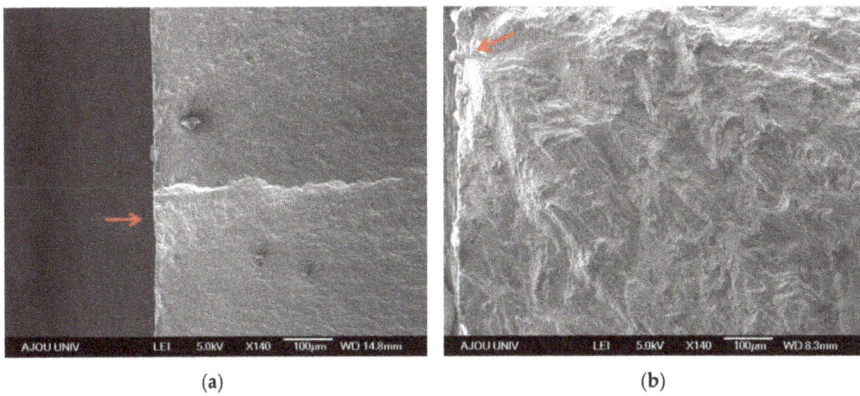

Figure 12. *Cont.*

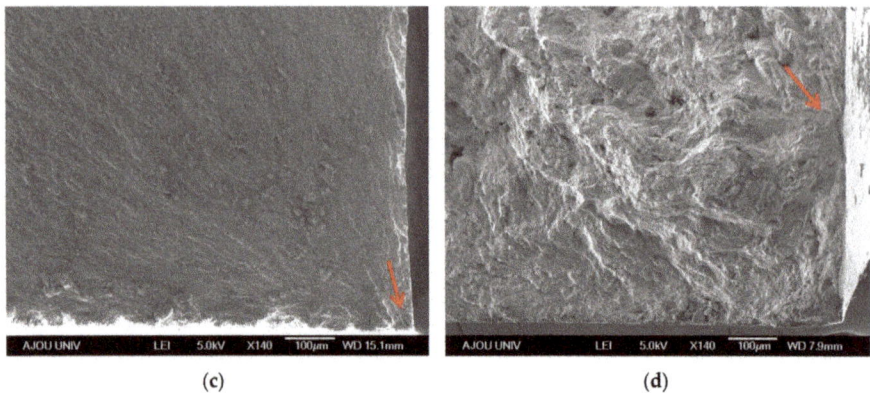

Figure 12. Crack initiations (×140): (a) 590 FB (490 MPa, 3.53 × 10^7 cycles); (b) welded 590 FB (450 MPa, 1.50 × 10^7 cycles); (c) 780 FB (700 MPa, 1.03 × 10^9 cycles); and (d) welded 780 FB (650 MPa, 5.08 × 10^7 cycles).

4. Conclusions

In this study, plate specimens for ultrasonic fatigue tests were successfully designed, and their designs were validated. A novel method to measure the dynamic Young's modulus was also proposed. The designed specimen confirmed that the resonant frequency corresponded to 19.927 kHz by FEM simulation. This value is very close to the target resonant frequency (20 kHz). Additionally, ultrasonic fatigue tests were performed using EDM-machined plate specimens at a test frequency of 19.914 kHz.

The fatigue strengths of the 590 FB steel and 780 FB steel specimens corresponded to 460 MPa and 667 MPa, respectively, during a gigacycle test. This indicated an approximate reduction of 7%–8% of the fatigue strength in the butt-weld specimens. This result verified the results obtained by previous studies, wherein it was observed that the heat of the welding process increased the grain size and decreased the fatigue life. Large grains were observed at the fractured surface and cross-section of the welded specimen. The tests also verified a single-point crack initiation at a very high cycle fatigue.

Hence, as indicated by the aforementioned results, this study confirmed a method to design ultrasonic fatigue test plate specimens. The fatigue behaviors of 590 FB steel, 780 FB steel, and welded plate specimens were examined by using an S–N curve based on gigacycle testing. It is expected that these results will benefit industries requiring parts that are manufactured and welded from high-strength steel materials. The results can also be applied to numerical fatigue simulations.

Acknowledgments: This research was supported by the Basic Science Research Program through the National Research Foundation of Korea (NRF) funded by the Ministry of Science, ICT and Future Planning (No. NRF-2015R1C1A1A02036547).

Author Contributions: Hyunho Yeom, Byungjoo Choi, and Taeho Seol performed the experiments; Hyunho Yeom, Moongu Lee, and Yongho Jeon analyzed the data; Hyunho Yeom and Yongho Jeon wrote the paper.

Conflicts of Interest: The authors declare no conflicts of interest.

References

1. Bathias, C.; Paris, P.C. *Gigacycle Fatigue in Mechanical Practice*; Marcel Dekker: New York, NY, USA, 2005; pp. 9–50.
2. Kazymyrovych, V. Very High Cycle Fatigue of Tool Steels. Ph.D. Thesis, Karlstad University, Karlstad, Sweden, 10 September 2010.
3. Crupi, V.; Epasto, G.; Guglielmino, E.; Risitano, G. Analysis of temperature and fracture surface of AISI4140 steel in very high cycle fatigue regime. *Theor. Appl. Fract. Mech.* **2015**, *80*, 22–30. [CrossRef]

4. Wang, Q.Y.; Bathias, C.; Kawagoishi, N.; Chen, Q. Effect of inclusion on subsurface crack initiation and gigacycle fatigue strength. *Int. J. Fatigue* **2002**, *24*, 1269–1274. [CrossRef]
5. Wang, Q.Y.; Berard, J.Y.; Dubarre, A.; Baudry, G.; Rathery, S.; Bathias, C. Gigacycle fatigue of ferrous alloys. *Fatigue Fract. Eng. Mater. Struct.* **1999**, *22*, 667–672. [CrossRef]
6. Gu, Y.; Tao, C.; He, Y.; Liu, C. The effect of frequency and sample shape on fatigue behaviors of DZ125 superalloy. *Theor. Appl. Mech. Lett.* **2012**, *2*. [CrossRef]
7. Jiang, Q.; Sun, C.; Liu, X.; Hong, Y. Very-high-cycle fatigue behavior of a structural steel with and without induced surface defects. *Int. J. Fatigue* **2016**, *93*, 352–362. [CrossRef]
8. Heinz, S.; Eifler, D. Crack initiation mechanisms of Ti6Al4V in the very high cycle fatigue regime. *Int. J. Fatigue* **2016**, *93*, 301–308. [CrossRef]
9. Müller-Bollenhagen, C.; Zimmermann, M.; Christ, H.J. Very high cycle fatigue behaviour of austenitic stainless steel and the effect of strain-induced martensite. *Int. J. Fatigue* **2010**, *32*, 936–942. [CrossRef]
10. Mayer, H.; Papakyriacou, M.; Zettl, B.; Stanzl-Tschegg, S.E. Influence of Porosity on the Fatigue Limit of Die Cast Magnesium and Aluminium Alloys. *Int. J. Fatigue* **2003**, *25*, 245–256. [CrossRef]
11. Majumdar, S.; Roy, S.; Ray, K.K. Fatigue performance of dual-phase steels for automotive wheel application. *Fatigue Fract. Eng. Mater. Struct.* **2016**, *40*, 315–332. [CrossRef]
12. Ellwood, R.D. Fatigue Performance of Downgauged High Strength Steel Automotive Suspension Component. Ph.D. Thesis, University of Wales Swansea, Swansea, UK, 2003.
13. Potukutchi, R.; Agrawal, H.; Perumalswami, P.; Dong, P. *Fatigue Analysis of Steel MIG Welds in Automotive Structures*; SAE International: Warrendale, PA, USA, 2004.
14. Takaoka, Y.; Shimoda, T.; Hara, J.; Seki, N.; Deguchi, T.; Koshio, K. Application of the latest technologies to fatigue strength improvement. In Proceedings of the TSCF2010 Shipbuilders' Meeting, Todyo, Japan, 28 October 2010.
15. Zhao, Z.P.; Qiao, G.Y.; Li, G.P.; Yang, W.W.; Liao, B.; Xiao, F.R. Fatigue properties of ferrite/bainite dual-phase X80 pipeline steel welded joints. *Sci. Technol. Weld. Join.* **2016**, *22*, 1–10. [CrossRef]
16. Shrama, Kadhum. Fatigue of Welded High Strength Steels for Automotive Chassis and Suspension Applications. Ph.D. Thesis, Cardiff University, Cardiff, UK, 2016.
17. Zhang, M.; Wang, W.; Wang, P.; Liu, Y.; Li, J. Fatigue behavior and mechanism of FV520B-I welding seams in a very high cycle regime. *Int. J. Fatigue* **2016**, *87*, 22–37. [CrossRef]
18. Zhu, M.L.; Liu, L.L.; Xuan, F.Z. Effect of frequency on very high cycle fatigue behavior of a low strength Cr-Ni-Mo-V steel welded joint. *Int. J. Fatigue* **2015**, *77*, 166–173. [CrossRef]
19. Zhao, X.; Dongpo, W.; Deng, C.; Liu, Y.; Zongxian, S. The fatigue behaviors of butt welds ground flush in the super-long life regime. *Int. J. Fatigue* **2012**, *36*, 1–8. [CrossRef]
20. Liu, Y.; Tian, R.; He, C. Gigacycle fatigue behaviors in fusion zone and heat affected zone of Q345 LA steel welded joints. In Proceedings of the ICF13, Beijing, China, 16–21 June 2013.
21. Matlock, D.K.; John, G.S. Third generation of AHSS: Microstructure design concepts. In *Microstructure and Texture in Steels*; Springer: London, UK, 2009; pp. 185–205.
22. Rojas, J.I.; Daniel, C. Onset Frequency of Fatigue Effects in Pure Aluminum and 7075 (AlZnMg) and 2024 (AlCuMg). *Metals* **2016**, *6*, 50. [CrossRef]
23. Buehler. *The Science behind Materials Preparation a Guide to Materials Preparation & Analysis*; Buehler Ltd.: Lake Bluff, IL, USA, 2007.
24. Westgate, S. The resistance spot welding of high and ultra-high strength steels. In Proceedings of the 3rd International Seminar on Advances in Resistance Welding, Berlin, Germany, 16–17 November 2004.
25. Seon, S.W.; Yi, W.; Park, H.D.; Hwang, Y.T. Characteristics of welds of pure titanium plate using ultrasonic attenuation. *J. Korean Soc. Nondestruct. Test.* **2013**, *33*, 205–211. [CrossRef]
26. Stille, S.; Tilmann, B.; Lorenz, S. Very high cycle fatigue (VHCF) behavior of structured Al 2024 thin sheets. In Proceedings of the 13th International Conference on Fracture, Beijing, China, 16–21 June 2013.

metals

MDPI

Article

High-Cycle Microscopic Severe Corrosion Fatigue Behavior and Life Prediction of 25CrMo Steel Used in Railway Axles

Yan-Ling Wang [1], Xi-Shu Wang [1,*], Sheng-Chuan Wu [2], Hui-Hui Yang [1] and Zhi-Hao Zhang [1]

[1] Department of Engineering Mechanics, Applied Mechanics Laboratory, Tsinghua University, Beijing 100084, China; vipsophia@126.com (Y.-L.W.); yanghh14@mails.tsinghua.edu.cn (H.-H.Y.); zzh6230@163.com (Z.-H.Z.)

[2] State Key Laboratory of Traction Power, Southwest Jiaotong University, Chengdu 610031, China; wusc@swjtu.edu.cn

* Correspondence: xshwang@tsinghua.edu.cn; Tel.: +86-10-6279-2972

Academic Editor: Filippo Berto
Received: 18 January 2017; Accepted: 7 April 2017; Published: 11 April 2017

Abstract: The effects of environmental media on the corrosion fatigue fracture behavior of 25CrMo steel were investigated. The media include air, and 3.5 wt % and 5.0 wt % NaCl solutions. Experimental results indicate that the media induces the initiation of corrosion fatigue cracks at multiple sites. The multi-cracking sites cause changes in the crack growth directions, the crack growth rate during the coupling action of the media, and the stress amplitude. The coupling effects are important for engineering applications and research. The probability and predictions of the corrosion fatigue characteristic life can be estimated using the three-parameter Weibull distribution function.

Keywords: corrosion fatigue; characteristic life prediction; 25CrMo steel; microscopic analysis; Weibull distribution

1. Introduction

CrMo steels, including low- and high-alloy chromium steels, are structural steels that are used in key components and parts of critical engineering structures such as gears and axles. In particular, 25CrMo steel has been used for high-speed axles because low-alloy chromium steels have the high strength and good ductility suitable for the high speed gears and axle components in engineering or transport applications. The mechanical requirements for components such as railway speeds axles are stricter because of their high-speed rotations and long service lives [1–4]. The literature has reported on the corrosion-fatigue [2], super-long life regime of railway axle steel [3,4] and fatigue properties of railway axles so far in the context of results of full-scale specimens [5]. The traditional fatigue strength design is based on a fatigue limit of materials of approximately 10^7 cycles or more for samples that do not fracture [6–8]. However, many engineering components (such as railway axles and vehicle wheels) are subjected to both environmental media and stress levels [2,9,10]. Therefore, there is always a discrepancy between traditional design and practical application. Accidents involving engineering components are not completely understood. In the past decades, a number of researchers studied the fatigue crack initiation and failure mechanisms [11,12] for very-high-cycle fatigue (VHCF) using fatigue life prediction methods [7,13] to avoid such accidents. For example, Li and Akid proposed a corrosion fatigue model including the stages of pitting and the pit-to-crack transition in order to predict the fatigue life of a structural material and the model showed good agreement with the experimental data at lower stress levels but predicted more conservative lifetimes as the stress increases [14]. Then, Beretta et al. modified Murtaza and Akid's model in order to obtain the description of corrosion–fatigue crack growth data thus allowing us to obtain a conservative prediction of the *S-N* diagram subjected to artificial rainwater. These works confirmed or verified that the present study is very important, especially the reliability analysis of corrosion fatigue data of high speed railway axle [2]. Pyttel et al. [15] determined that the surface fatigue strength during high cycle fatigue (10^5–10^7) (HCF) and the volume fatigue strength during VHCF (>10^7) should be studied in detail and presented a design or prediction processes for the entire fatigue life of metals. However, Pyttel et al. [15] also discussed the existence of a fatigue limit for metals. Miller et al. [6] also discussed the fatigue limit and methods for overcoming it. Therefore, some high strength CrMo steels with both good HCF performance and good corrosion fatigue performance [8,16,17] have been reported recently. Sakai et al. [18] and Huang et al. [19] previously reported on the HCF or VHCF issues of high carbon chromium bearing steel and low alloy chromium steel, respectively. Akita et al. [20] determined the effects of sensitization on the corrosion fatigue behavior of type 304 stainless steel annealed in nitrogen gas. The interest in the application of low-alloy CrMo steels to railways [1] along with other modified methods and effect factors [21,22] have led to substantial research efforts focused on the fatigue, corrosive fatigue, high-temperature fatigue and enhanced fatigue resistance in the past decade. Nevertheless, these results indicated that the fatigue behavior of some low-alloy CrMo steels in different environmental media was not sufficient for the intended applications. Additionally, some corrosive fatigue fracture mechanisms are unclear. For example, the probabilistic fatigue *S-N* curves for different media are based on the quantification of scattered fatigue data which require a certain number of specimens tested at various stress amplitudes [23]. The two-parameter (2-P) Weibull distribution-based modeling was preferred for the statistical analysis [24]. In this paper, we investigate the corrosion fatigue behavior of 25CrMo steel over 5×10^5–5×10^7 cycles in air and over 10^5–2×10^6 cycles in 3.5 wt % NaCl and 5.0 wt % NaCl aqueous solutions. For some of the scatter corrosion fatigue data, we use three-parameter (3-P) Weibull plots and the characteristic life of corrosion fatigue is predicted using detail fatigue rating (DFR). In addition, the microscopic fracture behavior is determined based on observations of the cross-sections of the fracture surfaces.

2. Materials and Experimental Method

2.1. Material or Specimen

25CrMo railway axle steel was prepared by vacuum inductive melting an ingot with a chemical composition (wt %) of 1.130 Cr, 0.600 Mn, 0.260 C, 0.240 Si, 0.210 Mo, 0.160 Cu, 0.150 Ni, 0.041 V, 0.007 P, and a balance of Fe. The room temperature mechanical properties of the 25CrMo steel used in this study are listed in Table 1.

Table 1. Mechanical properties of 25CrMo steel at room temperature.

Material	Ultimate Tensile Strength	Offset Yield Strength	Elongation	Young's Modulus
25CrMo steel	678 MPa	520 MPa	21.2%	205 GPa

The sample for the rotating bending fatigue tests in the environmental media shown in Figure 1 was designed with a circular center notch to control the stress concentration factor. The radius of the notch is 7 mm, and the minimum diameter is 4 mm [25–27]. All environmental fatigue test samples were machined from the surface layer of the railway axle by turning and grinding to the required dimensions and then abraded using abrasive paper to achieve a surface roughness of approximately $R_a = 0.6$ to 0.8 μm prior to the environmental fatigue testing.

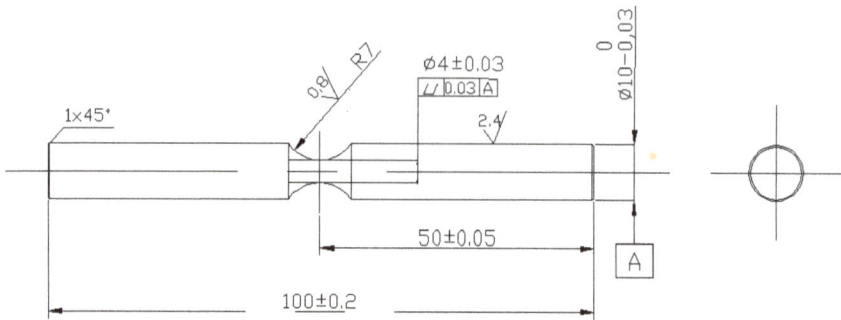

Figure 1. Sketch of the specimens indicating shape and size. The dimensions are in mm.

2.2. Mechanical Testing

The rotation bending environmental fatigue testing system used to investigate the high-cycle corrosion fatigue behavior of 25CrMo steel is shown in Figure 2a–c. Figure 2c shows the different corrosion media (air, the 3.5 wt % and 5.0 wt % NaCl aqueous solutions). To estimate the effective corrosive fatigue life 3–10 samples were tested at every stress level in the *S-N* curves. The applied stress amplitude is estimated as follows:

$$\sigma_a = \frac{32g\alpha LW}{\pi d^3} \ (\text{MPa}) \tag{1}$$

where d is the diameter of the gauge section (i.e., 4 mm), g is the acceleration due to gravity (9.8 m/s^2), α is the stress concentration factor (1.08), L is the distance from the gauge section end at which the load is applied (40.5 mm for a standard sample) and W is the applied load (kgf). All rotation bending fatigue tests were controlled by the load at a stress ratio R= −1. All rotation bending fatigue tests were controlled by the load at a stress ratio $R = -1$ and a rotating rate of about 3300 r/min (rotating frequency 55 Hz) according to ASTM E468-90.24 [28]. The fatigue test was stopped manually when the number of cycles exceeded 5×10^7 in air or the number of cycles exceeded 2×10^6 in 3.5 wt % NaCl and 5.0 wt % NaCl aqueous solutions. The corrosive liquid is dispensed by the corrosive fatigue system at a rate of 1.6 mL/min [26]. The pH values of the 3.5 wt % and 5.0 wt % NaCl aqueous solutions are 7.47 and 7.24, respectively.

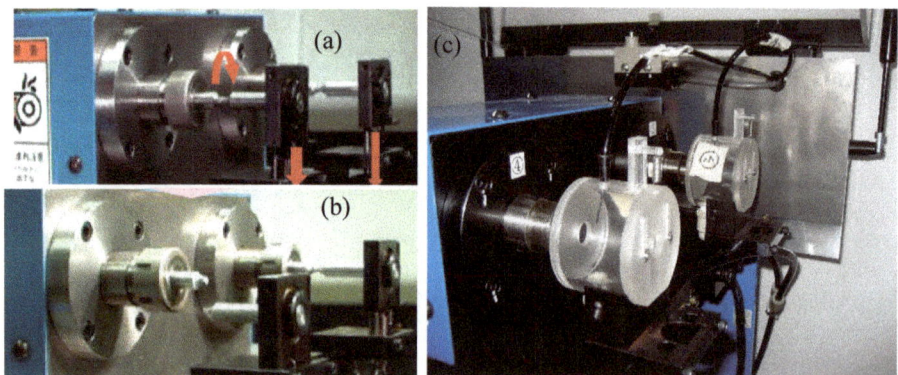

Figure 2. Fatigue testing loading modes for air or NaCl solutions. (**a**,**b**) Mechanical loading only; (**c**) mechanical loading in corrosive media.

2.3. Three-Parameter Weibull Distribution Function

Weibull analysis is a statistical method [21,24]. To further confirm the reliability of the corrosion fatigue data for the aforementioned testing conditions, a probabilistic analysis method was proposed using the 3-P Weibull distribution model, in which the failure probability function is expressed as follows:

$$F(x) = \begin{cases} 1 - e^{-(\frac{x-c}{b})^m} & x \geq c \\ 0 & x < c \end{cases} \tag{2}$$

where m, b, and c are the shape, scale, and location parameters, respectively, in this model and F is the failure probability of the corrosion fatigue data for 25CrMo steel up to a given value of x, the number of cycles to failure.

When the number of test specimens, n, is less than 20 ($n = 10$ for stress levels of 233.3 MPa and 273.0 MPa) the function F can be estimated as follows [29]:

$$F(x_i) = \frac{i - 0.3}{n + 0.4} \tag{3}$$

The values of m, b and c in Equation (2) can be estimated using Equation (3).

The equation $Q = \sum_{i=1}^{n} (X_i - \hat{X})^2$, in which X_i, $\hat{X}$ are the number of cycles to failure and the expected value of the number of cycles to failure under the same applied load and corrosive media, respectively, should be a minimum. Thus:

$$X - c - b \sqrt[m]{-\ln(1 - F(x))} = 0,$$
$$X_i - c - b \sqrt[m]{-\ln(1 - F_i)} = 0, i = 1, 2, 3, ...n \tag{4}$$

$$Q = \sum_{i=1}^{n} \left[X_i - c - b \sqrt[m]{-\ln(1 - F_i)} \right]^2 \to \min\{Q\}$$

and the estimated location parameter, $\hat{c}$, in the 3-P Weibull distribution model can be determined. The following is assumed:

$$\frac{1}{m} \ln \ln [1 - F(x)]^{-1} = \ln(x - \hat{c}) - \ln(b - \hat{c})$$
$$X = \ln \ln [1 - F(x)]^{-1}$$
$$Y = \ln(x - \hat{c}) \tag{5}$$
$$\zeta = 1/m, n = \ln(b - \hat{c})$$

Therefore, a linear equation can be obtained as follows:

$$Y = \zeta X + n \tag{6}$$

The corresponding correlation coefficient (ρ') for the linear regression is as follows:

$$\rho' = \frac{[\sum_{i=1}^{n} X_i Y_i - (\sum_{i=1}^{n} X_i \sum_{i=1}^{n} Y_i)/n]^2}{[\sum_{i=1}^{n} X_i^2 - (\sum_{i=1}^{n} X_i)^2/n] \cdot [\sum_{i=1}^{n} Y_i^2 - (\sum_{i=1}^{n} Y_i)^2/n]} \tag{7}$$

In this statistical analysis, the mean value or expected value, $E(x)$, (or $\hat{X}$) and the standard deviation, $S(x)$, can also be expressed as follows [23,30]:

$$E(x) = \hat{X} = (b - \hat{c})\Gamma(1 + \frac{1}{\zeta}), \Gamma(a) = \int_0^\infty x^{m-1} e^{-x} dx \tag{8}$$

$$S(x) = \sqrt{D(x)} = \sqrt{(b - \hat{c})^2 [\Gamma(1 + \frac{2}{m}) - \Gamma^2(1 + \frac{1}{m})]} \tag{9}$$

In addition, to estimate the reliability level of the statistical distribution function for these experimental data, the reliability index, β ($\beta = \frac{E(x)}{S(x)}$), of the probability distribution function was used. In general, the higher the $\beta = E(x)/S(x)$ value is, the smaller the degree of scatter for fatigue data is.

All of the aforementioned parameters in the Weibull distribution model were estimated using in-house procedure, which is developed by our team using Microsoft visual C++. In addition, to compare the reliability level of the number of cycles to failure at stress levels of 233.3 MPa and 273.0 MPa, all of the estimated parameters in the 3-P Weibull distribution function are listed in Table 2, in which the number of samples to failure is $n = 10$.

Table 2. Reliability parameters of fatigue fracture life under different environmental conditions.

Condition	m	b ($\times 10^5$)	$\hat{b}$ ($\times 10^5$)	c ($\times 10^5$)	P'	$E(x)$ ($\times 10^5$)	$S(x)$ ($\times 10^5$)	β
273.0 MPa 3.5 wt % NaCl	3.44	5.47	5.10	1.81	0.96	4.92	1.58	3.11
233.3 MPa 5.0 wt % NaCl	2.08	6.51	6.00	2.42	0.84	5.77	2.91	1.98

β: Reliability index ($\beta = E(x)/S(x)$). A higher value of β indicates a higher reliability.

3. Results

All fatigue data for the different environmental media are plotted in Figure 3. The *S-N* curves indicate that the effects of the environmental media on the fatigue fracture of 25CrMo steel are different. Based on the *S-N* curves, a fatigue limit of 25CrMo steel in the corrosive media was not reached. The differences in the corrosion fatigue behavior for the different test conditions are evident in the decreased gradient of the *S-N* curves. The *S-N* curves of 25CrMo steel in the corrosive media can be fitted using the following flow functions (the *S-N* curve for the 25CrMo in air is difficult to fit):

$$\sigma_{3.5wt\%Nacl} N_f^{0.21} = 4230.18 \tag{10}$$

$$\sigma_{5.0wt\%Nacl} N_f^{0.42} = 61083.68 \tag{11}$$

The scope of N_f of the *S-N* curve for the samples tested in the 5.0 wt % NaCl aqueous solution changes from 10^5 to 10^6, which is lower than that of *S-N* curves for the samples tested in air or a 3.5 wt % NaCl aqueous solution. For this range of number of cycles to failure, the difference between the number of cycles to failure for the two solutions is not significant (1.09×10^{-4} vs. 1.32×10^{-4}) considering the variability of the fatigue data. Thus, the corrosion fatigue mechanisms for 25CrMo steel in a 3.5 wt % and a 5.0 wt % NaCl aqueous solutions should be similar. However, when the number of cycles to failure is greater than 5×10^5, the corrosion fatigue life of 25CrMo steel in a 5.0 wt % NaCl aqueous solution decreased. That is, the environmental effect on the HCF fracture of this steel becomes increasingly important. Therefore, the environmental fatigue fracture of 25CrMo steel for 5×10^5 to 5×10^6 cycles warrants further investigation. The environmental effect on the fatigue fracture of 25CrMo axle steel not only depends on the formation of corrosive pits and solution exposure time but also depends on the local deformation of the metal (stress level). The former involves the transformation of corrosion pits to stress corrosion cracks and the latter involves a electrochemical process (anodic dissolution) and hydrogen embrittlement due to the cyclic deformation on the surface [26,31–33]. In addition, there is some scatter in the corrosive fatigue data, including the fatigue data for the samples tested in air, as shown in Figure 3. The scatter characteristics of the fatigue fracture data for the samples tested in air are more apparent, especially for $N_f > 10^7$. Therefore, for the complex environmental fatigue fracture mechanism of 25CrMo steel in aqueous solutions with varying concentrations of NaCl, reliability analysis of the *S-N* curves is necessary. Multiple samples were tested in the 3.5 wt % NaCl and 5.0 wt % NaCl aqueous solutions at the average stress levels shown in Table 3. Even if the stress ratio ($R = -1$) in the rotating fatigue test is different from the stress ratio ($R = 0.06$) in a push-pull electro-hydraulic servo, the estimated corrosion fatigue life at any stress level based on the typical environmental fatigue data can be determined for the region between 10^4 and 10^6 cycles to failure with the aid of detail fatigue rating (DFR) [34]. Although the fatigue fracture data have different decentralizations, the scatter of corrosion fatigue data (from 3.942×10^5 to 9.386×10^5) in the 5.0 wt % NaCl aqueous solution is greater than the scatter (from 3.588×10^5 to 6.911×10^5) for the 3.5 wt % NaCl aqueous solution. This may be because the stress level of the former (233.3 MPa) is lower than that of the latter (273.0 MPa). The reasons for this scatter and the mechanisms for environmental fatigue under the same testing condition are also validated by the microscopic analysis discussed in the next section.

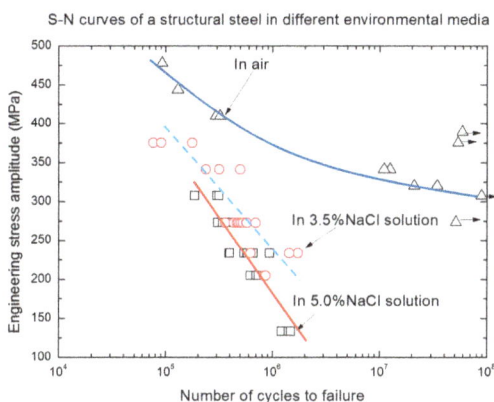

Figure 3. *S-N* curves for 25CrMo steel tested under different environmental conditions.

Table 3. Repeated fatigue testing data for 25CrMo steel under different environmental conditions.

Testing Conditions	No. 1 $\times 10^5$	No. 2 $\times 10^5$	No. 3 $\times 10^5$	No. 4 $\times 10^5$	No. 5 $\times 10^5$	No. 6 $\times 10^5$	No. 7 $\times 10^5$	No. 8 $\times 10^5$	No. 9 $\times 10^5$	No. 10 $\times 10^5$
273.0 MPa 3.5 wt % NaCl	3.588	4.351	4.676	4.819	4.830	5.068	5.310	5.690	5.711	6.911
233.3 MPa 5.0 wt % NaCl	3.942	3.987	5.440	5.779	5.799	5.811	6.200	6.446	6.590	9.386

4. Discussion

4.1. Fractography Analysis

In addition to the statistical analysis described above, observations of the fracture surfaces can also be used to characterize the scatter of the corrosion fatigue fracture for the same loading condition. For example, the fatigue fracture analyses use the scatter case under a stress level of 233.3 MPa as shown in Figure 3. The cross-sections of the fracture surfaces after different numbers of cycles are shown in Figures 4–6. Since the scatter in the data exists at other stress levels, the corrosion fatigue fracture characteristics are similar to those for the case mentioned above.

Figure 4 shows the corrosion fatigue fracture characteristics of 25CrMo steel in a 5.0 wt % NaCl aqueous solution ($N_f = 3.942 \times 10^5$). The two different fracture regions (regions A and C) can be divided into a corrosion fatigue crack initiation and propagation region (A) and a static fracture or instantaneous fracture region (C), as shown in Figure 4a. The former (A) is a relatively smooth region and the latter (C) is a relatively rough region at the macroscopic scale, as shown in Figure 4a. Region B, shown in Figure 4a, is the closest interface region between the corrosion fatigue crack propagation region and the static fracture region. The fatigue fracture morphology of region A includes many more fatigue crack propagation vestiges that have the typical concave-convex fatigue behavior shown in Figure 4b. This is because the 25CrMo steel has good ductility (the elongation is approximately 21.2%, as shown in Table 1) so that the concave-convex vestiges are visible in the fatigue crack propagation region. Additionally, the surface crack propagation length (the circumference is approximately $3a$) is approximately three times the inside fatigue crack propagation length (a) in region A, as shown in Figure 4a. The ratio between the inside fatigue crack propagation length and the surface fatigue crack propagation is approximately 0.3–0.5 for steels with good ductility. At the closest interface between the fatigue crack propagation region and the static fracture region, the concave-convex vestiges gradually disappear and the multi-secondary-cracks form, as shown in Figure 4c. Some plastic dimples are present at the closest interface between the fatigue crack propagation region and the static fracture region as shown in Figure 4c,d. The plastic dimples formed at the closest crack tips due to the higher stress concentrations. The maximum open displacement of these secondary cracks is approximately 3–4 μm as shown in Figure 4e,f. This suggests that the fracture toughness (K_{IC}) is relatively higher, but the corrosion fatigue strength of this steel in a 5.0 wt % NaCl aqueous solution is relatively low, as shown in Figure 3. This is because the secondary cracks rarely appear in the regions affected by corrosion in which the corrosive solution accelerates the hydrogen embrittlement at the fatigue crack tip.

Figures 5 and 6 show the other fracture characteristics for samples with $N_f = 6.590 \times 10^5$ and $N_f = 9.386 \times 10^5$ under the testing conditions of $\sigma = 233.3$ MPa in a 5.0 wt % NaCl aqueous solution, respectively. There are two different fatigue fracture characteristics due to the differences in the number of cycles to failure. That is, there are the different corrosion fatigue crack initiation sites and relationships between the corrosion fatigue crack propagation directions and paths. This is because there are different multi-cracking sites at the free surfaces of the round samples. The different crack growth directions from the surface to the inside of a circle cause the slip sidesteps and the different crack propagation paths, as shown in Figures 5a–d and 6a. The fatigue secondary cracks are present in the regions close to the fatigue crack propagation areas (A and B), as shown in Figure 6b,c. At the same time, the multi-cracks and the plastic dimples are also present in the static fracture region, as shown in Figure 6d.

To further describe the reason for the differences in the corrosion fatigue life for the 25CrMo steel samples tested under the same conditions, the proportional differences between the corrosion fatigue crack propagation area and the static fracture area are illustrated in Figure 7. These area fractions are 37.36% (Figure 7a), 38.12% (Figure 7b), and 55.22% (Figure 7c). Thus, a larger crack propagation area results in a longer fatigue life. Due to the multi-crack initiation sites on the free surface of the round samples, the fatigue crack propagation length (a) for multi-crack initiation sites is less than that for single fatigue crack initiation sites. This is because the fatigue fracture life depends primarily on the inside crack propagation length (a) in the environmental fatigue condition. A greater inside crack propagation length indicates a stronger stress-corrosion coupling effect [26,31–33,35]. Therefore, the probability of the corrosion solution acting on the surface crack at the multi-crack initiation sites is lower than it acting on the single crack initiation sites. Thus, the inside crack propagation length (a) after $N_f = 6.590 \times 10^5$ or $N_f = 9.386 \times 10^5$ (Figure 7b or Figure 7c) is much smaller than that after $N_f = 3.942 \times 10^5$ (Figure 7a). This is because the hydrogen atoms formed by the cathode partial reaction embrittle the material when the surface fatigue crack length is much greater, which results in a shorter fatigue life. Wittke et al. [33] described the interaction between strain and corrosive media, suggesting that the active straining at the crack tip during fatigue crack propagation can enhance the corrosion reactivity at the metal surface by increasing the area fraction and electrochemical reactivity of the fresh metal surfaces. This suggests that the effect of corrosion on fatigue damage inside the crack tip is greater than that at the crack tip surface after the corrosion liquid infiltrates into the inside of the newly-fractured metal surface. Therefore, the corrosion fatigue life is not completely dependent on the corrosion time, but depends primarily on the coupling effects of the electrochemical reaction and the plastic deformation. Subsequently, considering the effects of a stress gradient on the fatigue fracture process, if the inside crack length is greater for the rotating bending fatigue tests, the bending fracture of the sample occurs more easily, as shown in Figure 7a. Therefore, the reasons discussed above will result in the differences in the corrosion fatigue life for the same testing condition. Improving the surface quality of the sample and avoiding stress concentrations or heterogeneous defects on the surface can help decrease these differences.

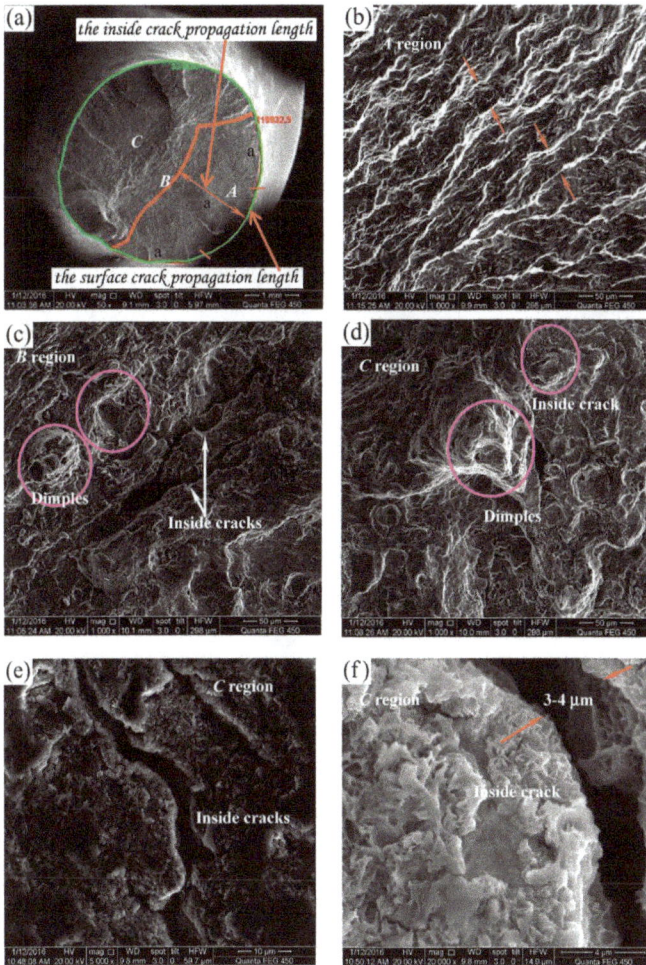

Figure 4. Typical fracture characteristics for $N_f = 3.942 \times 10^5$, $\sigma = 233.3$ MPa in a 5.0 wt % NaCl solution. (**a**) Macro-scale view of the fracture surface; (**b**) region labeled A; (**c**) region labeled B; (**d**) region labeled C; (**e**,**f**) crack patterns in the static fracture regions.

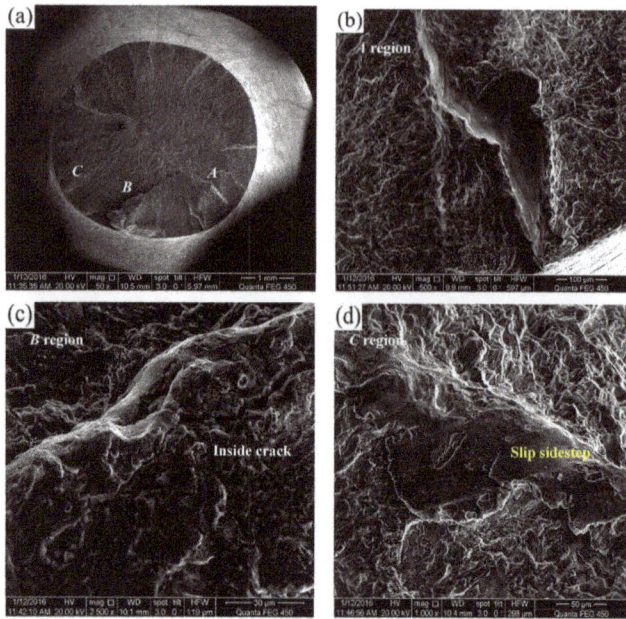

Figure 5. Typical fracture characteristics for $N_f = 6.590 \times 10^5$, $\sigma = 233.3$ MPa in a 5.0 wt % NaCl solution. (**a**) Macro-scale view of the fracture surface; (**b**) region labeled A; (**c**) region labeled B; (**d**) region labeled C.

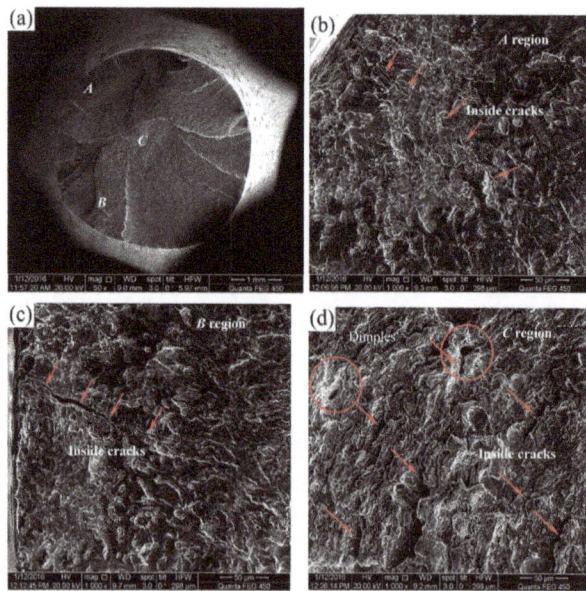

Figure 6. Typical fracture characteristics for $N_f = 9.386 \times 10^5$, $\sigma = 233.3$ MPa in a 5.0 wt % NaCl solution. (**a**) Macro-scale view of the fracture surface; (**b**) region labeled A; (**c**) region labeled B; (**d**) region labeled C.

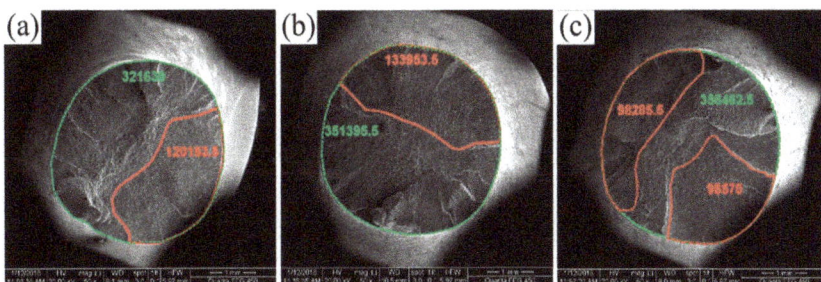

Figure 7. Typical fracture analysis of 25CrMo steel in the same testing condition (233.3 MPa, 5.0 wt % NaCl). Macro-scale fracture surface characteristics for (**a**) $N_f = 3.942 \times 10^5$; (**b**) $N_f = 6.590 \times 10^5$; (**c**) $N_f = 9.386 \times 10^5$.

4.2. Fatigue Life Prediction and Reliability Analysis

Since the corrosion fatigue data have a degree of scattering, the Weibull distribution is the most appropriate statistical analysis tool. Based on the statistical distribution parameters in Table 2, can the fatigue lives for the same material be predicted for other stress levels? To answer this question, we carefully analyzed each parameter in Table 2. The values of the shape parameter (m) and reliability index, β, at 273.0 MPa in a 3.5 wt % NaCl aqueous solution are greater than the values at 233.3 MPa in a 5.0 wt % NaCl aqueous solution. This means that the scatter of the former ($m = 3.44$, $\beta = 3.11$) is lower than that of the latter ($m = 2.08$, $\beta = 1.98$). Another important parameter in this reliability analysis is the scale parameter, b, which reflects the characteristic corrosion fatigue life. The two distributions of the characteristic lives are $N_f = 5.47 \times 10^5$ and $N_f = 6.51 \times 10^5$, which are slightly greater than the average values ($N_f = 5.095 \times 10^5$ and $N_f = 5.938 \times 10^5$, respectively). When n is less than ten, the difference becomes greater between the characteristic life and the arithmetic mean values of the fatigue lives. Therefore, using the characteristic life value ($\hat{b}$) of corrosion fatigue data is better than using the arithmetic mean value in the statistical analysis of the small sample number. Most S-N curves for the samples were tested under the same environmental condition but at different stress levels have approximately the same shape parameter (m). Therefore, other characteristic lives ($\hat{b}$) at additional stress levels for small sample numbers (such as $n = 3$) in the same S-N curve can be estimated based on the m parameter for the large sample numbers (such as $n = 10$). The estimation method is as follows [34,36]:

$$\hat{b} = (\frac{1}{n} \sum_{i=1}^{n} N_{fi}^m)^{1/m} \tag{12}$$

where n is the sample number, N_{fi} is the number of cycle to failure in the ith fatigue test, and m is the shape parameter in the Weibull distribution function. For example, the relationship between the experimental data and all characteristic lives ($\hat{b}$) at different stress levels determined using Equation (12) is shown in Figure 8. According to the aforementioned estimation method (Equation (12)) [36], the predication of additional characteristic fatigue lives ($\hat{b}$) at different stresses is possible if the shape parameter (m) is known. In Figure 8, the characteristic corrosion fatigue data (labeled with ●) is in good agreement with the experimental data (labeled with ○). Therefore, this estimation method based on the statistical distribution is simple and effective. That is, we can estimate the characteristic life by using the shape parameter (m) for the experimental data with the same testing condition to eliminate the invalid data. In addition, as a key parameter in reliability design, Sakin [37] obtained the fatigue life distribution diagrams for glass-fiber reinforced polyester composites by using a two-parameter Weibull distribution function, from which the reliability percentage (%) corresponding to any life (cycle) or stress amplitude could be found easily. At the same time, Sivapragash et al. [38] systematically analyzed the fatigue life prediction of ZE41A magnesium alloys using the Weibull distribution and obtained a probability distribution based on the failure of the material. However, reports of corrosion fatigue data of 25CrMo steel based on the 3-P Weibull distribution analysis are rare.

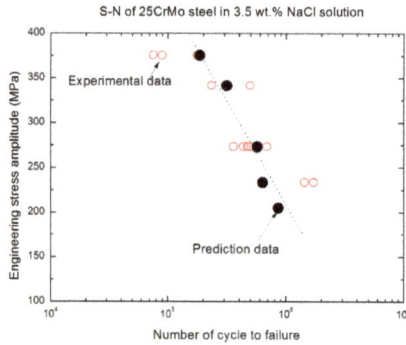

Figure 8. Experimental data and estimated value of the corrosion fatigue life for different stresses in a 3.5 wt % NaCl solution. The solid circles are the predicted values for 25CrMo steel in a 3.5 wt % NaCl solution, and the open circles are the experimental values for 25CrMo steel in a 3.5 wt % NaCl solution. The dotted line was the liner fit of the black solid circles.

Figure 9 gives the failure probability for corrosion fatigue of 25CrMo axle steel for two typical corrosion cases. The estimated curves shown in Figure 9 indicate that the relationship between the failure probability value of the corrosion fatigue fracture and N_f suggests that the service life of 25CrMo axle steel can be calculated for the different environmental conditions. This will assist with the design of safe structures or materials. For example, when the number of cycles to failure is 5×10^5, the probability values for corrosion fatigue fracture are 55% at 273.0 MPa, in a 3.5 wt % NaCl aqueous solution and 25% at 233.3 MPa, in a 5.0 wt % NaCl aqueous solution. Additionally, the failure probability of 25CrMo axle steel is not sensitive to the change in the number of cycles to failure in either the low or the high failure probability region, as shown in Figure 9. The critical value of the cyclic number is different for the changes in both the environmental condition and the stress level because the plateau values are different for each curve.

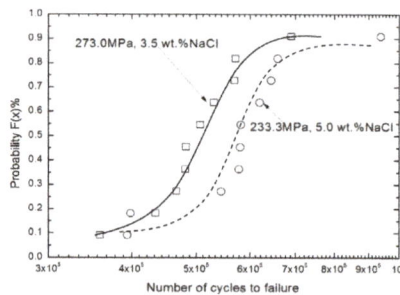

Figure 9. $F(x)$ vs. N_f for 25CrMo steel under different testing conditions. The circles represent the failure probability at 233.3 MPa in a 5.0 wt % NaCl solution, and the squares represent the failure probability at 273.0 MPa in a 3.5 wt % NaCl solution.

5. Conclusions

Through detailed corrosion fatigue tests, the fatigue fracture characteristics, reliability analysis, and characteristic life cycle predictions for a typical 25CrMo axle steel in three different environmental media were determined. The main conclusions are as follows:

1.   The experimental results indicate that, the higher the number of cycles to failure, the stronger the effect of the corrosive media on the fatigue damage. In this HCF regime, the corrosion fatigue behavior of 25CrMo steel is not completely dependent on the corrosion time (exposure time). It is also dependent on the coupling action of the electrochemical reaction and the plastic deformation. When the number of cycles to failure is greater than 10^6, the corrosion fatigue data must be analyzed using a statistical analysis, such as the Weibull distribution model.
2.   For the same environmental media and applied load, the amount of scatter in the corrosion fatigue data is caused primarily by the number of fatigue crack initiation sites and the direction of the fatigue crack propagation in the early stage. A lower number of fatigue crack initiation sites indicates a shorter fatigue crack initiation life, and a faster corrosion fatigue crack growth rate indicates that the total life of the sample is shorter. Improving the surface quality of the sample by reducing the surface defects or the stress concentration sites helps reduce the dispersion in the corrosion fatigue data.
3.   The corrosion fatigue fracture characteristics indicate that there are fatigue crack propagation vestiges in the crack propagation regions, the second cracks, the plastic dimples, and the slip sidesteps near the interface region between the fatigue crack propagation region and the static fracture region.
4.   The results of the 3-P Weibull distribution analysis indicate that the corrosion fatigue data are more reliable as the value of the shape parameter (m) increases. At the same time, the characteristic corrosion fatigue life at any stress level for a small sample tested under the same environmental condition can be estimated simply and effectively using Equation (12).

Acknowledgments: The present research is supported by the Natural Science Foundation of China (Nos. 11272173, 11572170), the Foundation of the State Key Laboratory of Traction Power Southwest Jiaotong University (No. TPL1503), China, and the Nantong Key Laboratory of New Materials Industrial Technology, and the Science and Technology and Research and Development Application Project of Sichuan Province (2017JY0216).

Author Contributions: Yanling Wang finished the fatigue experiments and microstructure observation and wrote part of the paper. Xishu Wang wrote the paper and completed the data analysis. Shengchuan Wu provided all materials for fatigue tests and checked a manuscript. Huihui Yang finished the fracture analysis of paper. Zhihao Zhang participated in a part experiment.

Conflicts of Interest: The authors declare no conflict of interest.

Nomenclature

a	crack length
b	scale parameter in Weibull distribution function
$\hat{b}$	predication value of characteristic life
c	location parameter in Weibull distribution function
α	the stress concentration factor (1.08)
d, L	the geometry size, respectively
g	the acceleration of gravity (9.8 m/s^2)
F(x)	Weibull distribution function
m	shape parameter in Weibull distribution function
n	experimental number of corrosion fatigue life
P'	corresponding correlation coefficient
E(x)	expected value of statistics
S(x)	standard deviation value
β	reliability index
N_f	number of cycles to failure
σ	stress amplitude (MPa)
R	stress ratio

References

1. Zhang, Y.T.; Bian, S.; Han, W.X. Effect of Mn on hardenability of 25CrMo axle steel by an improved end-quench test. *China Foundry* **2012**, *9*, 318–321.
2. Beretta, S.; Carboni, M.; Fiore, G.; Conte, A.L. Corrosion-fatigue of A1N railway axle steel exposed to rainwater. *Int. J. Fatigue* **2010**, *32*, 952–961. [CrossRef]
3. Zhao, Y.X.; Yang, B.; Feng, M.F.; Wang, H. Probabilistic fatigue S-N curves including the super-long life regime of a railway axle steel. *Int. J. Fatigue* **2009**, *31*, 1550–1558. [CrossRef]
4. Zhao, Y.X. A fatigue reliability analysis method including super long life regime. *Int. J. Fatigue* **2012**, *35*, 79–90. [CrossRef]
5. Cervello, S. Fatigue properties of railway axles: New results of full-scale specimens from Euraxles project. *Int. J. Fatigue* **2016**, *861*, 2–12. [CrossRef]
6. Miller, K.J.; O'Donnell, W.J. The fatigue limit and its elimination. *Fatigue Fract. Eng. Mater. Struct.* **1999**, *22*, 545–557. [CrossRef]
7. Wang, Q.Y.; Berard, J.Y.; Dubarre, A.; Baudry, G.; Rathery, S.; Bathias, C. Gigacycle fatigue of ferrous alloys. *Fatigue Fract. Eng. Mater. Struct.* **1999**, *22*, 667–672. [CrossRef]
8. Li, S.X.; Zhang, P.Y.; Yu, S.R. Experimental study on very high cycle fatigue of martensitic steel of 2Cr13 under corrosive environment. *Fatigue Fract. Eng. Mater. Struct.* **2014**, *37*, 1146–1152. [CrossRef]
9. Kang, D.H.; Kim, S.H.; Lee, C.H. Corrosion fatigue behaviors of HSB800 and its HAZs in air and seawater environments. *Mater. Sci. Eng. A Struct. Mater. Prop. Microstruct. Process.* **2013**, *559*, 751–758. [CrossRef]
10. Jones, R.H.; Simonen, E.P. Early stages in the development of stress-corrosion cracks. *Mater. Sci. Eng. A Struct. Mater. Prop. Microstruct. Process.* **1994**, *176*, 211–218. [CrossRef]
11. Lai, C.L.; Tsay, L.W.; Chen, C. Effect of microstructure on hydrogen embrittlement of various stainless steels. *Mater. Sci. Eng. A Struct. Mater. Prop. Microstruct. Process.* **2013**, *584*, 14–20. [CrossRef]
12. Bayraktar, E.; Garcias, I.M.; Bathias, C. Failure mechanisms of automotive alloys in very high cycle fatigue range. *Int. J. Fatigue* **2006**, *28*, 1590–1602. [CrossRef]
13. Poting, S.; Zenner, H. Parameter C lifetime calculation for the high cycle regime. *Fatigue Fract. Eng. Mater. Struct.* **2002**, *25*, 877–885. [CrossRef]
14. Li, S.X.; Akid, R. Corrosion fatigue life prediction of a steel shaft material in seawater. *Eng. Fail. Anal.* **2013**, *34*, 324–334. [CrossRef]
15. Pyttel, B.; Schwerdt, D.; Berger, C. Very high cycle fatigue—Is there a fatigue limit? *Int. J. Fatigue* **2011**, *33*, 49–58. [CrossRef]
16. Qian, G.A.; Zhou, C.G.; Hong, Y.S. Experimental and theoretical investigation of environmental media on very-high-cycle fatigue behavior for a structural steel. *Acta Mater.* **2011**, *59*, 1321–1327. [CrossRef]
17. Guo, Q.; Guo, X. Research on high-cycle fatigue behavior of FV520B stainless steel based on intrinsic dissipation. *Mater. Des.* **2016**, *90*, 248–255. [CrossRef]
18. Sakai, T.; Lian, B.; Takeda, M.; Shiozawa, K.; Oguma, N.; Ochi, Y.; Nakajima, M.; Nakamura, T. Statistical duplex S-N characteristics of high carbon chromium bearing steel in rotating bending in very high cycle regime. *Int. J. Fatigue* **2010**, *32*, 497–504. [CrossRef]
19. Huang, Z.Y.; Wagner, D.; Wang, Q.Y.; Bathias, C. Effect of carburizing treatment on the "fish eye" crack growth for a low alloyed chromium steel in very high cycle fatigue. *Mater. Sci. Eng. A Struct. Mater. Prop. Microstruct. Process.* **2013**, *559*, 790–797. [CrossRef]
20. Akita, M.; Uematsu, Y.; Kakiuchi, T.; Nakajima, M.; Tsuchiyama, T.; Bai, Y.; Isono, K. Effect of sensitization on corrosion fatigue behavior of type 304 stainless steel annealed in nitrogen gas. *Mater. Sci. Eng. A Struct. Mater. Prop. Microstruct. Process.* **2015**, *640*, 33–41. [CrossRef]
21. Mohammad, M.; Abdullah, S.; Jamaludin, N.; Innayatullah, O. Predicting the fatigue life of the SAE 1045 steel using an empirical Weibull-based model associated to acoustic emission parameters. *Mater. Des.* **2014**, *54*, 1039–1048. [CrossRef]
22. Ebara, R.C. Corrosion fatigue crack initiation in 12% chromium stainless steel. *Mater. Sci. Eng. A Struct. Mater. Prop. Microstruct. Process.* **2007**, *468*, 109–113. [CrossRef]
23. Zhao, Y.X.; Liu, H.B. Weibull modeling of the probabilistic S-N curves for rolling contact fatigue. *Int. J. Fatigue* **2014**, *66*, 47–54. [CrossRef]

24. Weibull, W. A statistical distribution function of wide applicability. *J. Appl. Mech. Trans. ASME* **1951**, *18*, 293–297.
25. Wang, X.S.; Guo, X.W.; Li, X.D.; Ge, D.Y. Improvement on the fatigue performance of 2024-T4 alloy by synergistic coating technology. *Materials* **2014**, *7*, 3533–3546. [CrossRef]
26. Wang, X.S.; Li, X.D.; Yang, H.H.; Kawagoishi, N.; Pan, P. Environment-induced fatigue cracking behavior of aluminum alloys and modification methods. *Corros. Rev.* **2015**, *33*, 119–137. [CrossRef]
27. Yang, H.H.; Wang, Y.L.; Wang, X.S.; Pan, P.; Jia, D.W. Synergistic effect of corrosion environment and stress on the fatigue damage behavior of Al alloys. *Fatigue Fract. Eng. Mater. Struct.* **2016**, *39*, 1309–1316. [CrossRef]
28. ASTM. *Standard Practice for Presentation of Constant Amplitude Fatigue Test Results for Metallic Materials Annual Book of ASTM Standards (2004) USA*; ASTM E468-90; ASTM: West Conshohocken, PA, USA, 2004.
29. Zheng, R.Y.; Yan, J.S. New estimation method of three-parameter Weibull distribution. *J. Mech. Strength* **2002**, *24*, 599–601. (In Chinese).
30. Lu, Z.J.; Zhang, S.L.; Liu, X.F.; Liu, C.H. Some problems of Weibull distribution. *Qual. Reliab.* **2007**, *32*, 9–13.
31. Cooper, K.R.; Kelly, R.G. Crack tip chemistry and electrochemistry of environmental cracks in AA 7075. *Corros. Sci.* **2007**, *49*, 2636–2662. [CrossRef]
32. Li, X.D.; Wang, X.S.; Ren, H.H.; Chen, Y.L.; Mu, Z.T. Effect of prior corrosion state on the fatigue small cracking behavior of 6151-T6 aluminum alloy. *Corros. Sci.* **2012**, *55*, 26–33. [CrossRef]
33. Wittke, P.; Klein, M.; Walther, F. Chemical-Mechanical Characterization of the Creep-Resistant Mg-Al-Ca Alloy DieMag422 Containing Barium-Quasistatic and Cyclic Deformation Behavior in Different Defined Corrosion Conditions. *Mater. Test.* **2014**, *56*, 16–23. [CrossRef]
34. Goranson, U.G. Elements of structural integrity assurance. *Int. J. Fatigue* **1994**, *16*, 43–65. [CrossRef]
35. Dhinakaran, S.; Prakash, R.V. Effect of low cyclic frequency on fatigue crack growth behavior of Mn-Ni-Cr steel in air and 3.5% NaCl solution. *Mater. Sci. Eng. A Struct. Mater. Prop. Microstruct. Process.* **2014**, *609*, 204–208. [CrossRef]
36. Huang, X.; Liu, J.Z.; Ma, S.J.; Hu, B.R. Sensitivity analysis of the parameters in detail fatigue rating equation. *Acta Aeronaut. ET Astronaut. Sin.* **2012**, *33*, 863–870. (In Chinese).
37. Sakin, R.; Ay, İ. Statistical analysis of bending fatigue life data using Weibull distribution in glass-fiber reinforced polyester composites. *Mater. Des.* **2008**, *29*, 1170–1181. [CrossRef]
38. Sivapragash, M.; Lakshminarayanan, P.R.; Karthikeyan, R.; Raghukandan, K.; Hanumantha, M. Fatigue life prediction of ZE41A magnesium alloy using Weibull distribution. *Mater. Des.* **2008**, *29*, 1549–1553. [CrossRef]

![metals logo] *metals*

MDPI

Article

Experimental Investigation on the Fatigue Life of Ti-6Al-4V Treated by Vibratory Stress Relief

Han-Jun Gao [1], Yi-Du Zhang [1], Qiong Wu [1,*] and Jing Song [2]

[1]   State Key Laboratory of Virtual Reality Technology and Systems,
      School of Mechanical Engineering and Automation, Beijing University of Aeronautics and Astronautics,
      Beijing 100191, China; gao.hanjun@buaa.edu.cn (H.-J.G.); ydzhang@buaa.edu.cn (Y.-D.Z.)
[2]   Beijing Institute of Aerospace Systems Engineering, Beijing 100076, China; songjing@buaa.edu.cn
*   Correspondence: wuqiong@buaa.edu.cn; Tel.: +86-10-8231-7756

Academic Editor: Filippo Berto
Received: 1 April 2017; Accepted: 28 April 2017; Published: 3 May 2017

Abstract: Vibratory stress relief (VSR) is a highly efficient and low-energy consumption method to relieve and homogenize residual stresses in materials. Thus, the effect of VSR on the fatigue life should be determined. Standard fatigue specimens are fabricated to investigate the fatigue life of Ti-6Al-4V titanium alloy treated by VSR. The dynamic stresses generated under different VSR amplitudes are measured, and then the relationship between the dynamic stress and vibration amplitude is obtained. Different specimen groups are subjected to VSRs with different amplitudes and annealing treatment with typical process parameters. Residual stresses are measured to evaluate the stress relieving effects. Finally, the fatigue behavior under different states is determined by uniaxial tension–compression fatigue experiments. Results show that VSR and annealing treatment have negative effects on the fatigue life of Ti-6Al-4V. The fatigue life is decreased with the increase in VSR amplitude. When the VSR amplitude is less than 0.1 mm, the decrease in fatigue limit is less than 2%. Compared with specimens without VSR or annealing treatment, the fatigue limit of the specimens treated by VSR with 0.2 mm amplitude and annealing treatment decreases by 10.60% and 8.52%, respectively. Although the stress relieving effect is better, high amplitude VSR will lead to the decrease of Ti-6Al-4V fatigue life due to the defects generated during vibration. Low amplitude VSR can effectively relieve the stress with little decrease in fatigue life.

Keywords: fatigue life; vibratory stress relief; titanium alloy Ti-6Al-4V; residual stress; annealing treatment

---

1. Introduction

Residual stresses exist in many fabricated structures due to plastic deformation from thermal and mechanical operations during manufacturing [1]. The presence of residual stresses in engineering components and structures significantly affect the fatigue behavior [2], strength [3], and dimensional stability [4]. Many studies have been conducted to investigate the effects of residual stress on the fatigue life of different materials [5–7]. The effects of residual stress and surface hardness on the fatigue life under different cutting conditions of 0.45%C steel were studied by Sasahara [8]. Their experimental results show that the fatigue life of machined components can be increased if compressive residual stress and high hardness within surface layer can be induced by a cutting process. Fatigue crack growth from a hole with a pre-existing compressive residual stress was simulated utilizing the 2D elastic-plastic finite element (FE) model by LaRue and Daniewic [9]. Pouget and Reynolds [10] determined that fatigue crack propagation in friction stir-welded AA2050 was strongly linked with the presence of residual stresses, and compressive residual stresses in the heat affected zone were responsible for the apparent improvement of fatigue behavior when the crack approached the weld.

Surface compressive stress can inhibit the initiation and propagation of fatigue cracks. Some processes, such as rolling, laser peening and shot peening, are utilized to improve the fatigue life by inducing surface compressive residual stress. The effect of different shot peening treatments on the reverse bending fatigue behavior of 7075-T651 aluminum alloy was investigated by Benedetti et al. [11]. They demonstrated that controlled shot peening that employed ceramic beads determined a remarkable increment of the high-cycle fatigue resistance, which ranges between 15% and 50%. Nikitin et al. [12] found that laser shock peening produced similar amounts of lifetime enhancements as deep rolling. The cycle, stress amplitude and temperature-dependent relaxation of compressive residual stresses was more pronounced than the decrease of near-surface work hardening. Bagherifard et al. [13] concluded that severe shot peening induced near surface grain refinement to nano and sub-micron range and transformed the austenite phase into strain-induced α'-martensite in a layered deformation band structure. The effect of small defects on fatigue threshold of different series of nitride and nitride-shot peened low alloy steel specimens was investigated by Fernández-Pariente et al. [14] through experiments.

Some researchers show interest in the fatigue behavior of Ti-6Al-4V. Zabeen et al. [15] evaluated the changing tendency of the Ti-6Al-4V titanium alloy residual stress field with fatigue crack growth after laser shock peening. Yamashita et al. [16] investigated the method of estimating the fatigue strength of small notched Ti-6Al-4V specimen using the theory of critical distance that employed the stress distribution in the vicinity of the notch root. Bourassa et al. [17] explored surface thermal/mechanical processing of Ti-6Al-4V to improve the fatigue strength of microknurled specimens via the production of a Ti-6Al-4V dual microstructure. Golden et al. [18] investigated the fatigue variability of an alpha + beta processed Ti-6Al-4V turbine engine alloy by conducting a statistically significant number of repeated tests at a few conditions.

The stress relieving process is generally conducted to improve the fatigue life and dimensional accuracy [19]. Natural stress relief, thermal stress relief (TSR), and vibratory stress relief (VSR) [20] are the most common methods. VSR is a general term used to refer to the reduction of residual stress by means of cyclic loading treatments. When the superposition of the dynamic stress and initial residual stress reaches the elastic limit of the material, plastic deformations occur in local positions, which leads to the residual stress release. Low cyclic stress amplitude can also release the stress after enough cycle times due to the generation of dislocation slip and multiplication at the micro level [21]. As a highly efficient and low-energy consumption method, VSR has received much attention from scholars and engineers in recent years. Sun et al. [22] concluded that the tensile properties of a marine shafting of 35# bar steel bar changed slightly before and after vibration, whereas the macro residual stress decreased notably by approximately 48%. Results in literature [23] show that the macro-residual stresses of the welded steel plates of D6AC and D406A decreased to 0 ± 25 MPa after VSR. Wang et al. [24] believed that the relaxation effect of compressive residual stress after VSR was better than that of tensile residual stress; the compressive residual stress and texture density increased first and then decreased with an increase in the vibration time. The said researchers also determined in another study that the strong basal textures of AZ31 Mg alloys would be weaken when the vibration time was more than 10 min [25].

Moreover, a residual stress decrease model of 304L under cyclic loading was established by Rao et al. [26] to evaluate the VSR effectiveness. Kwofie [27] proposed a theoretical plasticity model to study the residual stress relief mechanism under mechanical vibration. A combined method of TSR and VSR (TVSR) to extend the VSR effects was proposed by Lv and Zhang [28]. Their experimental results show that the maximum stress of the 7075 aluminum alloy plate decreased by 55.9% and 13.4% after TVSR and VSR, respectively. A mathematical model based on plasticity theorem with linear kinematic hardening was presented by Vardanjani et al. [29] to explain the reduction mechanism of residual stresses caused by VSR.

Nowadays, researchers have divergent opinions about the VSR effects on the fatigue life. Several of them believe that VSR can decrease the fatigue life, but others hold the opposite view. Djuric et al. [30] evaluated the fractal dimension of welded high-strength martensitic steel utilizing Barkhausen Noise

Analysis and observed an increase in the fatigue damage to the microstructure due to the applied VSR treatment. This study indicates that VSR can have a negative effect on the fatigue life. Wozney and Crawmer [31] indicated that stress reduction did not occur uniformly throughout vibrated structures, and the possibility of fatigue damage during VSR treatment couldn't be neglected. On the contrary, the VSR and TSR effects on the fatigue lives of welded specimens made of 0.18% carbon steel were investigated by Munsi et al. [32]. Their experimental results show that the fatigue lives of the specimens treated by TSR decreased by 43%, whereas those treated by VSR increased by 17%. Song and Zhang [33] determined that vibratory stress relief could improve the fatigue life of 7075-T651 aluminum alloy when the dynamic stress was less than 8% of the yield limit.

Therefore, the VSR effect on the fatigue life is still ambiguous. VSR can have different effects on the fatigue life of different materials. The present study manufactures a batch of standard fatigue specimens to investigate the fatigue life of Ti-6Al-4V titanium alloy treated by VSR. Dynamic stresses generated under different VSR amplitudes are initially measured, and then the relationship between dynamic stress and vibration amplitude is obtained. VSRs with 0.03, 0.05, 0.1, and 0.2 mm amplitude and annealing treatment with typical process parameters are then conducted on five specimen groups. Another specimen group without any treatment is employed for comparison. The residual stresses of each group are subsequently measured to study the stress relieving effect. Finally, the fatigue limits and cyclic stress amplitude-logarithm of cycle numbers (S-lgN) curves under different states are obtained by uniaxial tension-compression fatigue experiments.

2. Experiments

Standard fatigue specimens are manufactured from ϕ30 mm $\times$ 200 mm (diameter $\times$ length) extrusion Ti-6Al-4V titanium alloy bars. The specimen shape and size (Figure 1) are determined based on the China National Standard GB/T 3075-2008. Threads on both ends are utilized for clamping in the experiments.

A cylindrical counterweight with 40 mm diameter and 30 mm length, which is also made of Ti-6Al-4V titanium alloy, is added to one end of the specimen to lower the natural frequency and increase the dynamic stress (Figure 2a). The counterweight is fixed to the specimen by a thread in the dynamic stress measurement and VSR tests.

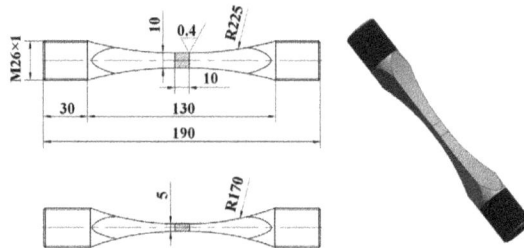

Figure 1. Ti-6Al-4V titanium alloy fatigue specimen.

2.1. Dynamic Stress Measurement Tests

Dynamic stress, which is the key factor that affects the stress relieving effect and fatigue life, is generated in the specimen during the VSR process. The dynamic stress magnitude can be controlled by changing the vibration magnitude. Therefore, dynamic stress measurement tests are conducted to obtain the relationship between the vibration magnitude and generated dynamic stress before the VSR tests.

A modal analysis is performed to calculate the vibration modes utilizing ANSYS Finite Element Analysis (FEA) software [34,35], as well as determine the strain measuring point and exciting frequency in the dynamic stress measurement and VSR tests. The FEA modal analysis steps are described as follows: (1) a 3D geometry model is imported; (2) the element type is selected as Solid 185; (3) the material properties are defined as shown in Table 1, which is provided by the manufacturer (Hangzhou Hengchao Metallic Materials Co., Ltd., Hangzhou, China); (4) the FE model is built (Figure 2a); (5) the boundary conditions are defined; (6) a solution is generated; and (7) the results are outputted.

Table 1. Mechanical properties of the Ti-6Al-4V titanium alloy.

Density/kg·m³	Elastic Modulus/GPa	Poisson's Ratio	Yield Strength/MPa
4440	110	0.33	980

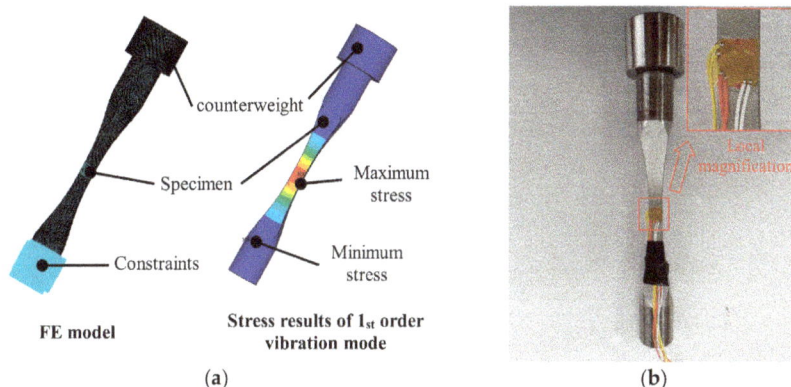

Figure 2. Finite element (FE) modal analysis and sticking position of the strain rosette: (**a**) FE model and first-order vibration mode result; and (**b**) sticking position of the strain rosette.

Simulation results show that the natural frequency of the first vibration mode is 52.07 Hz, and the maximum stress occurs in the middle of the specimen (Figure 2a). Thus, a strain rosette that consists of 0° (axial direction of the specimen), 45°, and 90° (perpendicular to the axial direction of the specimen) direction strain gauges is attached onto the middle of the specimen (Figure 2b).

A V8-440 HBT 900C vibration table (LDS Test and Measurement Ltd., Royston, England) is employed as the excitation equipment, and the dynamic strain gauge and digital signal-processing system produced by China Orient Institute of Noise & Vibration are utilized for the dynamic strain measurement and signal acquisition. The experimental setup is shown in Figure 3.

One end of the specimen without the counterweight is fixed to the vibration table, whereas the other end with a counterweight is free. The exciting frequency is set at 52 Hz according to the FEM simulation results. A total of 20 experimental sets are performed with 20 different vibration amplitudes. The vibration amplitudes (peak-to-peak value) range from 0.01 mm to 0.2 mm. The vibration direction is parallel to the normal direction of the strain rosette plane and perpendicular to the direction of the specimen's axial direction.

The vibration time at each amplitude is 1 min, and the effective values of dynamic strains at the 0°, 45°, and 90° directions can be read from the digital signal-processing system. Dynamic stresses can then be calculated utilizing Equations (1)–(3) [36]:

$$\sigma_1 = \frac{E}{2(1-\mu)}(\varepsilon_a + \varepsilon_c) + \frac{E}{\sqrt{2}(1+\mu)}\sqrt{(\varepsilon_a - \varepsilon_b)^2 + (\varepsilon_b - \varepsilon_c)^2}, \tag{1}$$

$$\sigma_2 = \frac{E}{2(1-\mu)}(\varepsilon_a + \varepsilon_c) - \frac{E}{\sqrt{2}(1+\mu)}\sqrt{(\varepsilon_a - \varepsilon_b)^2 + (\varepsilon_b - \varepsilon_c)^2}, \tag{2}$$

$$\sigma_{eqv} = \sqrt{\frac{1}{2}\left[(\sigma_1 - \sigma_2)^2 + (\sigma_2 - \sigma_3)^2 + (\sigma_3 - \sigma_1)^2\right]}, \tag{3}$$

where ε_a, ε_b, and ε_c are the measured strains at the $0°$, $45°$, and $90°$ directions, respectively; σ_1, σ_2, and σ_3 are the first, second, and third principal stresses, respectively (σ_3 is zero in the plane problems); σ_{eqv} is the equivalent stress (von Mises stress); E is the elastic modulus; and μ is Poisson's ratio. Thus, the generated dynamic stresses under different amplitudes can be obtained.

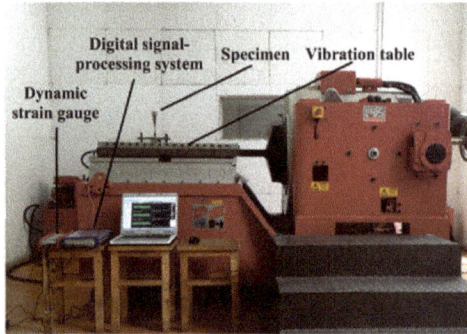

Figure 3. Dynamic stress measurement and vibratory stress relief (VSR) test setup.

2.2. VSR Tests

VSR tests are performed utilizing the same vibration table and fixture as the dynamic stress measurement tests. A total of 240 specimens are evenly divided into six groups (Figure 4). The specimens in Group A are in their original state without VSR or heat treatment. Four typical vibration amplitudes, which correspond to four typical dynamic stress levels, are selected based on the dynamic stress measurement test results. Thus, VSRs with 0.03, 0.05, 0.1, and 0.2 mm vibration amplitudes are performed on the specimens in Groups B, C, D, and E, respectively. The vibration time for each specimen is 10 min.

Moreover, annealing treatment is conducted on the specimens in Group F for comparison. The specimens are placed in a WZH-60 vacuum heat treatment furnace (produced by Beijing Research Institute of Mechanical & Electrical Technology, Beijing, China) and heated to 600 °C. The temperature is maintained at 600 °C for 6 h [37]. The specimens are then removed from the furnace and cooled to room temperature in open air.

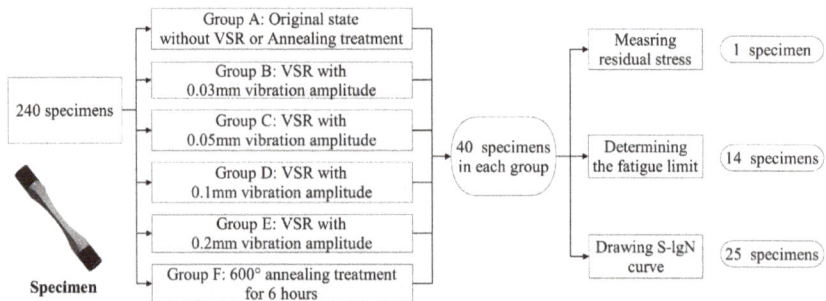

Figure 4. Grouping conditions of the vibratory stress relief(VSR) tests.

2.3. Residual Stress Measurement Tests

One of the 40 specimens in each group is selected for the residual stress measurement tests after VSR and annealing treatment. In particular, the measured stresses of the specimens in Group A reflect the original stresses without VSR or annealing treatment. Hence, the stresses before and after VSR and annealing treatment can be obtained.

The Prism System (produced by Stresstech Group, Vaajakoski, Finland, which is a residual stress measurement device based on the hole-drilling method and electronic speckle pattern interferometry technique, and a ϕ1.6 mm cemented carbide micro-milling cutter with titanium nitride coating (Richards Micro-Tool, Inc., Plymouth, MA, USA) are employed for the residual stress measurement (Figure 5). The measuring point is the same as the position where the strain rosette is attached (Figure 2b) in the dynamic stress measurement tests. A small hole with 1.6 mm diameter and 1 mm depth is drilled on the specimen by the micro-milling cutter. Hence, the stresses from the surface to 1 mm depth at the measuring point can be obtained.

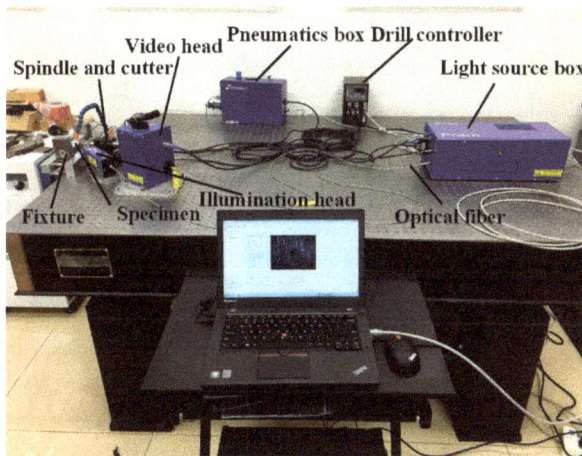

Figure 5. Setup of the residual stress measurement tests.

2.4. Uniaxial Tension-Compression Fatigue Experiments

The high-frequency fatigue testing machine QBG 200 (Changchun Qianbang Test Equipment Co., Ltd., Changchun, China) is adopted to perform the uniaxial tension-compression fatigue experiments (Figure 6). The stress ratio R is -1, and the load frequency is set at 81.7 Hz with a 5 Hz drop protection. The alternating and average load protections are both set at $\pm$5 KN, ambient temperature is 25 °C, and relative humidity is 39.2%.

A total of 14 specimens from each group are selected to determine the fatigue limit. The up and down method [38] is applied, whose procedure is described in Figure 7. The cyclic stress amplitude of the first specimen for Group A is set at 520 MPa, which is slightly higher than half of the titanium alloy yield limit (490 MPa). The measured fatigue limit of Group A (original state) indicates that the cyclic stress amplitude of the first specimen for Groups B, C, D, E, and F are set at 500, 500, 490, 450, and 470 MPa, respectively.

Figure 6. Setup of the fatigue experiments.

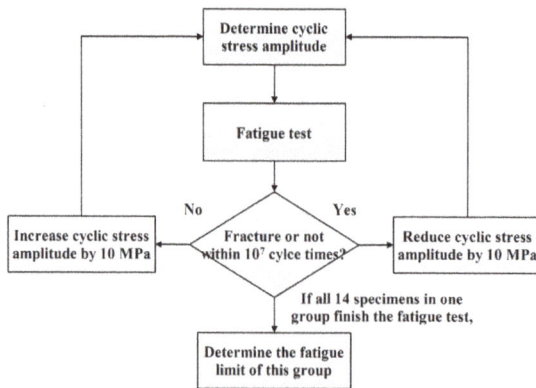

Figure 7. Procedure of the up and down method.

Fracture commonly occurs to the first specimen due to the high cyclic stress amplitude. No fracture occurs to one specimen within 10^7 cycle times for the first time with the decrease in the cyclic stress amplitude. The first appearance of the no-fractured specimen and its previous fractured specimen are considered as effective specimens. If two specimens have the opposite fracture results (i.e., one is fractured, whereas the other is unfractured) and adjacent cyclic stress amplitude (±10 MPa), then both are regarded as effective specimens. The fatigue limit can then be calculated by Equation (4):

$$\sigma_{R=-1} = \frac{1}{m}\sum_{i=1}^{n} v_i\sigma_i,$$ (4)

where m is the total number of effective specimens; n is the number of cyclic stress amplitudes; σ_i is the ith cyclic stress amplitude; and v_i is the number of specimens at σ_i cyclic stress amplitude.

After the fatigue limit is determined, 25 (5 × 5) of the 40 specimens in each group are selected to obtain the S-lgN curve with the grouping method [38]. Five cyclic stress amplitudes of each group are selected based on the fatigue limit. The fatigue cycles of 5 specimens under each cyclic stress amplitude are measured. Chauvenet's criterion [39] is employed to filter the experimental data and calculate the average fatigue cycle under each stress amplitude. Thus, the S-lgN curve of each group can be drawn.

3. Results and Discussion

3.1. Dynamic Stress Measurement Tests

The dynamic stress measurement test results are shown in Figure 8a,b. The first principal, second principal, and equivalent stresses under different amplitudes are calculated using Equations (1)–(3). This study utilizes the equivalent stress to reflect the dynamic stress magnitude. Figure 8b shows that the equivalent stress generated in the VSR can be expressed by VSR amplitude (denoted by A) utilizing the linear fitting method in Equation (5):

$$\sigma_{eqv} = 404.33 \cdot A + 0.18 \tag{5}$$

The dynamic stress increases linearly with the amplitude increase. The generated dynamic stresses under 0.03, 0.05, 0.1, and 0.2 mm amplitude VSRs are 8.68, 14.43, 28.76, and 57.22 MPa, respectively.

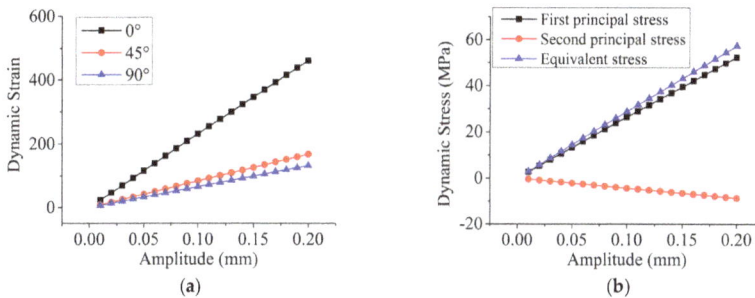

Figure 8. Dynamic stress and strain under different amplitudes: (**a**) dynamic strain; and (**b**) dynamic stress.

3.2. Residual Stress Measurement Tests

Plane stresses, including normal stresses in two mutually perpendicular directions and shear stress, are measured in the residual stress measurement tests. The first principal stress is calculated to compare the residual stress change before and after VSR or annealing treatment. The residual stresses of Groups A to F are shown in Figure 9a,b.

Figure 9a shows that the magnitude of the first principal stress on the surface layer (0.02 mm–0.1 mm depth) of each group is much larger than those in the deeper layer (0.1 mm–1 mm) because of the machining-induced residual stress [40,41]. The peak and average stresses of Group A are significantly larger than those of Groups B to F. This scenario indicates that VSR and annealing treatment largely contribute to the stress relief and homogenization of Ti-6Al-4V titanium alloy. Comparing the results of Groups B to E shows that the stress magnitude decreased when the VSR amplitude increased. Therefore, a higher amplitude that generates a higher dynamic stress during VSR is more conducive to the relief and homogenization of the residual stress than the lower amplitude.

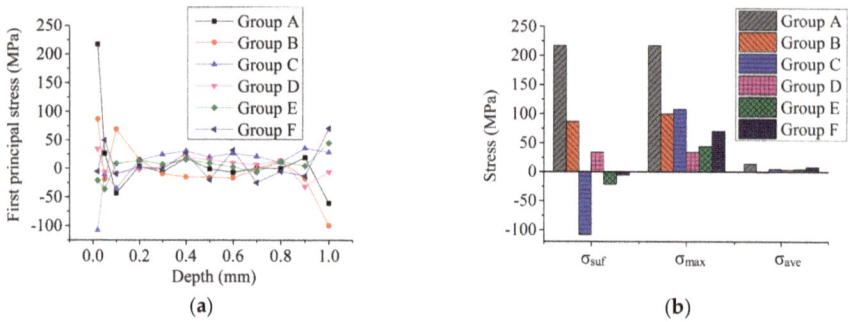

Figure 9. Measured stresses of Groups A to F after the VSR tests: (**a**) first principal stresses along the depth direction; and (**b**) comparison of the measured stress among the different groups (i.e., σ_{surf}, σ_{max}, and σ_{ave} denote the stresses at 0.02 mm depth, absolute value of the peak stress, and average stress along the depth direction, respectively).

3.3. Uniaxial Tension-Compression Fatigue Experiments

The fatigue limits of the specimens in Groups A to F are shown in Figure 10, whereas the S-lgN curves of Groups A to F are shown in Figure 11. The fatigue limit, surface residual stress, average stress, and dynamic stress of each group are compared in Table 2. Figures 10 and 11, as well as Table 2, show that the fatigue limits of Groups B to F are smaller than that of Group A. The fatigue limit decreased with the increase in the VSR vibration amplitude. The fatigue limit of Group F is smaller than those of Groups B, C, and D, but slightly larger than that of Group E. The S-lgN curve also decreased with the vibration amplitude increase. The S-lgN curve of Group F is generally lower than those of Groups B to E.

The fatigue life of Ti-6Al-4V titanium alloy decreased with the increase in the VSR vibration amplitude. Although the fatigue limit of Ti-6Al-4V titanium alloy decreased by 1.25%, 1.25%, and 1.87% after 0.03, 0.05, and 0.1 mm amplitude VSR, the surface residual stress (at 0.02 mm depth) decreased by 60.05%, 50.23%, and 84.29%, respectively. The average residual stress within 1 mm depth also decreased by 90.99%, 67.02%, and 74.68%, respectively. When the amplitude reaches 0.2 mm, the fatigue limit decreased by 10.06%, the surface stress decreased by 90.18%, and the average stress decreased by 68.68%. The fatigue limit after the annealing treatment decreased by 8.52%, the surface stress decreased by 97.28%, and the average stress decreased by 39.87%. These results indicate that low-amplitude VSR can significantly lower and homogenize the residual stress of Ti-6Al-4V titanium alloy with an extremely low cost to the fatigue life. By contrast, the high-amplitude VSR and annealing treatment have several negative effects on the fatigue life.

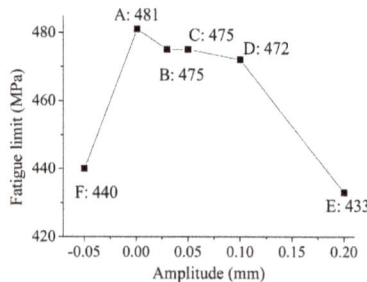

Figure 10. Fatigue limits under different amplitudes (−0.05 mm amplitude stands for the annealing treatment).

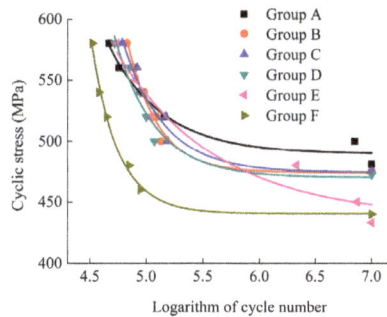

Figure 11. Cyclic stress amplitude-logarithm of cycle numbers (S-lgN) curves for Ti-6Al-4V titanium alloy.

Table 2. Fatigue limit and residual stress change after the VSR and annealing treatment.

Results	Group A	Group B	Group C	Group D	Group E	Group F
Fatigue limit (MPa)	481	475	475	472	430	440
Percentage change	/	−1.25%	−1.25%	−1.87%	−10.60%	−8.52%
Surface stress (MPa)	217	86.7	108	34.1	21.3	5.91
Percentage change	/	−60.05%	−50.23%	−84.29%	−90.18%	−97.28%
Average stress (MPa)	13.55	1.22	4.47	3.43	4.24	8.15
Percentage change	/	−90.99%	−67.02%	−74.68%	−68.68%	−39.87%
Dynamic stress (MPa)	/	8.68	14.43	28.76	57.22	/

When $\sigma_d + \sigma_r$ (σ_r is the residual stress, whereas σ_d is the dynamic stress in the VSR) is larger than the yield limit of the material in the local position, micro-plastic deformation occurs, and the internal residual stress is redistributed. The residual stress magnitude then decreases with the formation of a new stress equilibrium [42]. Therefore, micro-plastic deformation and stress redistribution that occur during the VSR process are the main reasons for the decrease and homogenization of residual stress.

The formation of fatigue damage can be divided into three stages: crack initiation, crack propagation, and instantaneous fracture. The crack is commonly generated in high stress areas under a cyclic load. The measured value of the strain gauge reflects the average strain in a certain area. Although the dynamic stress measured by the strain rosette is less than 60 MPa, the dynamic stress generated at the local position can be much larger than the measured value during the VSR process. The dynamic stress coupled with the initial residual stress can lead to local stress concentration, which then generates micro-cracks. This scenario explains the decrease in the fatigue life after VSR. Fewer cracks are produced after lower-amplitude VSR because lower dynamic stress is generated, and fewer stress concentration areas emerge during vibration. When the amplitude is less than a certain value, only a handful of cracks occur. Thus, the fatigue life slightly decreases. More cracks occur with the increases in the amplitude and dynamic stress. A larger decrease in the fatigue life is then observed after high-amplitude VSR.

Moreover, Figure 9a shows that the tensile stresses on the specimen surfaces in the original state, which are conducive to the initiation and propagation of fatigue cracks, are largely eliminated by VSR and annealing treatment. The decrease in the fatigue life demonstrates that the negative effects caused by the micro-crack generation during VSR are greater than the positive effects caused by the elimination of tensile stresses on the surface. If the surface residual stresses of the specimens in the original state is compressive, then a larger decrease in the fatigue life after VSR could be observed.

It should be noted that only five cyclic stress amplitudes of each state are selected to draw the S-lgN curve due to the high cost of time and raw material in experiments. The fatigue behavior can be better presented if more cyclic stress amplitudes are investigated. Besides, the results in this paper are basically effective on Ti-6Al-4V. The VSR effects on fatigue life vary with different metallic materials.

4. Conclusions

This study investigates the VSR effect on the fatigue life of Ti-6Al-4V. The dynamic stresses generated under different VSR amplitudes are initially measured, and then the relationship between the dynamic stress and vibration amplitude is obtained. VSRs with 0.03, 0.05, 0.1, and 0.2 mm amplitude and annealing treatment under typical process parameters are conducted on five specimen groups. Another specimen group without any treatment is employed for comparison. The residual stresses of each group are measured to study the stress relieving effects. Finally, the fatigue limits and S-lgN curves under different states are obtained by fatigue experiments. The following conclusions are drawn:

(1) The dynamic stress magnitude generated during the VSR process increases linearly with the amplitude increase. A higher amplitude is more conducive than the lower amplitude to the relief and homogenization of the residual stress of Ti-6Al-4V titanium alloy.

(2) VSR has a negative effect on the fatigue life of Ti-6Al-4V. The fatigue behavior decreased with the increase in the VSR vibration amplitude. When the VSR amplitude is less than 0.1 mm, the decrease in the fatigue limit is less than 2%. More than 50% of the surface residual stress and more than 67% of the average residual stress within 1 mm depth can be eliminated. The fatigue limit of the specimens treated by 0.2 mm amplitude VSR is decreased by 10.60%. Although the stress relieving effect is better, high amplitude VSR will lead to the decrease of Ti-6Al-4V fatigue life due to the defects generated during vibration. Low amplitude VSR can effectively relieve the stress with little decrease in fatigue life.

(3) Annealing treatment also decreases the fatigue life. The fatigue limit of the specimens that underwent annealing treatment is decreased by 8.52%. The S-lgN curve of the annealing treatment is generally lower than those of the VSRs with different amplitudes.

Acknowledgments: This work is supported by National Science and Technology Major Project (2014ZX04001011); State Key Laboratory of Virtual Reality Technology Independent Subject (BUAA-VR-16ZZ-07); Defense Industrial Technology Development Program (A0520110009); and Beijing Municipal Natural Science Foundation (3172021). The authors thank the referees of this paper for their valuable and very helpful comments.

Author Contributions: Han-Jun Gao, Yi-Du Zhang and Jing Song conceived and designed the experiments; Han-Jun Gao and Jing Song. performed the experiments; Han-Jun Gao and Qiong Wu analyzed the data; and Han-Jun Gao and Qiong Wu wrote the paper.

Conflicts of Interest: The authors declare no conflict of interest.

References

1. Barsoum, Z.; Barsoum, I. Residual stress effects on fatigue life of welded structures using LEFM. *Eng. Fail. Anal.* **2009**, *16*, 449–467. [CrossRef]
2. Schajer, G.S. Relaxation Methods for Measuring Residual Stresses: Techniques and Opportunities. *Exp. Mech.* **2010**, *50*, 1117–1127. [CrossRef]
3. Yeom, H.; Choi, B.; Seol, T.; Lee, M.; Jeon, Y. Very High Cycle Fatigue of Butt-Welded High-Strength Steel Plate. *Metals* **2017**, *7*, 103. [CrossRef]
4. Wu, Q.; Li, D.; Zhang, Y. Detecting Milling Deformation in 7075 Aluminum Alloy Aeronautical Monolithic Components Using the Quasi-Symmetric Machining Method. *Metals* **2016**, *6*, 80. [CrossRef]
5. Webster, G.A.; Ezeilo, A.N. Residual stress distributions and their influence on fatigue lifetimes. *Int. J. Fatigue* **2001**, *23*, 375–383. [CrossRef]
6. Cheng, X.; Fisher, J.W.; Prask, H.J.; Gnäupel-Herold, T.; Yen, B.T.; Roy, S. Residual stress modification by post-weld treatment and its beneficial effect on fatigue strength of welded structures. *Int. J. Fatigue* **2003**, *25*, 1259–1269. [CrossRef]
7. James, M.N.; Hughes, D.J.; Chen, Z.; Lombard, H.; Hattingh, D.G.; Asquith, D.; Yates, J.R.; Webster, P.J. Residual stresses and fatigue performance. *Eng. Fail. Anal.* **2007**, *14*, 384–395. [CrossRef]

8.   Sasahara, H. The effect on fatigue life of residual stress and surface hardness resulting from different cutting conditions of 0.45%C steel. *Int. J. Mach. Tools Manuf.* **2005**, *45*, 131–136. [CrossRef]

9.   LaRue, J.E.; Daniewicz, S.R. Predicting the effect of residual stress on fatigue crack growth. *Int. J. Fatigue* **2007**, *29*, 508–515. [CrossRef]

10. Pouget, G.; Reynolds, A.P. Residual stress and microstructure effects on fatigue crack growth in AA2050 friction stir welds. *Int. J. Fatigue* **2008**, *30*, 463–472. [CrossRef]

11. Benedetti, M.; Fontanari, V.; Scardi, P.; Ricardo, C.L.A.; Bandini, M. Reverse bending fatigue of shot peened 7075-T651 aluminium alloy: The role of residual stress relaxation. *Int. J. Fatigue* **2009**, *31*, 1225–1236. [CrossRef]

12. Nikitin, I.; Scholtes, B.; Maier, H.J.; Altenberger, I. High temperature fatigue behavior and residual stress stability of laser-shock peened and deep rolled austenitic steel AISI 304. *Scr. Mater.* **2004**, *50*, 1345–1350. [CrossRef]

13. Bagherifard, S.; Slawik, S.; Fernández-Pariente, I.; Pauly, C.; Mücklich, F.; Guaglianoa, M. Nanoscale surface modification of AISI 316L stainless steel by severe shot peening. *Mater. Des.* **2016**, *102*, 68–77. [CrossRef]

14. Fernández-Pariente, I.; Bagherifard, S.; Guagliano, M.; Ghelichi, R. Fatigue behavior of nitrided and shot peened steel with artificial small surface defects. *Eng. Fract. Mech.* **2013**, *103*, 2–9. [CrossRef]

15. Zabeen, S.; Preuss, M.; Withers, P.J. Evolution of a laser shock peened residual stress field locally with foreign object damage and subsequent fatigue crack growth. *Acta Mater.* **2015**, *83*, 216–226. [CrossRef]

16. Yoichi, Y.; Yusuke, U.; Hiroshi, K.; Shinozaki, M. Fatigue life prediction of small notched Ti-6Al-4V specimens using critical distance. *Eng. Fract. Mech.* **2010**, *77*, 1439–1453.

17. Bourassa, P.L.; Yue, S.; Bobyn, J.D. The effect of heat treatment on the fatigue strength of microknurled Ti-6Al-4V. *J. Biomed. Mater. Res. A* **1997**, *37*, 291–300. [CrossRef]

18. Golden, P.J.; John, R.; Porter, W.J. Variability in Room Temperature Fatigue Life of Alpha + Beta Processed Ti-6Al-4V (Preprint). *Int. J. Fatigue* **2009**, *31*, 1764–1770. [CrossRef]

19. Zhang, Y.; Li, J.; Shi, C.-B.; Qi, Y.-F.; Zhu, Q.-T. Effect of Heat Treatment on the Microstructure and Mechanical Properties of Nitrogen-Alloyed High-Mn Austenitic Hot Work Die Steel. *Metals* **2017**, *7*, 94. [CrossRef]

20. Dawson, R.; Moffat, D.G. Vibratory Stress Relief: A Fundamental Study of Its Effectiveness. *J. Eng. Mater. Technol.* **1980**, *102*, 169–176. [CrossRef]

21. Walker, C.A.; Waddell, A.J.; Johnston, D.J. Vibratory stress relief—An investigation of the underlying processes. *Proc. Inst. Mech. Eng. E-J. Process Mech. Eng.* **1995**, *209*, 51–58. [CrossRef]

22. Sun, M.C.; Sun, Y.H.; Wang, R.K. The vibratory stress relief of a marine shafting of 35# bar steel. *Mater. Lett.* **2004**, *58*, 299–303.

23. Sun, M.C.; Sun, Y.H.; Wang, R.K. Vibratory stress relieving of welded sheet steels of low alloy high strength steel. *Mater. Lett.* **2004**, *58*, 1396–1399. [CrossRef]

24. Wang, J.S.; Hsieh, C.C.; Lin, C.M.; Kuo, C.W.; Wu, W. Texture evolution and residual stress relaxation in a cold-rolled Al-Mg-Si-Cu alloy using vibratory stress relief technique. *Metall. Mater. Trans. A Phys. Metall. Mater. Sci.* **2013**, *44*, 806–818. [CrossRef]

25. Wang, J.S.; Hsieh, C.C.; Lai, H.H.; Kuo, C.W.; Wu, P.T.Y.; Wu, W. The relationships between residual stress relaxation and texture development in AZ31 Mg alloys via the vibratory stress relief technique. *Mater. Charact.* **2015**, *99*, 248–253. [CrossRef]

26. Rao, D.; Wang, D.; Chen, L.; Ni, C. The effectiveness evaluation of 314L stainless steel vibratory stress relief by dynamic stress. *Int. J. Fatigue* **2007**, *29*, 192–196. [CrossRef]

27. Kwofie, S. Plasticity model for simulation, description and evaluation of vibratory stress relief. *Mater. Sci. Eng. A* **2009**, *516*, 154–161. [CrossRef]

28. Lv, T.; Zhang, Y. A combined method of thermal and vibratory stress relief. *J. Vibroeng.* **2015**, *17*, 2837–2845.

29. Vardanjani, M.J.; Ghayour, M.; Homami, R.M. Analysis of the Vibrational Stress Relief for Reducing the Residual Stresses Caused by Machining. *Exp. Tech.* **2016**, *40*, 705–713. [CrossRef]

30. Djuric, D.; Vallant, R.; Kerschbaumer, K.; Enzinger, N. Vibration stress relief treatment of welded high-strength martensitic steel. *Weld. World* **2011**, *55*, 86–93. [CrossRef]

31. Wozney, G.P.; Crawmer, G.R. An investigation of vibrational stress relief in steel. *Weld. Res. Suppl.* **1968**, *23*, 411–419.

32. Munsi, S.M.Y.; Waddell, J.; Walker, C. The Influence of Vibratory Treatment on the Fatigue Life of Welds: A Comparison with Thermal Stress Relief. *Strain* **2001**, *37*, 141–149. [CrossRef]

33. Song, J.; Zhang, Y. Effect of vibratory stress relief on fatigue life of aluminum alloy 7075-T651. *Adv. Mech. Eng.* **2016**, *8*, 1–9. [CrossRef]
34. Wu, Q.; Zhang, Y.; Zhang, H. Dynamic characteristic analysis and experiment for integral impeller based on cyclic symmetry analysis method. *Chin. J. Aeronaut.* **2012**, *25*, 804–810. [CrossRef]
35. Baumeister, J.; Monno, M.; Goletti, M.; Mussi, V.; Weise, J. Dynamic Behavior of Hybrid APM (Advanced Pore Morphology Foam) and Aluminum Foam Filled Structures. *Metals* **2012**, *2*, 211–218. [CrossRef]
36. Gere, J.M.; Goodno, B.J. Mechanics of materials. In *Mechanics of Materials*; Van Nostrand Reinhold Co.: New York, NY, USA, 1972; pp. 211–291.
37. Fan, D. Heat treatment technology data manual. In *Heat Treatment Technology Data Manual*; China Machine Press: Beijing, China, 2000; pp. 497–499. (In Chinese)
38. Gao, Z.; Jiang, X.; Xiong, J.; Guo, G.; Gan, W.; Xia, Y.; Wang, S.; Zeng, B. Fatigue performance test design and data processing. In *Fatigue Performance Test Design and Data Processing*; Beihang University Press: Beijing, China, 1995. (In Chinese)
39. Chauvenet, W. A Manual of Spherical and Practical Astronomy. In *A Manual of Spherical and Practical Astronomy*; J. B. Lippincott & Co.: Philadelphia, PA, USA, 1960.
40. Fu, W.E.; Cohen, P.H.; Ruud, C.O. Experimental investigation of the machining induced residual stress tensor under mechanical loading. *J. Manuf. Process.* **2009**, *11*, 88–96. [CrossRef]
41. Lia, B.; Jiang, X.; Yang, J.; Liang, Y.S. Effects of depth of cut on the redistribution of residual stress and distortion during the milling of thin-walled part. *Thin-Walled Struct.* **2015**, *216*, 223–233. [CrossRef]
42. Yildiz, K.; Eken, S.; Kaya, M.O. Simulation of vibration stress relief after welding based on FEM. *Acta Metall. Sin. (Engl. Lett.)* **2008**, *21*, 289–294.

metals

MDPI

Article

Effects of Solution Treatment on Microstructure and High-Cycle Fatigue Properties of 7075 Aluminum Alloy

Chi Liu [1], Yilun Liu [1,2,*], Liyong Ma [3] and Jiuhuo Yi [1]

1   College of Mechanical and Electrical Engineering, Central South University, Changsha 410083, China; liuchi001@csu.edu.cn (C.L.); yijiuhuo@csu.edu.cn (J.Y.)
2   Light Alloy Research Institute, Central South University, Changsha 410083, China
3   School of Mechanical Engineering, Hebei University of Architecture, Zhangjiakou 075051, China; maliyong0001@163.com
*   Correspondence: ylliu@csu.edu.cn; Tel.: +86-731-88877991

Academic Editor: Filippo Berto
Received: 28 March 2017; Accepted: 22 May 2017; Published: 26 May 2017

Abstract: This research mainly focused on the effects of solution treatment on high-cycle fatigue properties, microstructure evolution, and fatigue fracture morphology of the high strength aluminum alloy (7075 aluminum alloy). The S-N curves and fatigue performance parameters of the alloy were obtained. We found that longer solution treatment time significantly influences the high-cycle ($N \geq 10^5$) fatigue properties of the Al-Zn-Mg-Cu alloy. Under the loading stress of 240 MPa, and the solution treatment of 2 h compared to 1.5 h, 1 h, and 0.5 h, the fatigue life was respectively improved by about 95.7%, 149%, and 359%. The microstructure observations conducted with a scanning electron microscope (SEM) and transmission electron microscope (TEM) are as follows: recrystallization occurs in the grains of the 7075 aluminum alloy under solution treatment, and the grains become large with the length of the solution treatment time. Cracks mainly initiate from the undissolved large phases, and prolonging the solution time can effectively promote the dissolution of the T phase and S phase, decrease the number of dislocations, and lower the rate of the initiation of fatigue cracks at the undissolved large phases due to dislocation glide and dislocation pile-up. In the second stage of crack propagation, the secondary cracks reduce the driving force and the rate of crack propagation, promoting the fatigue properties of the 7075 aluminum alloy, which can be verified by the observation result that fatigue striation widths become narrower with longer solution treatment times.

Keywords: 7075 aluminum alloy; solution treatment; high-cycle fatigue property; microstructure

1. Introduction

Solution treatment refers to the process of dissolving the coarse second phase in the alloy at a certain temperature and time, and then using rapid cooling to obtain a supersaturated solid solution [1]. In solution treatment, the time and the temperature are the core parameters [2], which have direct influences on the degree of the solid solution and the quality of the subsequent precipitated phase during the ageing and strengthening stage.

The effect of solution treatment on the microstructure and the performance of the 7050 aluminum alloy has been discussed in the literature [3–6]. The results show that solution treatment can have a significant effect on the grain size and the morphology of the second phase. While the second phases are dissolving into the base, the proportion of recrystallization and the size of the sub-grain increase, causing the weakening of the mechanical properties of the alloy [3]. In these studies, Li [4] found that the ideal solution treatment for the 7050 high strength aluminum is 470 °C × 1 h; the results of Deng [5] showed that the hardening layer of the 7050 aluminum alloy with a heat preservation solution under 490 °C tended to be 1.36 times deeper than that under 475 °C. In addition, Wang [6] pointed out that the ideal solution treatment for casting-rolling 7050 high strength aluminum is 480 °C × 1 h; References [7–9] obtained similar conclusions for the 7050 high strength aluminum alloy. Senkov et al. focused on the effect of parameter changes of the solution treatment on the micromorphology and volume fraction of the Al_3Sc phases and Al_3Zr phases [9]. Davies et al. [10] found that in the solution treatment, the second phase would dissolve in the solid solution, and the recrystallization occurred simultaneously. The recrystallized grain and the remaining large second phase significantly influenced the fracture toughness properties of the aluminum alloy. After the solution treatment, the second phase of the aluminum alloy, the recrystallization fraction, and the change of the grain morphology also have a great influence on the material's fatigue property [11–15]. The dissolution of the large second phase is beneficial to the increase of the supersaturation degree of a supersaturated solid solution before ageing, thus reducing the source of micro-cracks and promoting the fatigue performance of the alloy [16]. Fan et al. [17,18] found that grain boundaries inhibit the propagation of the fatigue cracks. The smaller the aluminum alloy grains are, the more boundaries exist and the more severe the inhibition is, correspondingly leading to a higher fatigue life.

However, the current studies mostly focus on the effect of solution treatment on the structure and mechanical properties, rather than on the fatigue properties of materials, especially the high-cycle fatigue properties. In practical applications, aluminum alloy materials usually work under relatively low cyclic stress loads, and the fracture patterns are often high-cycle fatigue fractures [19,20], in which there is usually no obvious plastic deformation, and which is difficult to detect and prevent. In this paper, the aviation-used 7075-T651 high strength aluminum alloy is taken as the object of study. Under a certain solution temperature—480 °C, the materials are solution treated with different times, and then the effects of the solution treatment time on the high-cycle fatigue properties are observed. According to the rules of alloy microstructure evolution and the analysis of fracture morphology, the influence mechanism is discussed in order to provide a theoretical guidance for the optimization of parameters in solution treatment and for the enhancement of the fatigue properties of aluminum alloys.

2. Material and Experiments

The material used in this study is a typical Al-Zn-Mg-Cu alloy (7075 aluminum alloy). The heat treatment processing for this material is T651, which means conducting the solution treatment at 470 °C for 1 h, quenching in cold water, artificial aging treatment at 120 °C for 24 h, and then 5% pre-stretching to release residual stress. The measured chemical composition of the studied aluminum alloy is listed in Table 1.

Table 1. Chemical composition of the 7075 alloy (wt %).

Zn	Mg	Cu	Cr	Fe	Si	Mn	Ti	Al
5.71	2.45	1.5	0.18	0.17	0.06	0.034	0.019	Bal.

Metallographic samples (with the dimensions 14 mm × 14 mm × 4 mm) and fatigue samples were prepared from the 7075 aluminum alloy plate (Southwest Aluminum Group Co., Ltd., Chongqing, China). The fatigue samples were cut with wire electrical discharge machining (WEDM) to reduce the residual stress and tissue damage [21], the shape and dimensions of which are shown in Figure 1, which met American society for testing and materials standard ASTM E466-2007. In order to avoid notches and micro-cracks in the specimens' surface that may form during the machining processing [22], surfaces were polished to an average surface roughness of less than 0.4 mm with metallographic abrasive paper (1600#) and nylon cloth.

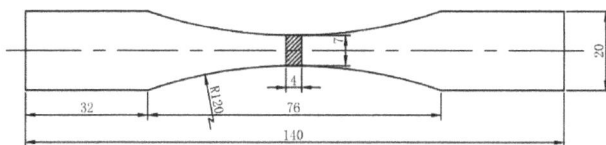

Figure 1. Smooth rectangular specimen of the fatigue sample (mm).

The solution treatment was conducted in a SX-12-10 resistance furnace (Cangyue Company, Shanghai, China), in which the solution temperature was 480 °C and the solution treatment times were 0.5 h, 1.0 h, 1.5 h, and 2.0 h, respectively. After quenching in the water at 25 °C, the artificial aging was performed at 120 °C for 24 h.

The fatigue test was conducted on a MTS810 fatigue machine (MTS Systems Corporation, Eden Prairie, MN, USA). The stress levels for axial stress fatigue performance were set between 240 MPa and 320 MPa, with a loading frequency of 40 Hz and a stress ratio of 0.1. Three horizontal samples were taken for each stress level. In order to study the initiation and propagation of the cracks, the fatigue fractures were fully cut off and observed under a TESCAN scanning electron microscope (SEM) (Tescan Company, Brno, Czech Republic). Micrographs of the precipitated phase characteristics were observed under a TecnaiG220 transmission electron microscope (TEM) (United States FEI limited liability company, Hillsboro, OR, USA). Samples were cut from the solution treatment specimens, and first thinned to 0.8 mm, and then thinned again on a TenuPol-5 electro-polished double spraying thinner machine (Struers, Copenhagen, Denmark). The solution was a mixed liquid of 30% hydrogen nitrate and 70% methanol, with a double spraying temperature between −35 °C to −25 °C, and a voltage of 20 kV.

3. Results and Discussion

3.1. Effects of Solution Treatment on the Microstructure

The microstructure of the 7075 aluminum alloy after solution treatment under the same temperature (480 °C) and for different times (0.5 h, 1.0 h, 1.5 h and 2.0 h) are shown in Figure 2.

Figure 2. Microstructure of the 7075 aluminum alloy with a solution time of (**a**) 0.5 h; (**b**) 1.0 h; (**c**) 1.5 h; (**d**) 2.0 h.

In Figure 2, it can be seen that varying degrees of recrystallization and grain growth occur with the increase of the solution treatment time. When the solution treatment time is 0.5 h (Figure 2a) the average size of the alloy grains are relatively small, due to the short time. Among these grains, some are isometric crystals. When the solution treatment time is increased to 1 h (Figure 2b), the recrystallization and grain growth continue because of grain boundary migration. When the solution treatment time increases to 1.5 h (Figure 2c), the grains after recrystallization merge and extend out, and the boundaries become straight. When the solution treatment time is increased to 2 h (Figure 2d), due to the long solution holding time, there is enough time for the grains to grow. With the interruption and the vanishing of the boundaries among the grains, the recrystallization becomes highly visible, and the average size of the grain is relatively the largest.

3.2. Effects of Solution Treatment on the Second Phase

The distributions and the appearances of the large phases of the 7075 aluminum alloy, solution treated at 480 °C and for different times (0.5 h, 1.0 h, 1.5 h, and 2.0 h), are listed in Figure 3. In the dark grey Al-base, there are a large number of undissolved large-phase particles distributed along the rolling direction, which are mostly grey and black, and non-uniform in distribution and size. Some of the particles are broken in the rolling process, and separated from the Al base due to plastic deformation.

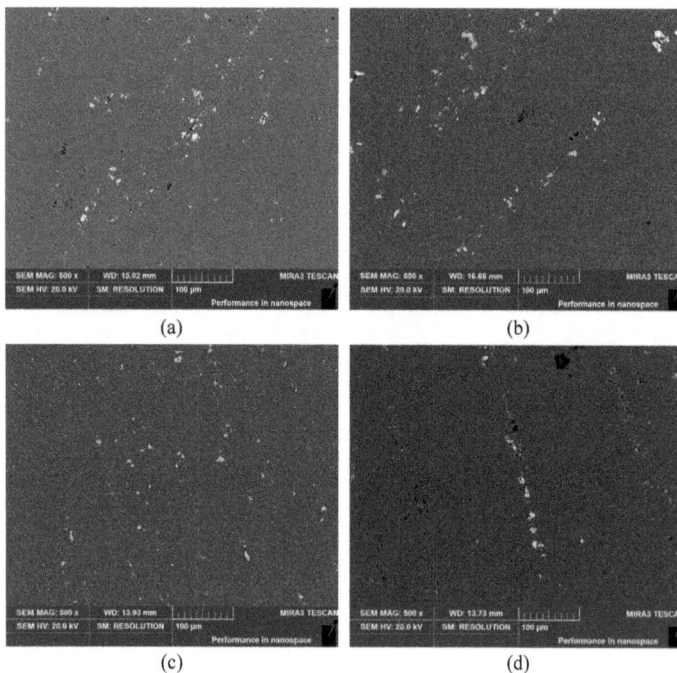

Figure 3. Distribution of the second phase in the 7075 aluminum alloy with a solution time of (a) 0.5 h; (b) 1.0 h; (c) 1.5 h; (d) 2.0 h.

In Figure 4, the grey and black large particles are analyzed by EDS (Energy Dispersive Spectrometer). The results show that there are four kinds of large second phases in the 7075 aluminum alloy after solution treatment: Fe-rich (Al_7Cu_4Fe) phase (Figure 4a), Si-rich (MgSi) phase (Figure 4c), S (Al_2CuMg) phase (Figure 4e), and T (AlZnMgCu) phase (Figure 4g). The Fe-rich phase, Si-rich phase, and S phase are larger primary phases of about 5–20 μm which are along the rolling direction, while the T phase is a dispersively distributed phase of about 0.5–3 μm.

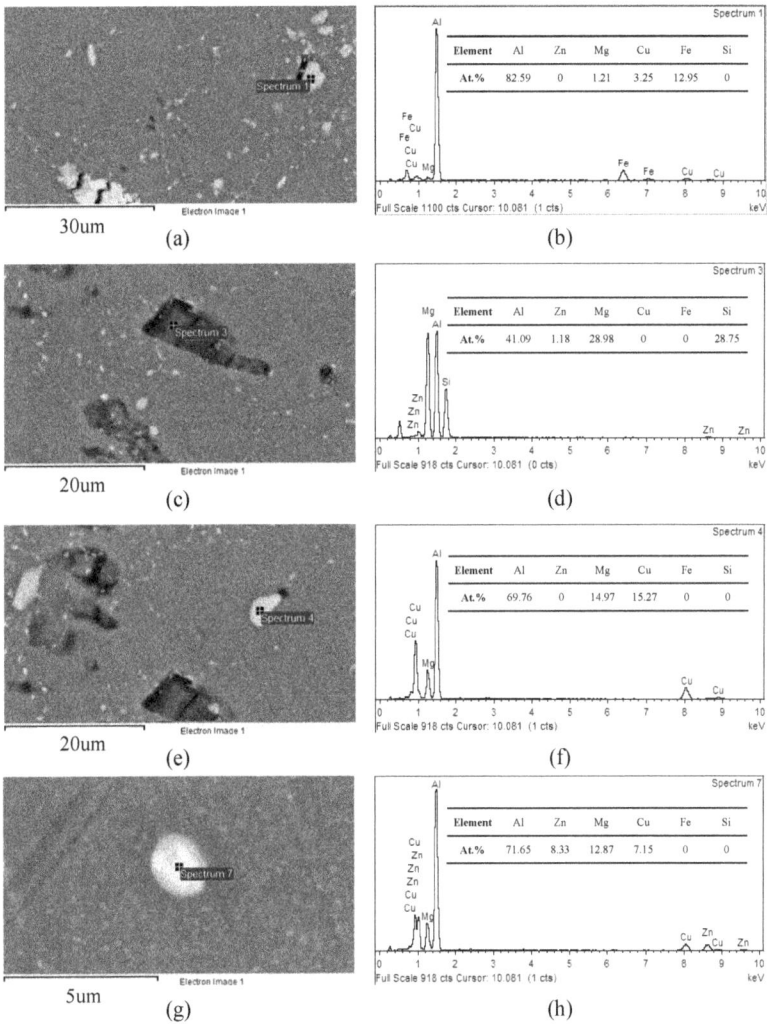

Figure 4. SEM (scanning electron microscope) images and EDS (Energy Dispersive Spectrometer) analysis results of the large second phase in the 7075 alloys with different solution times. (**a**,**b**) Fe-rich phase and its EDS spectrogram; (**c**,**d**) Si-rich phase and its EDS spectrogram; (**e**,**f**) S (Al_2CuMg) phase and its EDS spectrogram; (**g**,**h**) T (AlZnMgCu) phase and its EDS spectrogram.

In order to quantitatively express the effects of the solution time on the large second phase, the number and average size of the large second phases were measured and calculated by Image J software (Version 1.49, National Institutes of Health, Bethesda, MD, USA, 2015). Figure 5 shows the volume fraction of the large second phase with different solution times.

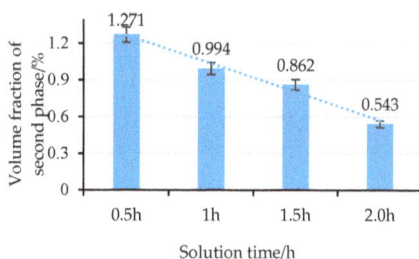

Figure 5. Volume fraction of the second phase with different solution times.

In Figure 5, under the 480 °C solution temperature, the volume fraction of the large second phase in the base decreases with the increase of the solution time, indicating that most of the remaining large second phases have dissolved in the base but not completely, and the undissolved phases with large size (Figure 4a,c,e) distribute along the rolling direction. After 0.5 h of solution time, the remaining phases are mostly large T phases, S phases, Fe-rich phases, and Si-rich phases. At that time, the volume fraction of the large second phases is about 1.27%. After 1.5 h, most of the T phases have dissolved in the base, and the volume fraction of the large second phases decreases to 0.86%, but still leaves a small portion of S phases, large Fe-rich phases, and large Si-rich phases. After 2 h, there are only impure phases such as large Fe-rich phases and large Si-rich phases in the base, and few T phases and S phases can be observed. At this time, the second phase is dissolved adequately, and the volume fraction is 0.54%, much less than before. Hence, the long solution time helps the dissolution of the second phases.

3.3. Effects of Solution Treatment on the Precipitated Phases

Figure 6 shows the results of the precipitated phases using TEM analysis of the 7075 aluminum alloy, that were solution treated with different times (0.5 h, 1.0 h, 1.5 h, and 2.0 h). After solution treatment, the dislocations in the base vanish in every picture. In Figure 6b, most of the precipitated phases dissolve in the base, and only a few second phases are left in the grains. With the increase of the solution treatment time, the density of the second phase decreases.

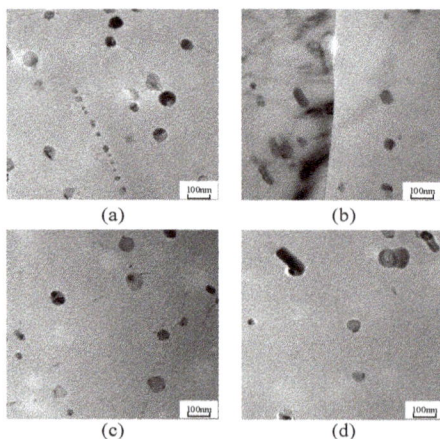

Figure 6. TEM micrographs of the precipitation characteristics of the 7075 aluminum alloy with solution times of (**a**) 0.5 h; (**b**) 1.0 h; (**c**) 1.5 h; (**d**) 2.0 h.

3.4. Effects of Solution Treatment on the High-Cycle Fatigue Life

The data of the uniaxial fatigue life of the 7075 aluminum alloy, solution treated in 480 °C for different times, are listed in Table 2. As is shown in Table 2, when the axial loading stress is between 280 MPa and 320 MPa, most of samples' fatigue life—N—is on the level of 10^4, while only the fatigue life of Case 4 (480 °C/2.0 h), at the stress level of 280 MPa, reaches the level of 10^5. When the axial loading stress decreases to 240 MPa, the fatigue life of most samples reach the level of 10^6. Only the fatigue life of the samples in Case 1 (480 °C/0.5 h) stays at the level of 10^5.

Table 2. Data of the uniaxial fatigue life of the 7075 aluminum alloy for different solution treatment times.

Spec No.	S_{max} (MPa)	Fatigue Life—N (Cycles)			
		Case 1 (480 °C/0.5 h)	Case 2 (480 °C/1 h)	Case 3 (480 °C/1.5 h)	Case 4 (480 °C/2 h)
1#		20,210	16,151	20,790	19,245
2#	320	18,410	18,050	15,990	20,095
3#		22,141	19,151	14,790	22,945
4#		34,157	41,151	37,956	59,410
5#	300	36,141	40,040	38,990	61,945
6#		39,520	42,642	40,247	53,921
7#		80,241	95,356	99,036	246,569
8#	280	71,451	86,547	95,701	226,569
9#		86,584	88,470	84,943	291,016
10#		193,039	261,477	372,882	699,403
11#	260	168,320	284,757	392,414	738,941
12#		181,241	224,646	420,440	751,814
13#		966,022	2,059,221	2,394,939	4,353,010
14#	240	904,514	1,824,641	2,560,780	4,568,122
15#		1,155,842	1,700,578	2,149,401	5,004,514

In addition, when the axial load stress is at a high stress level, greater than 300 MPa, the fatigue life of the specimens under the same loading stress is almost the same, less than 105 cycles. However, when the axial load stress is at a low stress level, less than 280 MPa, the high-cycle fatigue life is significantly different. For the average fatigue life data, see Figure 7.

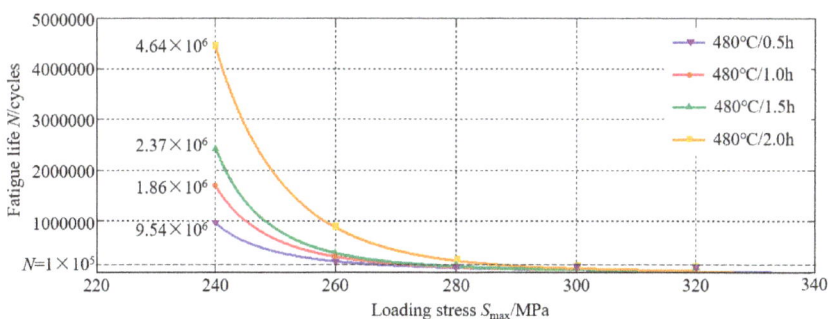

Figure 7. Fatigue life of the 7075 aluminum alloy under different solution times and loading stresses.

When the high-cycle fatigue life exceeds 10^5 cycles, it is significantly improved with the prolongation of the solution time. Under the loading stress of 240 MPa, compared to that of 1.5 h, 1 h, and 0.5 h, the fatigue life with the solution time of 2 h was respectively improved about 95.7%, 149%, and 359%.

In order to obtain the S-N curves of the 7075 aluminum alloy, the Weibull equation (Equation (1)) is used to fit the data:

$$\lg(N) = a + b\lg(S_{max} - A),\tag{1}$$

where a, b, and A are material constants, N is the fatigue life, and S_{max} is the maximum test stress. The fitting results are shown in Table 3.

Table 3. The fitting results of the 7075 aluminum alloy with different solution times.

Solution Time/h	Fitting Parameters			Fitting Formula
	a	b	A	
0.5	10.22	−2.90	210.7	$\lg(N) = 10.22 - 2.90\lg(S_{max} - 210.7)$
1.0	11.99	−3.75	205.4	$\lg(N) = 11.99 - 3.75\lg(S_{max} - 205.4)$
1.5	13.64	−4.53	199.4	$\lg(N) = 13.64 - 4.53\lg(S_{max} - 199.4)$
2.0	28.35	10.46	121.1	$\lg(N) = 28.35 - 10.46\lg(S_{max} - 121.1)$

Figure 8 shows the S-N curves of the 7075 aluminum alloy with different solution times. All the S-N curves have no horizontal asymptote, indicating that the 7075 aluminum alloy has no real fatigue limit. The fatigue life increases with the decrease of the stress level. Even though the stresses are at the same level, the fatigue life under different solution treatment times differs. As can be seen from the variation tendency of the S-N curve, when the stress is below a certain level, the fatigue life of all the samples can reach about 10^7 cycles. The data indicated by the arrows in Figure 8 illustrate that at those stress levels, when the loading cycles reached 10^7, the fatigue tests were stopped, while the specimens had not been fractured.

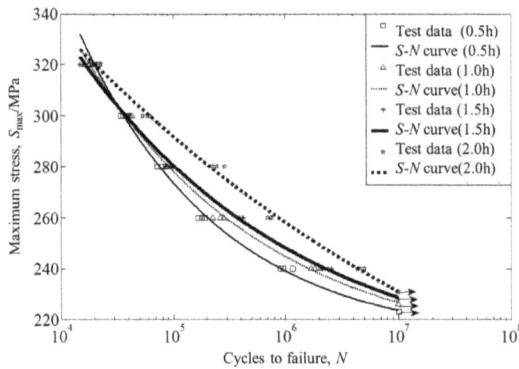

Figure 8. S-N curves of the 7075 aluminum alloy under different solution times.

Table 4 shows the fatigue endurance of 7075-T651 and 7075 Al alloys after different solution treatments. $\sigma_{1\times10^6}$ and $\sigma_{1\times10^7}$ describe the fatigue endurance, which are equal to the maximum stress σ_{max} at fatigue life $N = 1 \times 10^6$ and $N = 1 \times 10^6$, respectively [23]. From 0 to 2 h, the fatigue endurance $\sigma_{1\times10^7}$ at 0.5 h is 200 MPa, and at 2.0 h it is 231 MPa, indicating that an appropriate lengthening of the solution time is beneficial for the improvement of the fatigue properties of the 7075 aluminum alloy.

Table 4. Fatigue endurance of the 7075-T651 and 7075 Al alloys with solution treatment.

Material	7075 Al with Solution Treatment (480 °C)				7075-T651		
	0.5 h	1.0 h	1.5 h	2.0 h	Ref. [24]	Ref. [25]	Ref. [26]
R	0.1	0.1	0.1	0.1	-1	-1	0.1
$\sigma_{1 \times 10^6}$	239	245	248	258	248	230	250
$\sigma_{1 \times 10^7}$	200	226	228	231	223	193	180

In addition, the experimental results were compared with the fatigue endurance of 7075-T651 in references [24–26], and the effects of the stress ratio R were taken into account with the Walker equation (Equation (2)) [27].

$$N_f = [\sigma_{max}(\frac{1-R}{2})^\gamma \frac{1}{M}]^{1/n},$$ (2)

where M is the intercept constant at 1 cycle for a stress-life curve, γ is the fitting constant for the Walker method, and n is the exponent constant for a stress-life curve.

The value of $\sigma_{1 \times 10^6}$ and $\sigma_{1 \times 10^7}$ in references [24–26] are about 250 MPa and 200 MPa. The results show that there is a certain deviation in the value, which may be due to the machine conditions, fitting model, specimen surface treatment process differences, but the values are relatively close to our result, which also verifies the reliability of the fatigue test results.

3.5. Analysis of the Morphology of High-Cycle Fatigue Fractures

In Figure 7, at the applied stress level of 260 MPa, the fracture belongs to high-cycle fatigue fracture, with the fatigue life of all the samples exceeding 10^5. Therefore, the fatigue fracture under the applied stress of 260 MPa was observed using scanning electron microscopy.

Figure 9 shows the overall morphology of the high-cycle fatigue fractures. The fatigue fracture is made up of three regions, i.e., fatigue crack initiation region (A), crack propagation region (B), and fatigue final rupture region (C). The fatigue crack initiation region (A) is the smallest region near the surfaces of the samples. The crack propagation region (B), close to region (A), and with obvious characteristics of "radial cracks", is a smooth region. The arrows in the figure indicate the direction of crack propagation. With the continuous extension of fatigue cracks, the extension area constantly expands, the "radial cracks" gradually become sparse, and the crack propagation speed is also gradually accelerated. When exceeding the critical crack value, the specimen is broken simultaneously, forming a rupture region (C). The final rupture region (C) is a rough region, granulated and surrounded by shear lips of various sizes.

Figure 10 shows the SEM morphologies of the fatigue crack initiation region. The fatigue cracks in Figure 10a–c all initiate near the surfaces of the samples. The initiation of the cracks is the result of local shear stress. Under the cyclic loads, small slip bands are generated due to the dislocation movement of the alloy surface, and hence the initiation of the cracks in the slip band occurs. In Figure 10d, the cracks stem from the impure phases close to the alloy surface. Studies show that the large remaining phases in the 7075 aluminum alloy are mainly made up of undissolved T phases, S phases, Fe-rich phases, and Si-rich phases. The fatigue cyclic loads make dislocations slip to impure phases and pile up, and stress concentration is thus formed. When the stress is larger than the fracture strength and bonding strength with the base, the impure phases will be broken or separated from the bonding phases with the base, hence forming the cracks. Figure 10 also shows that the radial lines are centered around the fatigue initiation area (A). This is because in the process of expansion, the crack front is deviated in the direction of expansion due to the difference resistance, so that the crack begins to expand along a series of planes with a height difference, and the different fracture surfaces intersect to form a step. These steps constitute the characteristic morphology of the radial lines on the fatigue fracture.

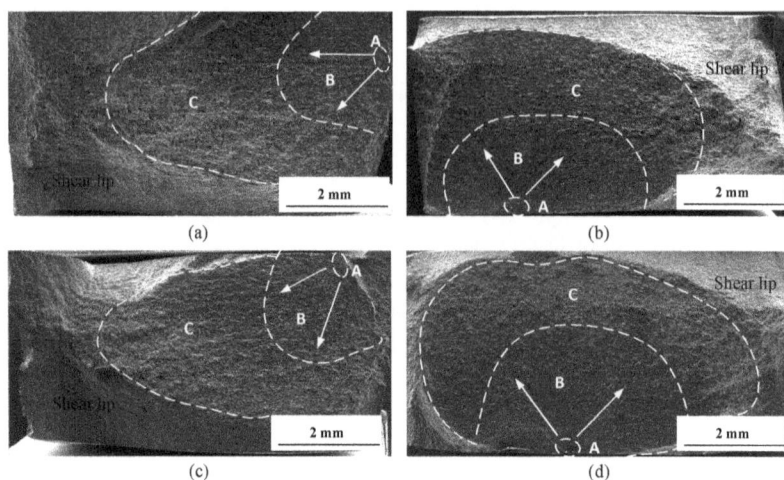

Figure 9. Fatigue fracture of the 7075 aluminum alloy under applied stresses of 260 MPa and solution times of (**a**) 0.5 h; (**b**) 1.0 h; (**c**) 1.5 h; (**d**) 2.0 h.

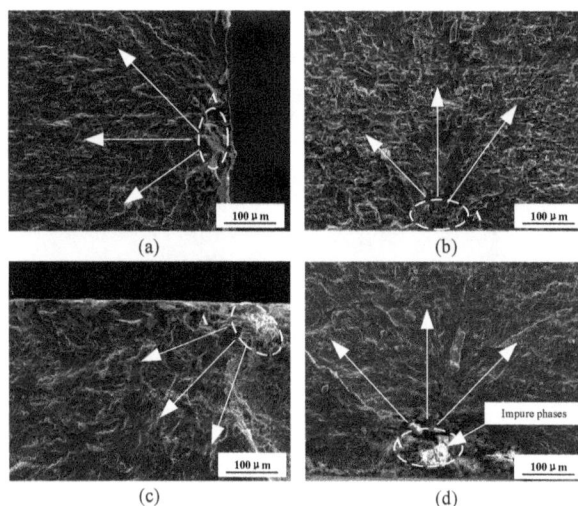

Figure 10. Fatigue crack initiation regions of the 7075 aluminum alloy under applied stresses of 260 MPa and solution times of (**a**) 0.5 h; (**b**) 1.0 h; (**c**) 1.5 h; (**d**) 2.0 h.

The process of crack propagation is divided into two stages [14]. In the first stage, the cracks propagate along the direction with the maximum shear stress, extending to the depth of one or several grains. In the second stage, due to the principal stress on the crack tips, the cracks propagate in the plane perpendicular to the maximum normal stress. This propagation plane is a large one with obvious fatigue striations.

Figure 11 shows the SEM morphologies of the first stage of crack propagation. The fracture in this stage presents a zigzag plane [28]; with no obvious fatigue striations, it belongs to the shear fracture. Additionally, around the shear zone there are some micro-voids which are formed by the undissolved phases and bases under the fatigue loads.

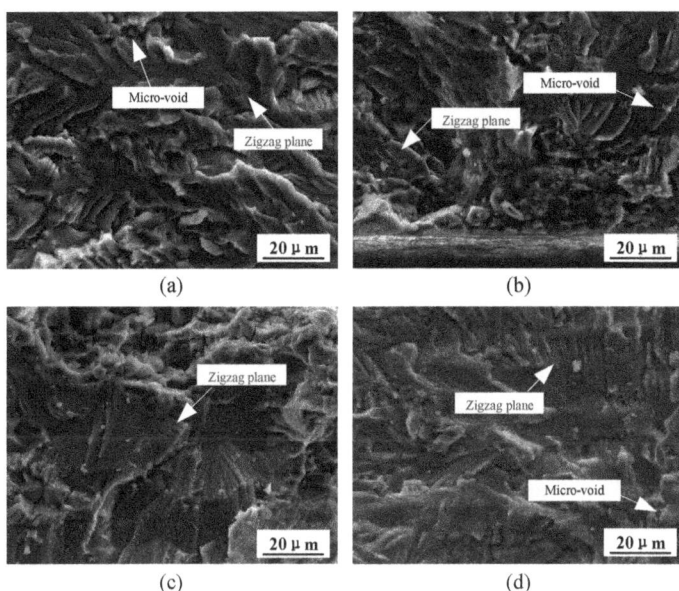

Figure 11. SEM morphologies of fatigue crack propagation (first stage) of the 7075 aluminum alloy under applied stresses of 260 MPa and solution times of (**a**) 0.5 h; (**b**) 1.0 h; (**c**) 1.5 h; (**d**) 2.0 h.

Figure 12 shows the SEM morphologies of the second stage of crack propagation. Distinct fatigue striations are highly visible. The intervals of the fatigue striations are stable, indicating a stable propagation of the cracks. Meanwhile, in the direction parallel to the fatigue striations, there are a number of secondary cracks propagating from the surface of the fracture to the interior, appearing in an interrupted distribution. The depth of the cracks is much longer than that on the fracture. Due to the secondary cracks, the paths of the crack propagation become more complex. The secondary cracks are able to lower the drivers of propagation and promote the resistance to propagation. With the 2 h solution treatment, there are more secondary cracks on the fracture, decreasing the crack propagation rate, and the fatigue life and fatigue limit are thus promoted.

In Figure 13, L is the distance from the fatigue crack initiation region to the observation point. At the observation point, five adjacent striation spacings were measured, and then the average value of these adjacent striation spacings were regarded as the width of the fatigue striation—D. Figure 13 shows that the interval of fatigue striations increases following the growth of the cracks, indicating a steady growth of the cracks in this region; the interval decreases with the increase of the solution treatment time. At the same position of 1.5 mm and 2.0 mm, the fatigue striation at a solution time of 2 h is only 0.38 μm and 0.24 μm, respectively, which were relatively small. Under the same conditions, the smaller the fatigue striations are, the higher the fatigue life is. The changing trends are consistent with the data shown in Figure 7.

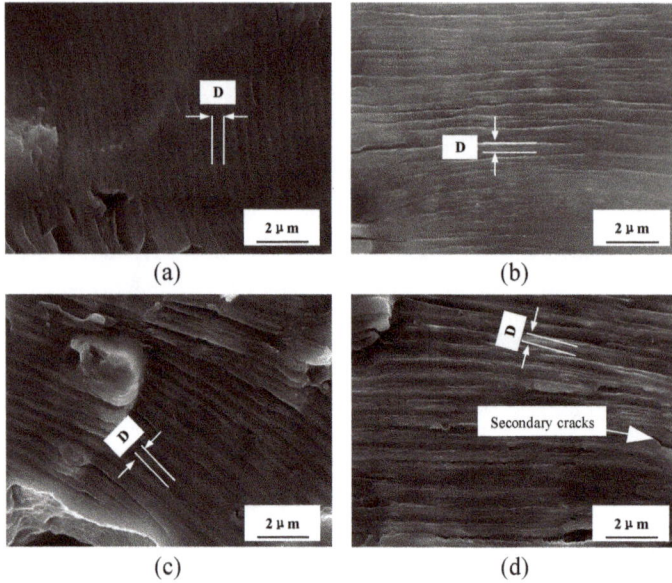

Figure 12. SEM morphologies of fatigue crack propagation (second stage) of the 7075 aluminum alloy under applied stresses of 260 MPa and solution times of (**a**) 0.5 h; (**b**) 1.0 h; (**c**) 1.5 h; (**d**) 2.0 h.

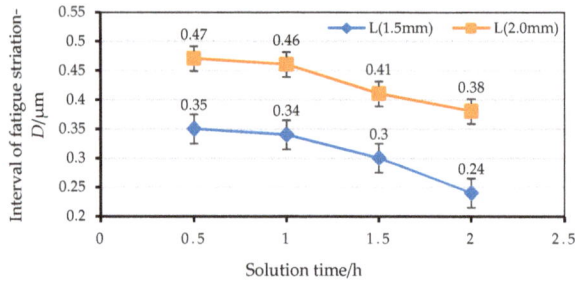

Figure 13. Intervals of fatigue striation—D in locations L = 1.5 mm/2.0 mm with different solution times.

Figure 14 presents the SEM images of the final rupture region. The fractures are mainly made up of dimples. In addition, in the fatigue final rupture region, there are a lot of tearing ridges tearing ridges similar to the fracture characteristics under static tension. With the 2 h solution treatment time, the distribution of the dimples is more uniform than the other fractures.

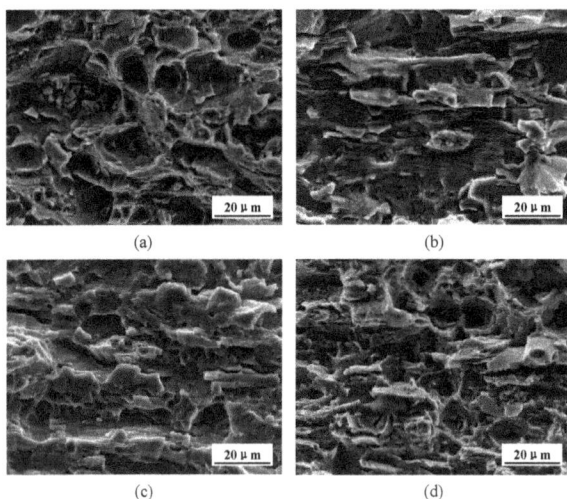

Figure 14. SEM morphologies of fatigue final rupture regions of the 7075 aluminum alloy under applied stresses of 260 MPa and solution times of (**a**) 0.5 h; (**b**) 1.0 h; (**c**) 1.5 h; (**d**) 2.0 h.

4. Conclusions

The effect on the high-cycle fatigue properties and the microstructure of a typical Al-Zn-Mg-Cu alloy after different solution treatments are discussed in this paper. Some important conclusions can be made as follows:

1. As the solution treatment time increases, different levels of recrystallization and grain growth occur in the studied alloy. The recrystallized grains merge and extend outward, flattening the crystal boundaries. When the solution treatment time is 2 h, most boundaries of adjacent grains disappear, becoming continuous rather than interrupted. Recrystallization is obvious and the average grain sizes are large.

2. The large secondary phases in the Al-Zn-Mg-Cu alloy are mainly undissolved T phase, S phase, Fe-rich phase, and Si-rich phase. Appropriate extension of the solution time can promote the dissolution of the second phase particles, reduce the number and size of the second phase particles, and improve the degree of solid solution of the alloy, so that the alloy microstructure is more uniform.

3. The solution time has a significant influence on the high-cycle fatigue properties of the Al-Zn-Mg-Cu alloy, especially when the fatigue life exceeds the 10^5 level at low stress levels. The large second phases gradually dissolve with the increase of the solution time, which can lower the probability of crack initiation. In the second stage of crack propagation, fatigue striations and secondary cracks are obviously observed. The number of secondary cracks increases with extending the solution time, which can reduce the drive force and crack propagation growth rate, promoting the high-cycle fatigue properties of the Al-Zn-Mg-Cu alloy. The changing trend of the fatigue striation interval validated this conclusion.

Acknowledgments: This work was supported by the National Natural Science Foundation of China (No. 51375500), the Fundamental Research Funds for the Central Universities of Central South University (No. 2015zzts038x), and Science and Technology Research Projects of Universities in Hebei China (No. ZC2016030).

Author Contributions: Chi Liu and Yilun Liu conceived and designed the experiment; Chi Liu and Liyong Ma performed the experiments; Chi Liu and Jiuhuo Yi analyzed the data; Jiuhuo Yi contributed reagents, materials, and analysis tools; Chi Liu wrote the paper.

Conflicts of Interest: The authors declare no conflicts of interest.

References

1. Andreatta, F.; Terryn, H.; de Wit, J.H.W. Effect of solution heat treatment on galvanic coupling between intermetallics and matrix in AA7075-T6. *Corros. Sci.* **2003**, *45*, 1733–1746. [CrossRef]
2. Fernandez Gutierrez, R.; Sket, F.; Maire, E.; Wilde, F.; Boller, E.; Requena, G. Effect of solution heat treatment on microstructure and damage accumulation in cast Al-Cu alloys. *J. Alloys Compd.* **2017**, *697*, 341–352. [CrossRef]
3. Xiong, C.X.; Deng, Y.L.; Wan, L.; Zhang, X.M. Evolutions of microstructures and textures of 7050 Al alloy plate during solution heat treatment. *Chin. J. Nonferr. Met.* **2010**, *20*, 427–434.
4. Jie, L.I.; Song, R.; Chen, X. Study on strengthening solution treatment for 7050 high strength aluminum alloy. *Hot Work. Technol.* **2009**, *38*, 125–128.
5. Deng, Y.L.; Wan, L.; Zhang, Y.; Zhang, X.R. Effect of solution treatment on quenched depth of 7050-T6 aluminum alloy. *Heat Treat. Met.* **2009**, *34*, 44–47.
6. Wang, H.B.; Meng, F.L.; Zhao, H.Y.; Cai, Y.H.; Zhang, J.S. Effects of solid solution treatment on microstructure and properties of cast-rolled 7050 aluminum alloy. *Trans. Mater. Heat Treat.* **2013**, *34*, 99–103.
7. Zeng, Z.; Peng, B.; Zeng, S.; Jiang, H.; Ning, A. Effects of progressive solution treatment on microstructure and mechanical properties of 7055 aluminum. *Spec. Cast. Nonferr. Alloys* **2009**, *29*, 672–674.
8. Chen, K.; Liu, H.; Zhang, Z.; Li, S.; Todd, R.I. The improvement of constituent dissolution and mechanical properties of 7055 aluminum alloy by stepped heat treatments. *J. Mater. Process. Technol.* **2003**, *142*, 190–196. [CrossRef]
9. Senkov, O.N.; Shagiev, M.R.; Senkova, S.V.; Miracle, D.B. Precipitation of $Al_3(Sc, Zr)$ particles in an Al-Zn-Mg-Cu-Sc-Zr alloy during conventional solution heat treatment and its effect on tensile properties. *Acta Mater.* **2008**, *56*, 3723–3738. [CrossRef]
10. Davies, R.K.; Randle, V.; Marshall, G.J. Continuous recrystallization related phenomena in a commercial Al-Fe-Si alloy. *Acta Mater.* **1998**, *46*, 6021–6032. [CrossRef]
11. Hailin, Y.; Shouxun, J.; Zhongyun, F. Effect of heat treatment and Fe content on the microstructure and mechanical properties of die-cast Al-Si-Cu alloys. *Mater. Des.* **2015**, *85*, 823–832.
12. Nayak, I.K.; Rao, S.V.R.; Kapoor, K. Effect of heat treatment and surface condition on inter-granular corrosion of alloy 800. *Mater. Perform. Charact.* **2016**, *5*, 239–252. [CrossRef]
13. Lazaro, J.; Solorzano, E.; Escudero, J.; de Saja, J.A.; Rodriguez-Perez, M.A. Applicability of solid solution heat treatments to aluminum foams. *Metals* **2012**, *2*, 508–528. [CrossRef]
14. Gang, L.; Jun, S.; Ce-Wen, N.; Kang-Hua, C. Experiment and multiscale modeling of the coupled influence of constituents and precipitates on the ductile fracture of heat-treatable aluminum alloys. *Acta Mater.* **2005**, *53*, 3459–3468.
15. Song, F.X.; Zhang, X.M.; Liu, S.D.; Bai, T.; Han, N.M.; Tan, J.B. Effects of solution heat treatment on microstructure and corrosion properties of 7050 Al alloy. *J. Aeronaut. Mater.* **2013**, *33*, 14–21.
16. Yan, L.; Du, F.S.; Dai, S.L.; Yang, S.J. Effect of microstructures on fatigue crack propagation in 2E12 aluminum alloy. *Chin. J. Nonferr. Met.* **2010**, *20*, 1275–1281.
17. Fan, X.; Jiang, D.; Zhong, L.; Wang, T.; Ren, S. Influence of microstructure on the crack propagation and corrosion resistance of Al-Zn-Mg-Cu alloy 7150. *Mater. Charact.* **2007**, *58*, 24–28. [CrossRef]
18. Zhao, T.; Zhang, J.; Jiang, Y. A study of fatigue crack growth of 7075-T651 aluminum alloy. *Int. J. Fatigue* **2008**, *30*, 1169–1180. [CrossRef]
19. Rong, L.; Yingping, J.; Shijie, W.; Shuzhen, L. Effect of microstructure on fracture toughness and fatigue crack growth behavior of Ti17 alloy. *Metals* **2016**, *6*, 186. [CrossRef]
20. Okazaki, Y. Comparison of fatigue properties and fatigue crack growth rates of various implantable metals. *Materials* **2012**, *5*, 2981–3005. [CrossRef]
21. Krahmer, D.M.; Polvorosa, R.; López de Lacalle, L.N.; Alonso-Pinillos, U.; Abate, G.; Riu, F. Alternatives for specimen manufacturing in tensile testing of steel plates. *Exp. Tech.* **2016**, *40*, 1555–1565. [CrossRef]
22. Avilés, R.; Albizuri, J.; Rodríguez, A.; López de Lacalle, L.N. Influence of low-plasticity ball burnishing on the high-cycle fatigue strength of medium carbon aisi 1045 steel. *Int. J. Fatigue* **2013**, *55*, 230–244. [CrossRef]
23. Benedetti, M.; Fontanari, V.; Bandini, M.; Taylor, D. Multiaxial fatigue resistance of shot peened high-strength aluminum alloys. *Int. J. Fatigue* **2014**, *61*, 271–282. [CrossRef]

24. Han, J.; Dai, Q.-X.; Zhao, Y.-T.; Li, G.-R. Study on fatigue performance of 7075-T651 aluminum alloys. *J. Aeronaut. Mater.* **2010**, *30*, 92–96.
25. Trško, L.; Guagliano, M.; Bokůvka, O.; Nový, F. Fatigue life of AW 7075 aluminium alloy after severe shot peening treatment with different intensities. *Procedia Eng.* **2014**, *74*, 246–252. [CrossRef]
26. Oskouei, R.H.; Ibrahim, R.N. The effect of a heat treatment on improving the fatigue properties of aluminium alloy 7075-T6 coated with TiN by PVD. *Procedia Eng.* **2011**, *10*, 1936–1942. [CrossRef]
27. Dowling, N.E.; Calhoun, C.A.; Arcari, A. Mean stress effects in stress-life fatigue and the walker equation. *Fatigue Fract. Eng. Mater. Struct.* **2009**, *32*, 163–179. [CrossRef]
28. Le Jolu, T.; Morgeneyer, T.F.; Denquin, A.; Gourgues-Lorenzon, A.F. Fatigue lifetime and tearing resistance of AA2198 Al-Cu-Li alloy friction stir welds: Effect of defects. *Int. J. Fatigue* **2015**, *70*, 463–472. [CrossRef]

![metals logo] *metals*

MDPI

Article

Experimental Investigation of Thermal Fatigue Die Casting Dies by Using Response Surface Modelling

Hassan Abdulrssoul Abdulhadi [1,3,*], Syarifah Nur Aqida Syed Ahmad [1], Izwan Ismail [2], Mahadzir Ishak [1] and Ghusoon Ridha Mohammed [1,3]

1   Faculty of Mechanical Engineering, University Malaysia Pahang, Pekan 26600, Malaysia;
    aqida@ump.edu.my (S.N.A.S.A.); mahadzir@ump.edu.my (M.I.); ghusoon_ridha@yahoo.com (G.R.M.)
2   Faculty of Manufacturing Engineering, University Malaysia Pahang, Pekan 26600, Malaysia;
    izwanismail@ump.edu.my
3   Baghdad-Institute, Foundation of Technical Education, Baghdad 10074, Iraq
*   Correspondence: PMM14001@stdmail.ump.edu.my; Tel.: +60-129-457-483

Academic Editor: Filippo Berto
Received: 7 March 2017; Accepted: 25 April 2017; Published: 26 May 2017

Abstract: Mechanical and thermal sequences impact largely on thermo-mechanical fatigue of dies in a die casting operations. Innovative techniques to optimize the thermo-mechanical conditions of samples are major focus of researchers. This study investigates the typical thermal fatigue in die steel. Die surface initiation and crack propagation were stimulated by thermal and hardness gradients, acting on the contact surface layer. A design of experiments (DOE) was developed to analyze the effect of as-machined surface roughness and die casting parameters on thermal fatigue properties. The experimental data were assessed on a thermo-mechanical fatigue life assessment model, being assisted by response surface methodology (RSM). The eminent valuation was grounded on the crack length, hardness properties and surface roughness due to thermal fatigue. The results were analyzed using analysis of variance method. Parameter optimization was conducted using response surface methodology (RSM). Based on the model, the optimal results of 26.5 µm crack length, 3.114 µm surface roughness, and 306 $HV_{0.5}$ hardness properties were produced.

Keywords: response surface methodology; machining parameters; design of experiments; thermal fatigue

---

1. Introduction

Reducing process lead time and design time are important aspect to reduce total cost of die casting process. Minimizing the trial-and-error stage of the production can further assist in reducing the cost. The shape of the cavity and die geometry is directly related to soldering tendency of a die casting. The more complicated geometry a die has, the more likely soldering can occur. The dies of complicated geometry usually procure sharp angles, core pins, nooks, and the part that acts as hot spots. These hot spots can induce soldering, due to their higher temperatures than other areas [1]. Moreover, prolonging die service life and preventing catastrophic die failures is essential because the die cost contributes considerably to the overall process cost. Die-casting dies are subjected to high mechanical and thermal loads, which can cause appalling or delayed die failure due to mount up damages. Most failures developed gradually and can be predicted. However, ongoing failures after few cycles can root for a larger economical losses due to wrecked tools, expensive down times, and disordered delivery schedules [2].

Thermal fatigue is one of the most common complications encountered in a die casting process. It is a result of the cyclic, rapid, non-uniform heating and cooling of dies. Thermal fatigue gets amplified by mechanical loadings and other damage mechanisms, such as erosion and corrosion.

Thermal fatigue is one of the major foundations of poor-quality castings and die failures. The initial phase of the thermal fatigue initiates with the formation of micro-crack networks denoted as heat checking. Heat checking results in the deterioration of die cavity surfaces. Further damage is due to heat checking leads to clasping, defective castings, and subsequently die damage and failure [3–8]. When an unconstrained object is slowly and uniformly heated or cooled, it will expand or contract proportionally to its coefficient of thermal expansion, and a temperature change will occur in relation to the mentioned temperature. However, if the body is constrained during the heating and cooling process, it will develop stresses and strains because it cannot reach its unconstrained dimensions properly. Similarly, if a body does not load and release through the same stress–strain path, repeated or cyclic heating and cooling might outcome in the build-up of inelastic deformation. If the time-varying stress within the body is tensile, accumulation may lead to a thermal fatigue. Even an unconstrained body may develop large thermal stresses if exposed to non-uniform or rapid heating and cooling.

Objects with sudden geometric and compositional variations retain internal constraints which are caused by temperature distribution. In a case of rapid surface temperature changes in a solid body, the temperature of a nearby surface layer increases and decreases promptly, whereas the rest of the body cannot react fast enough to the alterations in temperature. Different expansions and contractions within the body provide internal constraints, and various layers expand and contract at different degrees. Each region gets constrained by its neighbouring region, which results in a stress-strain field within the body [7–9]. Thermally induced stresses are not the only factors involved in the thermal fatigue phenomenon. Die material properties at elevated temperatures, mechanical loads from filling and locking, residual stresses, and cavity surface conditions also affect a material's response to heat checking and a thermal fatigue. Tempering, decarburization, and phase or structure changes due to an exposure to elevated temperatures accelerate thermal fatigue.

To predict and avoid die failures and their consequences, thermal management of dies, the interactions between thermal and stress/strain fields, material properties, and the influence of all these factors on die service life must be understood. By investigating the effects of these factors on die life during the design stage of dies and the casting process, prolonged die service life can be succeeded. A reliable die service life prediction allows for accurate estimates of actual die and production cost, reduces undesired machine down times, and helps to achieve good production management during production processes [2,9–11]. Furthermore, thermo-mechanical fatigue (TMF) experiments on die casting require costly equipment, time consuming, and often conducted through thermal testing under the same operating conditions as that for die-casting dies; however, in fact, the temperature and number of cycles are fixed and kept constant [9–12]. The damage on the die obtained by TMF experiments is less than that obtained in isothermal fatigue (IF) experiments [13]. The purpose of this study is to fit a model of the die life by using an experimental data. Experimental procedures were designed to mimic the commercial die casting process. The thermal fatigue process was conducted in a cyclic manner with constant sample temperature during the test [14–16]. Analysis of dies was performed by associating several of the previously mentioned factors that influence the temperature profile and the thermal gradients with the structural state, within the tool. The step was implemented by employing the temperature profiles obtained through a previous 1D thermal analysis and imposing them on the die cavity/casting interface via 2D thermal finite element models (FEMs), respectively. Thermal finite element analysis was conducted using ABAQUS software (Version 6.13.1, ABAQUS, Johnston, RI, USA, 2013), and this analysis was followed by a sequentially coupled structural analysis. The cooling effects of the lubricants were studied through experimental and numerical investigations. The effects of initial die cavity surface temperature and spray fluid density (spray volume per-unit surface area per-unit time as defined by Lee), have been a focus of several previous researchers [3–6,17–20].

In this study the effects on service life were investigated through response surface methodology (RSM), employing a three-factor Box-Behnken design. The thermal fatigue cycle parameters were evaluated with three levels for each factor. An experimental investigation was conducted to quantify the relationship between the input factors (as-machined surface roughness, R_{a1}, wall thickness, and immersion time) and responses (crack length, surface roughness due to thermal fatigue, R_{a2}, and hardness properties). Moreover, by combining existing and new data on heat transfer in die-casting dies and die material properties with RSM modelling, a method for predicting the onset of heat checking in H13 steel dies has been proposed.

2. Experimental Procedures

2.1. Materials

The samples in this study were made from H13 tool steel .In the fatigue testing, the sample was dipped into molten A356 aluminium alloy was used in the study. Chemical configuration for both H13 tool steel and A356 aluminium alloy are shown in Tables 1 and 2 respectively. The geometry of the samples is shown in Figure 1 with different sample wall thicknesses to analyze thermal gradient, determine efficient cooling in the central part and thermal fatigue cracks and varied surface roughness due to thermal fatigue.

Table 1. Chemical composition of American Iron and Steel Institute (AISI) H13.

Element	C	Si	Mn	Cr	Mo	V	W
wt (%)	0.51	1.26	0.413	5.5	1.52	1.0	0.02

Table 2. Chemical composition of Al 356.

Element	Cu	Mg	Mn	Si	Zn	Ti	Fe
wt (%)	0.25	0.45	0.35	7.5	0.35	0.25	0.2

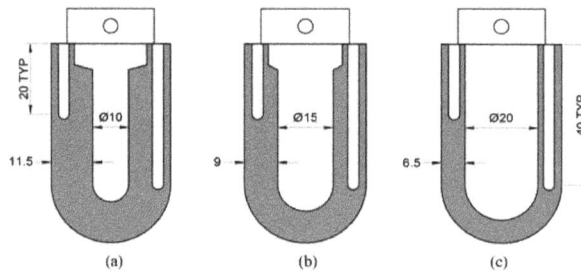

Figure 1. Samples with different wall thickness of (**a**) 11.5 mm (**b**) 9.0 mm and (**c**) 6.5 mm.

2.2. Experimental Apparatus and Work Procedure

A simulative laboratory experiment was conducted to generate controlled thermo-mechanical fatigue failures. Thermal wear set-up simulates an aluminium alloy die-casting conditions with testing cycle time of 24 s, 28 s and 32 s. The samples were dipped into molten aluminium at 700 °C and quenched in water at 32 °C for 7 s, 9 s and 11 s to generate thermal gradients during cooling process. In between the heating and water quenching, the samples were air-cooled at 28 °C for 5 s before repeating the next cycle. Thermal fatigue loading was achieved by 1850 cyclic movements. The inner wall of the samples was cooled by spraying water at room temperature for 3 s.

Samples surfaces were examined through optical microscope (MT8000 Series Metallurgical Microscopes, Microscope.com, Roanoke, VA, USA) and scanning electron microscope (SEM) (HITACHI Tabletop Microscope TM3030Plus, Hitachi High Technologies America, Inc., Schaumburg, IL, USA). For each immersion experiment scheme cycle, a characteristic testing sample from the final treated sample (when the sample was clamped on the equipment), was obtained. The transverse parts of the cycled samples were investigated for the exclusive evaluation of the detrimental penetration depth, microhardness, and microstructural amendments by using (SEM) and Vickers hardness tester (Instran ITW Test & Measurement Co., Ltd., Shanghai, China), respectively. The sample surface profile was measured with a two-dimensional surface profilometer MarSurf PS1 (Mahr, Göttingen, Germany). Surface roughness were measured using MAHR parameter MarSurf PS1 (Mahr, Göttingen, Germany), before and after the process. Thermally worn samples were investigated for crack analysis, metallographic study, and hardness properties. Metallographic study and crack analysis were performed along a longitudinal cross-section through SEM and Energy-dispersive X-ray spectroscopy (EDXS) analysis. The crack length on the sample surface was measured with ImageJ software (Version 2, National Institutes of Health, Bethesda, MD, USA, 2008); an image processing software available online [21]. The crack tips were then marked manually. The hardness properties were measured across the sample thickness with Vickers indenter.

2.3. Factors and Response Surface Modeling

Among the factors that may affect the response, several were selected to be kept constant during the experiment. The selected factors included are material, cooling fluid, and the sample heat treatment. In addition, several important factors related to thermal fatigue were considered. These factors included as-machined surface roughness, sample wall thickness, and immersion time cycle, which is a parameter that depends on the amount of cracks and hardness induced in the material (Table 3). The maximum crack length, gradient hardness, and roughness after the thermal fatigue test were observed as responses in the model. Thermal-mechanical and temperature cycles are phase variations, which means that the minimum load coincides with the minimum temperature, and the maximum tensile load is applied with the maximum temperature [22,23]. The investigated material was the tool steel used in the thermal fatigue wear test (H13 tool steel).

Table 3. Surface design factor levels.

Factor	Name	Low	High
A	Surface roughness (μm)	2.5	5.5
B	Wall Thickness (mm)	6.5	11.5
C	Immersion Time (s)	7	11

The experimental plan and results for the crack length (CL_s) of samples, surface roughness due to thermal fatigue (R_{a2}) and hardness properties obtained from the experimentation are presented in Table 4.

Table 4. Outcome of defect thermal fatigue cycles.

	Input Factors			Outcome (Response)		
Std	R_{a1} (μm)	Wall Thickness (*wt*) (mm)	Immersion Time (*T*) (s)	CL_s (μm)	R_{a2} (μm)	Hardness (HV$_{0.5}$)
1	2.5	6.5	9	30	2.8	293
2	5.5	6.5	9	46.8	6.3	239
3	2.5	11.5	9	32	2.7	297.4
4	5.5	11.5	9	46	5.9	257
5	2.5	9	7	26.5	3.5	294
6	5.5	9	7	43	6.53	242
7	2.5	9	11	30	2.9	288
8	5.5	9	11	52	5.9	235
9	4	6.5	7	32.3	4.6	291.7
10	4	11.5	7	35	4.6	265
11	4	6.5	11	41	4.6	250
12	4	11.5	11	46	4.35	291
13	4	9	9	38.5	4.3	285
14	5.5	9	9	47.3	6	246
15	2.5	9	9	30.6	2.8	297
16	5.5	11.5	7	47.6	6	234
17	2.5	6.5	11	33	2.9	267

3. Results and Discussion

3.1. Effect of Surface Roughness, Wall Thickness and Immersion Time on Crack Spherical Shape (SPH)

The second-order polynomial model was used to approximate the relationship between the three factors and response crack with Equation (1) suitably. The model showed that crack increased with an increase in the immersion time and decreased with an increase in the wall thickness. Another observation indicated that these factors (as-machined surface roughness, wall thickness, and immersion time) were "significant" to crack, respectively [11]. Table 5 shows the analysis of variance indicating that the model is adequate because the *p*-value of the square is much more than linear.

$$CL_s = +2.3622 + 0.41111 \times A + 0.086558B + 0.17442C + 7.7737 \times 10^{-4}AB + 6.1881 \times 10^{-3}AC - 5.1788 \times 10^{-3}BC, \quad (1)$$

where CL_s is crack length, A is as-machined surface roughness, B is wall thickness, and C is immersion time.

Table 5. Response surface 2FI model for crack length.

Source	Sum of Squares	Df	Mean Square	F Value	p-Value Prob > F
Model	6.42	6	1.07	41.91	<0.0001
A—surface roughness, R_{a1}	5.72	1	5.72	224.21	<0.0001
B—wall thickness	0.11	1	0.11	4.26	0.0660
C—immersion time	0.87	1	0.87	34.20	0.0002
$A \times B$	4.196×10^{-5}	1	4.196×10^{-5}	1.644×10^{-3}	0.9685
$A \times C$	1.701×10^{-3}	1	1.701×10^{-3}	0.067	0.8016
$B \times C$	3.311×10^{-3}	1	3.311×10^{-3}	0.13	0.7263
Residual	0.26	10	0.026		
Cor Total	6.67	16			

Crack SPH was obtained experimentally, and other values were predicted, in which the estimated regression coefficient for the second order predicted the crack and analysis of variance, as shown in Table 6. The small *p*-values for the linear term showed that their contribution was significant in the model. Moreover, the main effects were individually significant at the 0.05 significance level. Crack SPH was significant to the response model at $\alpha = 0.05$. From the value of R_1 (96.17%), the fit of data can be measured from the estimated model.

Table 6. Analysis of variance for crack.

Standard Deviation (Std. Dev)	0.16	Adeq Precision	19.815
Mean	6.19	Pred R-Squared	0.8678
Coefficient of Variation (C.V. %)	2.58	Adj R-Squared	0.9388
PRESS	0.88	R-Squared	0.9617

Figure 2 shows that the residuals follow a straight line, and the errors appear to be normally distributed. In such case, many parameters were carefully estimated to justify the test. Figure 3 shows that the response surface model determined above is relevant to the crack surface. The factors that are not on the plot are at their average level. The plots exhibited a square according to the surface 2FI model, and immersion time and R_{a1} contribute equally to reducing crack. Meanwhile, reduced crack length extends the lifetime of dies, but this factor cannot compensate for the reverse effect due to an increase in the other factors; wall thickness and R_{a1}.

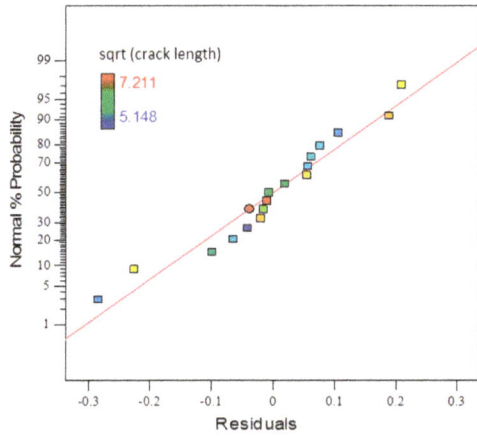

Figure 2. Normal probability plot of residuals.

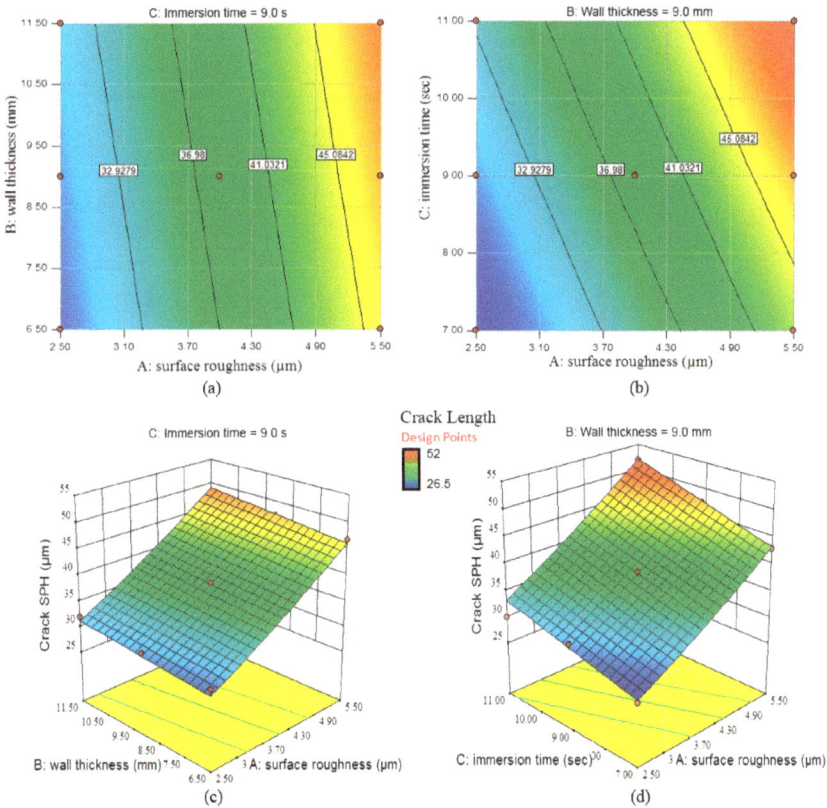

Figure 3. Contour plot of crack length responding to (**a**) R_{a1} and wall thickness, (**b**) R_{a1} and immersion time. (**c,d**) a 3D view of crack length interaction with the respective parameters.

3.2. Effect of As-Machined Surface Roughness, Wall Thickness and Immersion Time on R_{a2}

To describe the effective factors, a second-order equation was established under the conditions of surface roughness due to thermal fatigue, R_{a2}, is expressed as follows:

$$R_{a2} = 1.191 + 0.249A + 0.027B - 4.274 \times 10^{-3}AB + 4.564 \times 10^{-3}AC - 2.291 \times 10^{-3}BC, \quad (2)$$

where R_{a2} is the surface roughness due to thermal fatigue.

Table 7 indicates that the model is adequate because the p-value of the square is much more than linear. To assess the validity of the model (Equation (2)), a probability plot (Figure 6) was used to compare the measured and predicted y_2. The R_{a1}, wall thickness, and immersion time data are plotted in Figure 4. Figure 4 show that most value matched one another well, except that the difference between the measured and predicted values exceeded 0.6 μm. The smallest value of R_{a2} was measured for H13 tool steel after thermal fatigue cycles (1850 cycles) with a constant temperature of 700 °C. The data on additional factors were used to generate the probability plot, except for two data points where the model overpredicted the measured data by over 0.6 μm. The additional test data fit the model reasonably and supports the validity of the model in predicting the values of R_{a2}.

Table 7. Estimated regression coefficients for surface roughness due to thermal fatigue, R_{a2}.

Source	Sum of Squares	Df	Mean Square	F Value	p-Value Prob > F
Model	1.75	6	0.29	88.40	<0.0001
A—surface roughness, R_{a1}	1.61	1	1.61	488.86	<0.0001
B—wall thickness	6.229×10^{-3}	1	6.229×10^{-3}	1.89	0.1995
C—immersion time	0.015	1	0.015	4.47	0.0605
$A \times B$	1.268×10^{-3}	1	1.268×10^{-3}	0.38	0.5492
$A \times C$	9.257×10^{-4}	1	9.257×10^{-4}	0.28	0.6080
$B \times C$	6.479×10^{-4}	1	6.479×10^{-4}	0.20	0.6672
Residual	0.033	10	3.301×10^{-3}		
Cor Total	1.78	16			

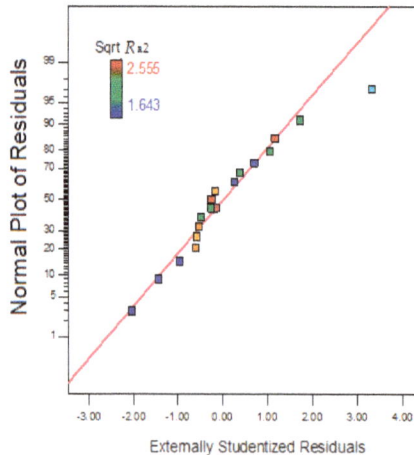

Figure 4. Normal residual probability.

Figure 5 shows a contour plot of R_{a2}, responding to three parameters. The increase in R_{a1}, and immersion time affected R_{a2} dramatically. In Figure 6, the R_{a2} reached the highest value when the R_{a1} increases.

The response surface plots (i.e., graphing of Equation (2)) for R_{a2} against R_{a1} and wall thickness ratio, and for R_{a2} against R_{a1} and immersion time are shown in Figure 5a–d respectively. In Figure 5b, the measured R_{a2} increased with R_{a1} and immersion time [24,25]. Generally, the predicted R_{a2} for the different samples was lower than that for steel B and C, which is consistent with the measured results (Figure 5a,b). The effect of the difference in time and wall thickness on y_2 varied for the different H13 samples (Figure 5b,d). With R_{a1} versus immersion time increasing from 0% to 24%, R_{a2} increased for parameters *A* and *C*. With a decrease in R_{a2}, the effect on tool steel also decreased.

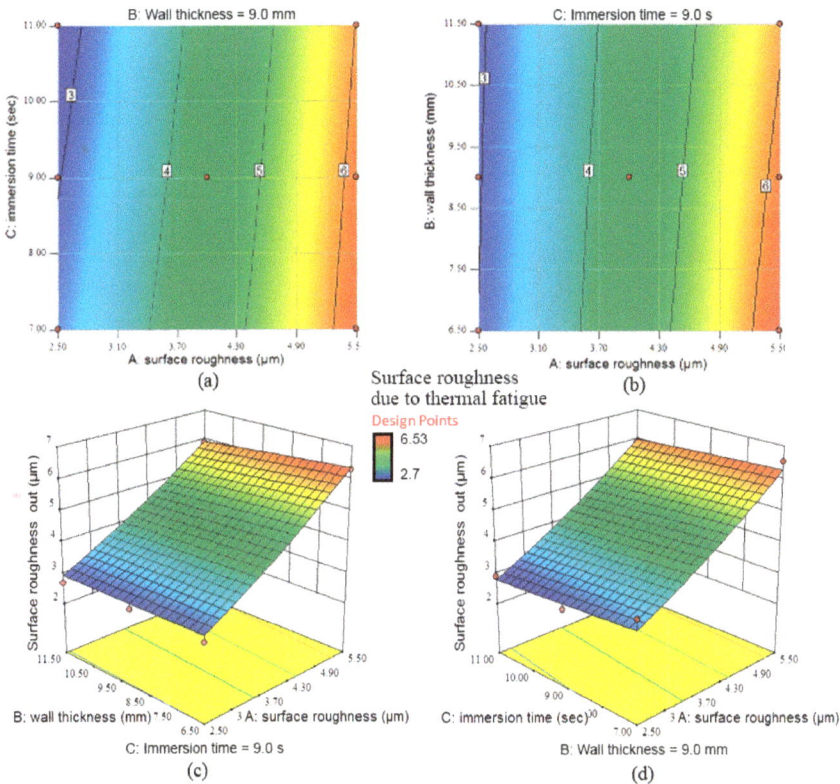

Figure 5. Contour plot of surface roughness due to thermal fatigue, R_{a2}, responding to (**a**) R_{a1} and immersion time, (**b**) R_{a1} and wall thickness. (**c**,**d**) a 3D view of R_{a2} interactions with the respective parameters.

3.3. *Effect of As-Machined Surface Roughness, Wall Thickness and Immersion Time on Hardness Properties*

The second-order equation was established to describe the influencing factors and conditions investigated in this study on the surface hardness. The second-order model can be expressed as

$$y_3 = 20.304 + 0.535A - 0.879B + 0.159C + 0.030AB - 4.244 \times 10^{-3}AC + 0.102BC$$
$$-0.159A^2 - 5.755 \times 10^{-3}B^2 - 0.062C^2,$$

(3)

where y_3 is hardness.

The model (Equation (3)) shows that surface hardness increased with an increase in surface roughness and wall thickness [26,27]. Furthermore, the square of surface roughness, R_{a1}, and immersion time provided a good indication that as-machined surface roughness was a major factor in R_{a2} changes. The analysis of variance as shown in Table 8 indicates that the model was adequate because the *p*-value of the response surface quadratic model is significant.

Table 8. Response Surface Quadratic model by the analysis of variance (ANOVA).

Source	Sum of Squares	df	Mean Square	F Value	p-Value Prob > F
Model	8.87	9	0.99	235.55	<0.0001
A—surface roughness, R_{a1}	6.66	1	6.66	1591.11	<0.0001
B—wall thickness	0.19	1	0.19	45.01	0.0003
C—immersion time	0.11	1	0.11	26.97	0.0013
$A \times B$	0.058	1	0.058	13.86	0.0074
$A \times C$	7.642×10^{-4}	1	7.642×10^{-4}	0.18	0.6820
$B \times C$	1.24	1	1.24	295.20	<0.0001
A^2	0.34	1	0.34	81.02	<0.0001
B^2	4.378×10^{-3}	1	4.378×10^{-3}	1.05	0.3404
C^2	0.21	1	0.21	49.97	0.0002
Residual	0.029	7	4.185×10^{-3}		
Cor Total	8.90	16			

The hardness values for all the samples at different through-the-wall thickness increased slightly with the increasing temperature. However, compared with hardness due to thermal fatigue cycles, the $HV_{0.5}$ values were much lower, when the R_{a1} and immersion time increased. These tests require the error term to be normally and independently distributed with mean zero and variances. Figure 6 shows the normal probability, fitted values, and histogram of residuals, respectively.

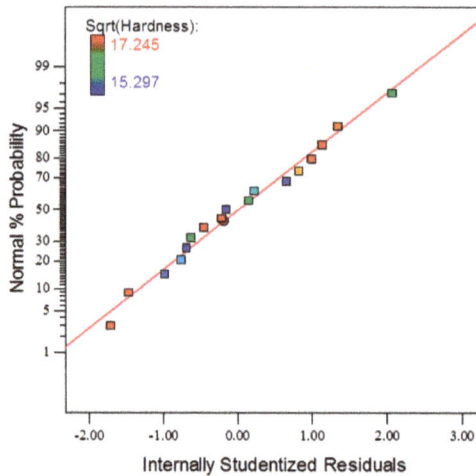

Figure 6. Normal probability plot of residuals.

Figure 7 shows the hardness contours at three different parameters. The increase in R_{a1} and immersion time clearly affected R_{a2} dramatically. In Figure 5, R_{a2} reached the highest value, when the R_{a1} increases. Hardness properties increased more rapidly with increasing temperature in comparison to increasing time. All the empirical equations predicting change in the properties are based on the three factors, namely, as-machined surface roughness, wall thickness, and immersion time.

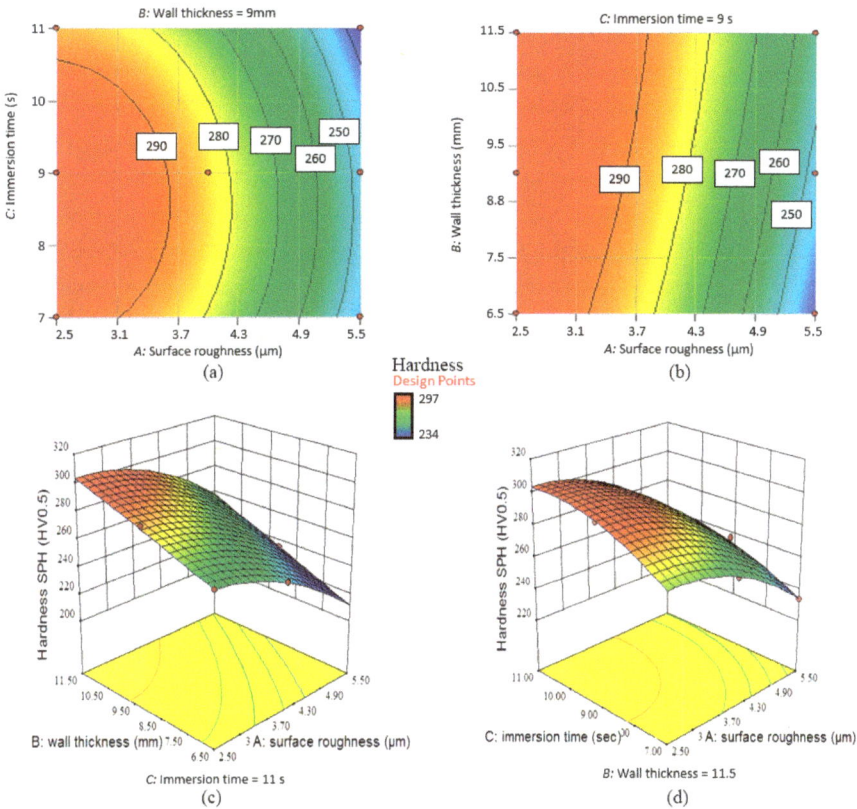

Figure 7. Contour plots of hardness properties responding to (**a**) R_{a1} and immersion time, (**b**) R_{a1} and wall thickness. (**c,d**) a 3D view of hardness properties, interaction with the respective parameters.

3.4. Influence of Cooling Rate on Hardness Properties and Crack Length

In the experiment, the maximum and the variety of temperatures reached at the surface and the difference inside the sample was investigated only as a function of the cooling system. A cooling system brought the maximum temperature at the surface. Therefore, the effect of cooling system on crack length as a function of maximum temperature at the surface of the sample. The value assessments of maximum crack length decreased with the increase in the maximum temperature. The identical trend was observed in the experiment conducted for the thermal fatigue cracking parameters. The relationship between the cracking and microhardness measured at a distance equal to the average maximum crack length, followed the temperature trend and confirmed the observation made for different immersion times, as shown in Figure 8. The longer crack is mainly because of the higher temperature at the surface, the lower is the hardness next to crack.

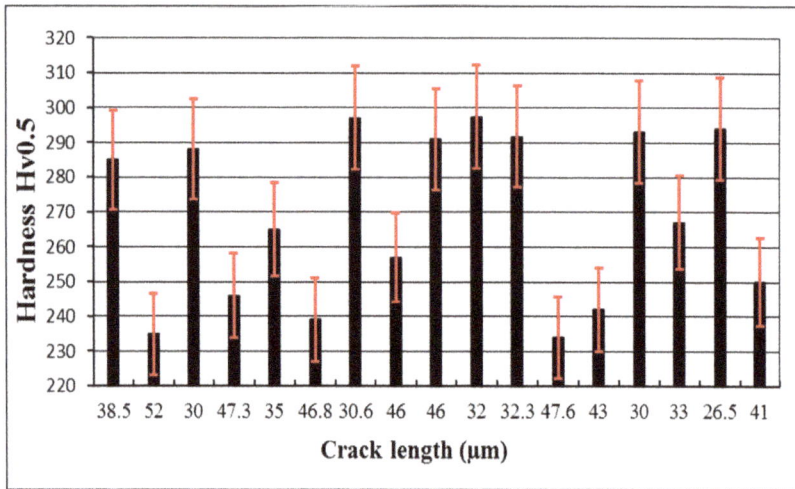

Figure 8. Relation between hardness HV$_{0.5}$ and crack length.

4. Conclusions

In the following research study, a mathematical model with second order has been employed to anticipate the crack parameters including the immersion time, surface roughness, R_{a1}, and wall thickness process by evaluating the surface roughness due to thermal fatigue, R_{a2}, and the temperature respectively, which is based on response surface methodology (RSM). The mathematical model outcomes were later equated with the experimental findings. It was observed that the immersion time, R_{a1}, and wall thickness affected R_{a2} and temperature distribution when milling the H13 tool steel. Some important conclusions derived from the study are concluded as following:

1. The response surface methodology (RSM) resulted in an advantageous procedure for the surface roughness and temperature analysis. In addition, designing experiments are essential to produce statistics, which in turn is beneficial in expanding the calculating equations for surface roughness, crack length, and hardness properties. The investigation of variance for the second order for both the studied model displays that the immersion time is the most affected parameter which afflicted the hardness and the crack lengths followed by wall thickness.
2. Both second order models were observed to be expedient in forecasting the key effects and the square effects of diverse dominant arrangements of the machining constraints. The process was found cost-effective in shaping the effect of several parameters in a methodical way. In addition, the process for the thermal fatigue cycle of H13 tool steel, the legitimacy of the process is typically restricted to the collection of factors measured during the investigation.
3. The RSM model could effectively relate the machining parameters with the responses, crack, surface roughness due to thermal fatigue, and hardness properties. The optimal parameter setting resulted crack length of 26.5 µm, surface roughness of 3.114 µm, and hardness properties of 306 HV$_{0.5}$.
4. The results generated by the predicted model are equated with the experimental results. The observed percentage error is very low, which is only 2% for both the predicted models.

Acknowledgments: The authors introduce their thanks to university Malaysia Pahang to support this project by RDU1403150.

Author Contributions: Hassan Abdulrssoul Abdulhadi designed and conducted the experiments, and analysed the results; Syarifah Nur Aqida Syed Ahmad and Izwan Ismail designed the thermal fatigue test setup; Mahadzir Ishak and Ghusoon Ridha Mohammed contributed materials and analysis tools.

Conflicts of Interest: The authors declare no conflict of interest. and the founding sponsors had no role in the design of the study; in the collection, analyses, or interpretation of data; in the writing of the manuscript, and in the decision to publish the results.

References

1. Matiskova, D.; Gaspar, S.; Mura, L. Thermal factors of die casting and their impact on the service life of moulds and the quality of castings. *Acta Polytech. Hung.* **2013**, *10*, 65–78.
2. Abdulhadi, H.A.; Aqida, S.N.; Ishak, M.; Mohammed, G.R. Thermal fatigue of die-casting dies: An overview. In Proceedings of the MATEC Web of Conferences: The 3rd International Conference on Mechanical Engineering Research (ICMER 2015), Kuantan, Malaysia, 18–19 August 2015; pp. 1–6.
3. Wang, G.; Zhao, G.; Wang, X. Effects of cavity surface temperature on reinforced plastic part surface appearance in rapid heat cycle moulding. *Mater. Des.* **2013**, *44*, 509–520. [CrossRef]
4. Ruby-Figueroa, R. Response surface methodology (RSM). In *Encyclopedia of Membranes*; Drioli, E., Giorno, L., Eds.; Springer: Berlin/Heidelberg, Germany, 2016; pp. 1729–1730.
5. Hidayanti, N.; Qadariyah, L.; Mahfud, M. Response surface methodology (RSM) modeling of microwave-assisted transesterification of coconut oil with K/γ-Al$_2$O$_3$ catalyst using box-behnken design method. In Proceedings of the AIP Conference, Brisbane, Australia, 4–8 December 2016.
6. Lee, J.-H.; Jang, J.-H.; Joo, B.-D.; Son, Y.-M.; Moon, Y.-H. Laser surface hardening of AISI H13 tool steel. *Trans. Nonferr. Met. Soc. China* **2009**, *19*, 917–920. [CrossRef]
7. Xu, W.; Li, W.; Wang, Y. Experimental and theoretical analysis of wear mechanism in hot-forging die and optimal design of die geometry. *Wear* **2014**, *318*, 78–88. [CrossRef]
8. Altan, T.; Knoerr, M. Application of the 2D finite element method to simulation of cold-forging processes. *J. Mater. Process. Technol.* **1992**, *35*, 275–302. [CrossRef]
9. Long, A.; Thornhill, D.; Armstrong, C.; Watson, D. Predicting die life from die temperature for high pressure dies casting aluminium alloy. *Appl. Therm. Eng.* **2012**, *44*, 100–107. [CrossRef]
10. Chen, C.; Wang, Y.; Ou, H.; Lin, Y.-J. Energy-based approach to thermal fatigue life of tool steels for die casting dies. *Int. J. Fatigue* **2016**, *92*, 166–178. [CrossRef]
11. Klobčar, D.; Kosec, L.; Kosec, B.; Tušek, J. Thermo fatigue cracking of die casting dies. *Eng. Fail. Anal.* **2012**, *20*, 43–53. [CrossRef]
12. Klobčar, D.; Tušek, J. Thermal stresses in aluminium alloy die casting dies. *Comput. Mater. Sci.* **2008**, *43*, 1147–1154. [CrossRef]
13. Fissolo, A.; Amiable, S.b.; Vincent, L.; Chapuliot, S.p.; Constantinescu, A.; Stelmaszyk, J.M. Thermal fatigue appears to be more damaging than uniaxial isothermal fatigue: A complete analysis of the results obtained on the CEA thermal fatigue device splash. In Proceedings of the ASME 2006 Pressure Vessels and Piping/ICPVT-11 Conference, Vancouver, BC, Canada, 23–27 July 2006; pp. 535–544.
14. Aqida, S.N.; Calosso, F.; Brabazon, D.; Naher, S.; Rosso, M. Thermal fatigue properties of laser treated steels. *Int. J. Mater. Form.* **2010**, *3*, 797–800. [CrossRef]
15. Auersperg, J.; Dudek, R.; Michel, B. Combined fracture, delamination risk and fatigue evaluation of advanced microelectronics applications towards RSM/DOE concepts. In Proceedings of the EuroSime 2006—7th International Conference on Thermal, Mechanical and Multiphysics Simulation and Experiments in Micro-Electronics and Micro-Systems, Como, Italy, 24–26 April 2006; pp. 1–6.
16. Mohammed, G.; Ishak, M.; Aqida, S.; Abdulhadi, H. Effects of heat input on microstructure, corrosion and mechanical characteristics of welded austenitic and duplex stainless steels: A review. *Metals* **2017**, *7*, 39. [CrossRef]
17. Box, G.E.; Draper, N.R. *Empirical Model-Building and Response Surfaces*; Wiley: New York, NY, USA, 1987.
18. Giddings, J.C.; Graff, K.A.; Caldwell, K.D.; Myers, M.N. *Field-Flow Fractionation*; ACS Publications: Salt Lake City, UT, USA, 1993.
19. Jayabal, S.; Natarajan, U.; Sekar, U. Regression modeling and optimization of machinability behavior of glass-coir-polyester hybrid composite using factorial design methodology. *Int. J. Adv. Manuf. Technol.* **2011**, *55*, 263–273. [CrossRef]
20. Paul, V.T.; Saroja, S.; Vijayalakshmi, M. Microstructural stability of modified 9Cr-1Mo steel during long term exposures at elevated temperatures. *J. Nuc. Mater.* **2008**, *378*, 273–281. [CrossRef]

21. Gallagher, S.R. Digital image processing and analysis with imagej. In *Current Protocols Essential Laboratory Techniques*; John Wiley & Sons, Inc.: New York, NY, USA, 2008.

22. Aslan, N.; Cebeci, Y. Application of box-behnken design and response surface methodology for modeling of some turkish coals. *Fuel* **2007**, *86*, 90–97. [CrossRef]

23. Dixson, R.; Fu, J.; Orji, N.; Guthrie, W.; Allen, R.; Cresswell, M. *CD-AFM Reference Metrology at NIST and Sematech*; The International Society for Optics and Photonics: Bellingham, WA, USA, 2005; pp. 324–336.

24. Patel G.C., M.; Krishna, P.; Parappagoudar, M.B. Squeeze casting process modeling by a conventional statistical regression analysis approach. *Appl. Math. Model.* **2016**, *40*, 6869–6888. [CrossRef]

25. Moverare, J.J.; Gustafsson, D. Hold-time effect on the thermo-mechanical fatigue crack growth behaviour of inconel 718. *Mater. Sci. Eng. A* **2011**, *528*, 8660–8670. [CrossRef]

26. Ammar, O.; Haddar, N.; Remy, L. Numerical computation of crack growth of low cycle fatigue in the 304l austenitic stainless steel. *Eng. Fract. Mech.* **2014**, *120*, 67–81. [CrossRef]

27. Hafiz, A.M.K.; Bordatchev, E.V.; Tutunea-Fatan, R.O. Influence of overlap between the laser beam tracks on surface quality in laser polishing of AISI H13 tool steel. *J. Manuf. Process.* **2012**, *14*, 425–434. [CrossRef]

![metals logo] *metals*

Article

Physical-Mechanism Exploration of the Low-Cycle Unified Creep-Fatigue Formulation

Dan Liu * and Dirk John Pons

Department of Mechanical Engineering, University of Canterbury, Christchurch 8140, New Zealand;
dirk.pons@canterbury.ac.nz
* Correspondence: dan.liu@pg.canterbury.ac.nz; Tel.: +64-021-113-9534

Received: 18 August 2017; Accepted: 15 September 2017; Published: 18 September 2017

Abstract: Background—Creep-fatigue behavior is identified as the incorporated effects of fatigue and creep. One class of constitutive-based models attempts to evaluate creep and fatigue separately, but the interaction of fatigue and creep is neglected. Other models treat the damage as a single component, but the complex numerical structures that result are inconvenient for engineering application. The models derived through a curve-fitting method avoid these problems. However, the method of curving fitting cannot translate the numerical formulation to underlying physical mechanisms. Need—Therefore, there is a need to develop a new creep-fatigue formulation for metal that accommodates all relevant variables and where the relationships between them are consistent with physical mechanisms of fatigue and creep. Method—In the present work, the main dependencies and relationships for the unified creep-fatigue equation were presented through exploring what the literature says about the mechanisms. Outcomes—This shows that temperature, cyclic time and grain size have significant influences on creep-fatigue behavior, and the relationships between them (such as linear relation, logarithmical relation and power-law relation) are consistent with phenomena of diffusion creep and crack growth. Significantly, the numerical form of "$1 - x$" is presented to show the consumption of creep effect on fatigue capacity, and the introduction of the reference condition gives the threshold of creep effect. Originality—By this means, the unified creep-fatigue equation is linked to physical phenomena, where the influence of different dependencies on creep fatigue was explored and relationships shown in this equation were investigated in a microstructural level. Particularly, a physical explanation of the grain-size exponent via consideration of crack-growth planes was proposed.

Keywords: creep fatigue; physical mechanism; temperature; cyclic time; grain size; fatigue capacity

1. Introduction

Creep-fatigue damage is physically explained as the combined effects of fatigue and creep, due to reversed loading and elevated temperature respectively. Historically, creep-fatigue models have been developed through either a constitutive-based method, or an empirically-based method.

The constitutive-based method is typically conducted by exploring the underlying physical mechanisms of fatigue and creep, and by including parameters representing the properties of the material. Some constitutive models have been extended from the conventional idea of continuum damage, including the linear damage rule [1,2] and the crack growth law [3]. Examples of such models are those proposed by Takahashi [4,5] and Warwick [6] which are based on ductility and energy exhaustion with the total damage being divided into fatigue damage and creep damage, the model proposed by Sehitoglu [7] that divides the total damage into three sub-damages caused by fatigue, oxidation and creep, and the model proposed by Ainsworth [8] that partitions the crack as a whole into sub-effects caused by fatigue and creep.

The constitutive models based on continuum damage are successful in evaluating creep and fatigue separately, but do not address the interaction of fatigue and creep, hence weakening the applicability. The deeper issue is that the interactive effect between fatigue and creep is complex and the numerical formulism is unknown. Thus, it is difficult to further improve the continuum-damage-based models by introducing an interactive effect. Some constitutive models try to avoid the need to evaluate the interactive component, by treating the total damage or the fatigue ability as a single component. Examples are the model [9,10] extended from the Chaboche model [11], the model [12,13] that described the crack-tip behavior, and the model [14,15] that evaluated dissipated energy. Although these models present a comprehensive numerical representation of creep fatigue, they provide mathematically complex structures. Therefore, these models are not convenient to be used by engineers who need to obtain fatigue life through simpler physical parameters such as applied loading, temperature and cyclic time.

The empirical-based method is typically conducted by fitting mathematical formula to the empirical data, with the emphasis on high quality of fit. The empirical-based models might be more favorable since these models provide simpler/clearer numerical representations. By the empirical-based method the creep-fatigue behavior is numerically represented—this is achieved by incorporating the creep-related parameters into one of the conventional fatigue models, such as the Basquin equation [16] for the high-cycle regime, or the Coffin-Manson equation [17,18] for the low-cycle regime.

The low-cycle creep-fatigue behavior has received much attention, and resulted in different strain-based creep-fatigue models. The first model developed by the empirical-based method was attempted by Coffin [19], who proposed the frequency-modified Coffin-Manson equation. Then, this formulation was further modified by Solomon [20] and Shi et al. [21] through directly introducing a temperature dependency. Discarding the family of equations based on the frequency-modified Coffin-Manson equation has given additional numerical representations of creep fatigue, such as the equations proposed by Jing et al. [22], Engelmaier [23] and Wong and Mai [24].

However, these creep-fatigue models have significant limitations. The issue is that there are multiple variables to model: failure mode (creep, creep-fatigue and fatigue), multiple temperatures, different cyclic times (frequency), and different materials (type of material and grade thereof). The existing curve-fitting models cover one or both of variable temperature and cyclic time, but are specific to one material grade. This is because these models were derived from one specific situation. In addition, these existing models cannot cover the full range of conditions from pure fatigue to pure creep, since they were derived from the empirical data of creep fatigue. In this case, the existing models cannot natively be extended to other materials or failure modes. Furthermore, these existing models cannot provide an economical method for engineering design because numerous creep-fatigue experiments are required to achieve high quality of curve fitting.

These three disadvantages arise as an intrinsic limitation of the method of curve fitting, since statistical fitting accuracy does not necessarily translate to a robust description of the underlying physical mechanisms. Therefore, these models are only numerical representations of creep-fatigue behavior, and no underlying physical mechanisms of fatigue and creep are provided. The fitting method and related existing creep-fatigue models do not provide a route to achieve a unified formulation across multiple materials, integrated representation of failure modes, and engineering economy. However, these disadvantages are recently improved by the low-cycle unified creep-fatigue equation [25–27] (the description of this new model will be presented in Section 2). While this new model does not formalize the interactive effect between fatigue and creep, nonetheless the error of life prediction caused by this is within an acceptable range (we assume the ratio of predicted fatigue life to experimental result is between 0.75 and 1.25). This is because the coefficients in this new model are not solely derived from the empirical data of pure fatigue or/and pure creep, even though the relationships between different variables in this model are based on the mechanisms of fatigue and creep.

The purpose of the current paper is to link the low-cycle unified creep-fatigue equation to physical phenomena. To do this we considered each of the main dependencies (including temperature, cyclic time and grain size) and relationships for the unified creep-fatigue equation, and explored what the literature says about the mechanisms of fatigue and creep. Specifically, the creep and fatigue mechanisms based on the underlying physical mechanisms in terms of temperature, cyclic time and grain size were discussed, then the consistency between the physical mechanisms and the structure of the unified creep-fatigue equation was investigated.

2. Brief Description of the Unified Model

The low-cycle unified creep-fatigue equation (Equation (1)) [25–27] is presented as follow:

$$\varepsilon_p = C_0 c(\sigma, T, t_c, d) N^{-\beta_0} \tag{1}$$

with

$$c(\sigma, T, t_c, d) = 1 - \left[c_1(\sigma)\left(T - T_{ref}\right) + c_2 \log\left(t_c/t_{ref}\right)\right] \cdot \left[A\left(d/d_{ref}\right)^m\right]$$

$$T - T_{ref} = \begin{cases} T - T_{ref} \text{ for } T > T_{ref} \\ 0 \text{ for } T \leq T_{ref} \end{cases}$$

$$t_c/t_{ref} = \begin{cases} t_c/t_{ref} \text{ for } t_c > t_{ref} \\ 1 \text{ for } t_c \leq t_{ref} \end{cases}$$

where ε_p is the plastic strain, C_0 is the fatigue ductility coefficient, β_0 is the fatigue ductility exponent, N is the creep-fatigue life, σ is the stress, T is the temperature, t_c is the cyclic time, d is the grain size, T_{ref} is the reference temperature, t_{ref} is the reference cyclic time, d_{ref} is the reference grain size, and A and m are constants. The derivation of this unified creep-fatigue formulation is based on the concept of fatigue capacity, which numerically shows the process of gradual consumption of full fatigue capacity by creep effect. Particularly, the relationships between different variables in this unified formulation can be explained by physical meaning. This is a remarkable characteristic for this new formulation, which results in several significant advantages on the creep-fatigue description. Specifically, the unified creep-fatigue equation was well validated at multiple temperatures and cyclic times for multiple materials [25–27], such as 63Sn37Pb, 96.5Sn3.5Ag, stainless steel 316 and Inconel 718, thus the unified characteristic is presented. In addition, this unified formulation can be restored to the Coffin-Manson equation at pure-fatigue condition and can be reorganized to the Manson-Haferd parameter [28] at pure-creep condition, thus the integrated characteristic is included. Furthermore, the coefficients of this unified formulation can be extracted with minimum experimental effort, which provides an effective and efficient method for engineering design. In general, function $c_1(\sigma)$ and constant c_2 are extracted from creep-rupture tests, and the constants C_0, β_0, A and m are derived from creep-fatigue tests. A brief description of the validation on material GP91 casting steel is presented below, based on [25]. Briefly, the pure-creep condition ($\varepsilon_p = 0$) gives the formula of the constant c_2 (Equation (2)) through letting $T = T_{ref}$, and suggests function $c_1(\sigma)$ (Equation (3)) through letting $t_c = t_{ref}$.

$$c_2 = \frac{1}{\log\left(t_a/t_{ref}\right)} \tag{2}$$

$$c_1(\sigma) = -\frac{c_2}{P_{MH}(\sigma)} \tag{3}$$

Then, creep-fatigue data are applied to obtain the constants of A, m, C_0 and β_0 through minimizing the error (Equation (4)) between the predicted creep-fatigue life ($N_{pre,ij}$) and experimental results ($N_{exp,ij}$):

$$error = \sum_{i,j} \left(\log N_{pre,ij} - \log N_{exp,ij}\right)^2 \tag{4}$$

The reference temperature for GP91 casting steel is chosen as 610 K, the reference cyclic time is defined as 1 s, and the reference grain size is selected as 25 μm. The creep-rupture data [29] gives the point of convergence $(T_{ref}, \log t_a)$ which is evaluated as (610 K, 18.281), and thus,

$$c_2 = \frac{1}{\log\left(t_a/t_{ref}\right)} = \frac{1}{\log(10^{18.281}/1)} = 0.0547 \tag{5}$$

And the relationship between stress and the Manson-Hafer parameter (P_{MH}) is given as:

$$-\frac{1}{P_{MH}(\sigma)} = 0.0174 + 2.20 \times 10^{-4}\sigma - 3.2 \times 10^{-7}\sigma^2 \tag{6}$$

Then, substituting Equations (5) and (6) into Equation (3), function $c_1(\sigma)$ is expressed as:

$$c_1(\sigma) = -\frac{c_2}{P_{MH}(\sigma)} = 9.51808 \times 10^{-4} + 1.20344 \times 10^{-5}f_m\sigma - 1.75045 \times 10^{-8}f_m^2\sigma^2 \tag{7}$$

where f_m is a moderating factor which is introduced to compress the constant stress into an equivalent creep damage under the cyclic situation. The magnitude of f_m is given as 0.6366 for the sinusoidal wave. The empirical data of creep fatigue (T = 673 K, t_c = 5 s, d = 25 μm; T = 823 K, t_c = 5 s, d = 25 μm; T = 873 K, t_c = 2 s, d = 35 μm) obtained from Ref. [30] give the constants of C_0, β_0, A and m through minimizing the difference between predicted life and experimental life (numerically optimization) (Equation (4)). Specifically, C_0 = 0.9532, β_0 = 0.669, A = 0.5588 and m = −0.4053. In addition, the ratios of predicted fatigue life to experimental fatigue life are plotted in Figure 1. This figure shows that all data points fall in a reasonable range (within the upper bound (+25%) and the lower bound (−25%)). This implies that the unified formulation has high accuracy of fatigue-life prediction.

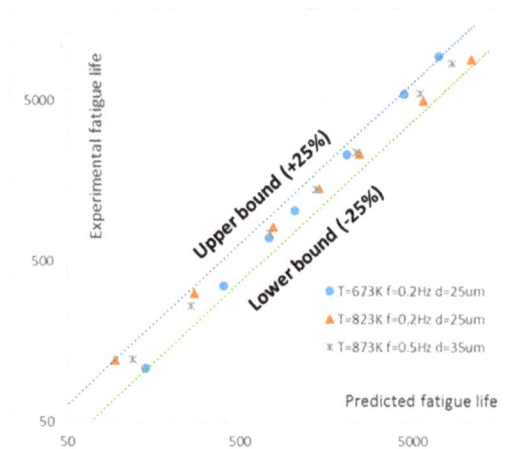

Figure 1. Application of the unified model to predict fatigue life vs. experimental fatigue life for GP91 casting steel, with raw data from physical tests from [29,30].

We also indicated that the unified formulation provides a more economical method of fatigue-life evaluation for engineering application. This was discussed and proved in [31], where the unified formulation was compared with Wong and Mai's equation. Specifically, taking 63Sn37Pb solder as an example, both Wong and Mai's equation and the unified formulation give high accuracy of fatigue-life prediction when all eight groups of creep-fatigue data are imposed. Then, three groups of creep-fatigue data were selected to extract the coefficients of these two formulations. When Wong and Mai's equation with the coefficients obtained at this stage is extended to predict fatigue life at total eight creep-fatigue situations, the average error dramatically worsens. However, only a slight reduction of average error was presented by the unified formulation. This implies that the unified formulation requires less creep-fatigue experiments to extract the coefficients, and thus a more economical method is provided for engineering application.

Overall, the new model presents a good balance between accuracy and economy, which is very important for engineering applications. More information of the advantages presented by the unified formulation is shown in Ref. [31].

3. Influence of Relevant Variables on Creep-Fatigue Behavior

Creep-fatigue process presents the fatigue behavior under elevated temperature, where creep effect is active. At the microstructural level, fatigue and creep show different underlying principles. Generally, fatigue effect occurs via cracks through the grains, while creep effect involves the grain boundary cracking [32]. The accumulation of creep-fatigue damage implies that full fatigue capacity is gradually consumed by creep effect, which is macroscopically influenced by temperature and cyclic time, and is related to grain size in the microstructural level. These relevant variables are accommodated in the unified creep-fatigue equation. We discuss these three main dependencies since they are frequently investigated and presented in the existing creep-fatigue models. Practically, the creep-fatigue behavior is also influenced by other effects, such as mean stress/strain and applied loading. The effect of mean stress/strain is not included into the unified equation, which is a limitation of this new model and will be considered in future research. In addition, the effect of applied loading (stress and strain) on creep is indirectly incorporated into function $c_1(\sigma)$, where this function is extracted through the relationship between applied loading and the Manson-Haferd parameter. It is generally accepted that the increased loading results in more vulnerable bonds between atoms due to change to inter-atomic spacing. This provides a favorable situation to initiate the atomic movements. In this case, creep damage is intensified with increased loading, and then fatigue capacity is reduced.

The influences of these three main variables (temperature, frequency/cyclic time and grain size) are shown below.

3.1. Temperature Dependency

Temperature has significant influence on creep, but presents negligible effect on pure fatigue [33] where creep is dormant. Normally, creep effect is activated when temperature is higher than 35% of the melting temperature [34], which may be attributed to the behavior of atomic vibrations [35]. To be specific, atomic vibration is accelerated (also the internal energy is increased) with rising temperature, where favorable conditions to break the bonds between atoms are provided. This threshold temperature is defined as the reference temperature, which is included in the unified formulation (Equation (1)). In this case, the discussion of temperature dependency focuses on creep behavior. Normally, creep mechanisms are divided into Nabarro-Herring creep, Coble creep, grain boundary sliding and dislocation creep [32,36]. Nabarro-Herring creep and Coble creep show a strong dependence on temperature, where diffusional flow of atoms occurs under conditions of relatively high temperature. Grain boundary sliding involves displacements of grains against each other. This is the main mechanism for the creep failure of ceramics at high temperature because of glassy-phase formation at grain boundary, which provides good sliding condition along grain boundary. Therefore, this creep behavior is not a major contributor to metals and hence is removed from further consideration here. Dislocation creep presents drastic dislocation through the crystal lattice, which results from both line defects and point defects at relatively low temperature. Therefore, high stress is needed and small diffusional flow is involved. This process shows a highly sensitive to the applied stress on material, but not temperature. This is indicated in Ref. [32], wherein a higher stress sensitive exponent is presented compared to the other three creep mechanisms. Based on the brief description of these four creep mechanisms shown above, the diffusion creep (Nabarro-Herring creep and Coble creep), which has strong temperature dependency is used to explain the influence of temperature on creep.

The temperature effect on creep-fatigue is attributed to elevated temperature leading to weaker bonding between atoms at the grain boundary. This is due to better conditions for diffusion. Then this causes the movements of vacancies [37–39]. This transfer finally results in the overall deformation of the material. Specifically, a vacancy is defined as a point defect in a crystal, where an atom is missing from its original lattice site. During this process, an atom needs to overcome the energy barrier to move from its current site to the nearby vacant site. By this means, the high temperature can provide atoms with enough energy to break their bonds with neighboring atoms, and then lead to the location transfer (motion) of atoms [40]. This process can be identified as a thermodynamic system with a strong driving force of temperature for diffusion [36]. In addition, diffusion basically is a net movement of atoms from high concentration region to a low concentration region. This reflects the initial driving force for the transfer of atoms. During this process, temperature is an important factor to determine the rate of diffusion, wherein the elevated temperature speeds up the random atom motion, which gives the atom access to a greater physical volume of space, and the new atomic configuration opportunities are provided.

Consequently, the increasing temperature accelerates the process of diffusion (more creep damage occurs), and then reduces the fatigue capacity for the creep-fatigue condition.

3.2. Frequency/Cyclic Time Dependence

Normally, cyclic time does not have significant influence on pure fatigue and at the same order of frequency magnitudes [33]. In our research, the frequency/cyclic time is limited to a range typical for general engineering situations. Thus, the influence of frequency/cyclic time on pure-fatigue life is ignored, and the discussion of frequency/cyclic time effect is based on the creep behavior.

Creep is normally defined as a time-dependent deformation under a constant loading, which indicates that creep damage is intensified with increasing time. The general influence of frequency/cyclic time on fatigue capacity could be explained through the transient-creep-plus-elastic model shown in Figure 2 [32], where σ is the applied stress. In this model, a high frequency load (which is related to elastic strain rate) primarily causes deformation of spring S_2. This is because of the dynamic resistance effect of dashpot (η_1(MPa)). Hence the slope of stress-strain curve is S_2 (N/m). However, low frequency loads (events over longer time) cause the deformation of both springs S_1 and S_2, and the dashpot is relatively inconsequential. In this case, the strain (ε (m/m)) may be expressed as:

$$\varepsilon = \frac{\sigma}{S_1} + \frac{\sigma - \eta_1}{S_2} \tag{8}$$

Equation (8) shows that the slope (S_e (N/m)) of stress-strain curve is $S_1 S_2 / (S_1 + S_2)$, which is smaller than S_2. This implies that large time can reduce stiffness, then increase strain and lead to more creep damage.

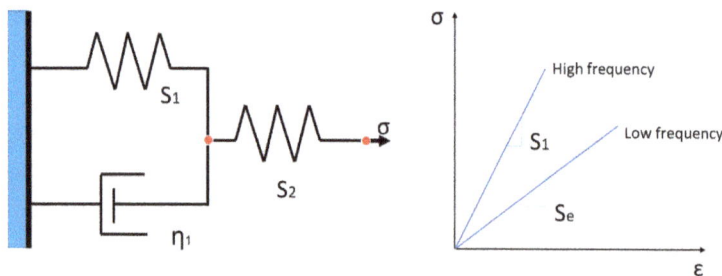

Figure 2. The transient-creep-plus-elastic model. Image adapted from Ref. [32].

Consequently, low frequency (large cyclic time) gives more accumulation of creep under a given temperature and loading. In other words, the longer time leads to more diffusion in microstructure when a material is exposed to high temperature under constant applied loading, and then produces more creep-fatigue damage.

3.3. Grain-Size Dependence

Grain size has significant influence on both fatigue damage and creep damage, but has contrary effects on creep and fatigue. Generally, smaller grain size is more positive for pure-fatigue resistance, while bigger grain size is more beneficial for pure-creep ductility [32]. This phenomenon results from the different failure mechanisms shown by fatigue and creep.

Fatigue failure is caused by the progressive accumulation of plastic deformation under cyclic loading. During fatigue process, cracks initiate at the early stage and then gradually propagate through grain boundary with the increasing number of cycles. It is easy to understand that the propagation of crack needs to penetrate the grain boundary to extend to the next grain, and this may also require a reorientation of the crack growth direction [41,42]. This means that the more grains the crack encounters the slower the progression of fatigue failure, and hence the greater loading (stress or strain) is required to make crack achieve the critical length of failure. In other words, the materials with finer grain size provide better performance of fatigue resistance.

As mentioned above, grain boundary is barrier for the propagation of crack under pure-fatigue condition. However, the grain boundary becomes the source of creep damage at pure-creep condition [36]. Since the stress concentration is intensive in the intersection point among three adjacent grains, the crack growth along grain boundary is promoted in these triple points [43–45]. Consequently the triple point provides an opportunity for further crack propagation under creep. Also relevant to note is that the triple point also contains crystalline defects, where multiple directional opportunities for crack propagation along the grain boundaries are provided. In this way, the finer the grain size, the greater the internal area of grain boundaries and volumetric density of triple points, hence enhanced opportunity for crack propagation under creep. In addition, the influence of grain size on creep also could be explained by the physical mechanism of diffusion creep. Specifically, the atomic diffusion causes the elongation of the grain along the stress axis, which implies finer grain size results in more significant deformation than coarse grain size under a given stress within a same area.

The discussion above shows that the influences of grain size on fatigue and creep are contrary, and thus combined effect between fatigue and creep should be determined by the proportion of fatigue contribution and creep contribution. The research conducted by Hatanaka and Yamada [46], Hattori et al. [47] and Pieraggi and Uginet [48] show that fatigue capacity reduces with increased grain size. This implies that the fatigue effect makes more contribution to creep-fatigue damage than creep effect under the zero-hold-time cyclic loading and relatively short cyclic time. This may be because the total failure time is too short to produce major creep damage in the low-cycle regime. We could image that, for the situation of cyclic loading with hold time or relatively long cyclic time, the contribution of creep effect would increase. Finally, if the creep effect becomes more significant than fatigue effect, the bigger grain size would have more benefits for creep-fatigue behavior [49].

4. Consistency between the Unified Formulation and Physical Phenomena

Section 3 shows that temperature, time and grain size have significant influences on creep and fatigue behaviors, thus the relationships between them can be ideally derived from the microstructural level. These relationships are well included in the unified equation, wherein the main relationships between different variables show a high consistency with physical mechanisms of fatigue and creep.

4.1. Linear Relationship between Temperature and Strain

The unified creep-fatigue equation can be reorganized to the form:

$$\varepsilon_p = C_0 N^{-\beta_0} - \left[c_1(\sigma)\left(T - T_{ref}\right) + c_2 \log\left(t_c/t_{ref}\right) \right] \cdot \left[A\left(d/d_{ref}\right)^m \right] C_0 N^{-\beta_0} \tag{9}$$

The first term in Equation (9) shows full fatigue capacity, and the second term reflects the strain caused by creep effect. Significantly, a linear relationship is presented between temperature and creep-related strain (also applied plastic strain ε_p), which is consistent with underlying creep mechanism. According to description of four different creep mechanisms (Nabarro-Herring creep, Coble creep, grain boundary sliding and dislocation creep) in Section 3.1, in the present work, diffusion creep (including Nabarro-Herring creep and Coble creep) is regarded as the main creep mechanism for creep fatigue since it has strong temperature dependency. Therefore, the discussion of temperature-strain relation is built on the mechanism of diffusion. As mentioned in Section 3.1, the process of diffusion is identified as a thermodynamic system. In this case, a piece of crystal containing n atoms is selected, wherein an atom inside is transferred to the surface due to diffusion (Figure 3), and thus a vacancy is formed. Normally, the creep process can be described by Gibbs free energy [50]. This parameter shows the thermodynamic potential to form this vacancy under the situation with a given pressure and a given temperature, is presented by Equation (10).

$$\Delta G_f = \Delta E_f + P\Delta V_f - T\Delta S_f \tag{10}$$

where ΔG_f is the Gibbs free energy for formation of a vacancy, ΔE_f is the change in internal energy due to formation of a vacancy, ΔV_f is the volume of a vacancy, ΔS_f is the entropy for formation of a vacancy, P is the pressure and T is the temperature. If n_v vacancies are formed during the process of diffusion, the total change in free energy is presented by Equation (11):

$$\Delta G = n_v \Delta G_f - T\Delta S_c \tag{11}$$

where ΔG is the total change in free energy, and ΔS_c is the configurational entropy (Equation (12)) which reflects W different ways of distribution of n_v vacancies among the n sites.

$$\Delta S_c = k \ln W \tag{12}$$

where k is the Boltzmann's constant. Then, the free energy shown by Equation (11) is reorganized as:

$$\Delta G = n_v \Delta G_f - kT \ln C_n^{n_v} = n_v \Delta G_f - kT \ln \left[\frac{n!}{(n - n_v)! n_v!} \right] \tag{13}$$

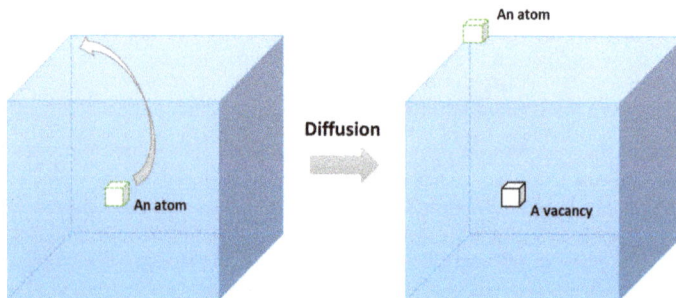

Figure 3. A movement of an atom.

By using Stirling's equation:

$$\ln(x!) = x \ln x - x + O(\ln x) \tag{14}$$

Equation (13) can be simplified and approximated as Equation (15) through assuming $n \ll n_v$:

$$\Delta G \approx n_v \Delta G_f - kTn_v \left(1 + \ln \frac{n}{n_v}\right) \tag{15}$$

This equation shows that the total free energy has a vacancy-number dependency, and thus this free energy varies during the process of transfer. Normally, the minimum value occurs at the two ends of movements, where an equilibrium situation is achieved and is numerically presented through letting $\partial \Delta G / \partial n_v = 0$. Then, this operation gives the equilibrium atomic fraction of vacancies:

$$N_v = \frac{n_v}{n} = exp\left(-\frac{\Delta G_f}{kT}\right) \tag{16}$$

Normally, diffusion is always described by Fick's law [3,37,39]:

$$J = -D\frac{d\varphi}{dx} \tag{17}$$

where J is the diffusion flux which shows the amount of substance flowing across a unit area, D is the diffusion coefficient, x is the position, and φ reflects the concentration of vacancies and is defined as the number of vacancies per unit volume (Equation (18)):

$$\varphi = \frac{N_v}{\Omega} \tag{18}$$

where Ω is the atomic volume. Therefore, a proportional relation between diffusion flux and temperature component can be presented:

$$J \propto exp(-1/T) \tag{19}$$

Since creep process indicates that diffusion finally leads to overall deformation, the strain caused by creep effect is proportional to temperature dependency:

$$\varepsilon \propto exp(-1/T) \tag{20}$$

The expression of $exp(1/T)$ can be simplified as a linear dependency when temperature is relatively high enough and within the application range (normally the range of experimental investigation), such as the temperature range from 650 K to 1000 K for the GP91 casting steel [30].

Therefore, we conclude that the underlying physical phenomenon is that diffusion of atoms is based on considerations of free energy and formation of vacancies. The literature shows that this mechanism is represented by an exponential dependency of the form $\varepsilon \propto exp(-1/T)$. We propose that the underlying physical explanation is that the ability of an atom to diffuse is based on the volume of space containing vacancies that it can recruit (hence an exp relationship), and on the rate at which vacancies form within that space. The latter requires activation energy, hence is dependent on temperature to achieve the necessary mobility at the atomic level. The formation of vacancies is therefore retarded at low temperatures, becomes active at intermediate temperatures, and is saturated at sufficient high temperature (all available vacancies have been formed), hence the $exp(-1/T)$ form of the relationship. At sufficiently high temperature the saturation causes this to simplify to a linear relation. This relation is also consistent with the empirical data on the materials of 63Sn37Pb solder [21] and GP91 casting steel [30]. The data on these two materials at the life of 5000 cycles are tabulated in Table 1. Then, the linear relationship between temperature and strain is presented in Figure 4. Results show good quality of linear fitting, with R^2 = 0.9932 for 63Sn37Pb solder and 0.9946 for GP91 casting steel.

Table 1. Data of temperature-strain relation for 63Sn37Pb [21] and GP91 casting steel [30].

Materials	Temperature (K)	Strain	Cyclic Time (s)	Life (Cycles)
63Sn37Pb	233	0.00412	1	5000
	298	0.00364		
	248	0.00332		
	398	0.00307		
GP91 Casting Steel	673	0.00317	20	5000
	823	0.00199		
	873	0.00174		

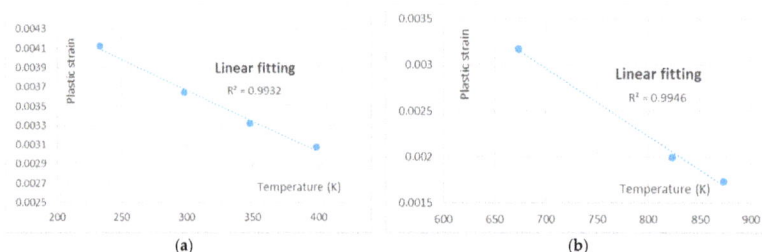

Figure 4. Curve fitting of temperature vs. strain for experimental data: (**a**) 63Sn37Pb; (**b**) GP91 casting steel. Raw data from [21,30] respectively.

Therefore, the creep-related strain is linearly proportional to temperature, and the overall fatigue capacity is reduced by a thermal effect, hence giving rise to the $c_1(\sigma)\left(T - T_{ref}\right)$ term in the unified formulation.

4.2. Logarithmical Relation between Temperature and Cyclic Time

A logarithmical relation between temperature and cyclic time is presented by the unified creep-fatigue equation, which is consistent with creep mechanism. As shown in Section 4.1, the diffusion behavior gives the logarithmical relationship between temperature and diffusion flux (Equation (19)). The definition of "diffusion flux" indicates that this term measures the amount of substance flowing through a cross sectional area during a unit time. In this context, it is a measure of the strain rate, at the microstructural level. Thus, a time dependency is included into this parameter in the form of a rate function. Then, Equation (19) can be presented as:

$$J = \frac{dD_v}{dt} \rightarrow \dot{\varepsilon} \propto exp(-1/T) \tag{21}$$

where D_v reflects the amount of substance following through a unit area. Significantly, Equation (21) gives a logarithmical relation between temperature and cyclic time (unite time), and this is consistent with the relationship shown in the unified creep-fatigue equation.

Theoretically, the temperature and cyclic time effects both decrease the fatigue capacity. In this case, we propose that to a first approximation they are independent of each other rather than convoluted with each other, hence the overall effect is additive. However, this statement is only reasonable when the creep damage is not specified. If creep is included then temperature can be related to time under one specific damage to show an inversely proportional relation. For example, for one specific creep damage, the effect caused by increasing temperature could be compensated through decreasing cyclic time. Consequently, considering these two situations, the numerical representation of creep effect is given as the sum of the temperature and cyclic time effects, hence of the form: $\left[c_1(\sigma)\left(T - T_{ref}\right) + c_2 \log\left(t_c/t_{ref}\right)\right]$.

The logarithmical relation for cyclic time is also represented by the conventional time-temperature parameter, wherein the time dependence is addressed as rupture time. This is based on the integrated characteristic shown by the unified creep-fatigue model. Specifically, the integrated characteristic shows that the unified creep-fatigue equation can be reorganized to the Manson-Haferd parameter at the pure-creep condition, where a logarithmical temperature-time relation is accommodated. Although this time-temperature parameter was entirely derived from empirical data (no physical basis) [32], it has been successfully validated on different materials, and thus is believed to have ability to describe creep behaviors. In addition, the pure-creep condition could be regarded as the idealization of a creep-fatigue situation with extremely prolonged cyclic time. In this case, it is reasonable for the unified formulation to present creep mechanism in a logarithmical relation between temperature and cyclic time.

In the specific case of cyclic creep, the load fluctuates between tension and compression. The cyclic time is a measure of the duration of time to which the material is exposed to the diffusion flow under tension and compression. Under tensile loading, we propose that the flow is, in practice, limited by dislocation pinning, grain boundaries, and other flow-limiting effects at the microstructural level. This causes the rate of diffusion to be reduced over longer periods, hence the total plastic strain due to this component has the form $c_2 \log\left(t_c/t_{ref}\right)$. This logarithmical relation between plastic strain and cyclic time is also consistent with empirical data on the materials of 63Sn37Pb solder [21] and stainless steel 316 [51]. The creep-fatigue data for the materials at the life of 5000 cycles are tabulated in Table 2, and the curve-fitting results are then presented by Figure 5. The result illustrates a good linear fit between strain and log of cyclic time, which implies a logarithmical relation between cyclic time and plastic strain.

Table 2. Data of cyclic time/strain rate-strain relation for 63Sn37Pb [21] and SS316 [51].

Materials	Cyclic Time (s)	Strain	Temperature (K)	Life (Cycles)
	10	0.00344		
63Sn37Pb	100	0.00298	298	5000
	1000	0.00243		

Materials	Strain Rate (%/min)	Strain	Temperature (K)	Life (Cycles)
	0.4	0.00218		
Stainless Steel 316	4	0.00253	973	5000
	40	0.00309		

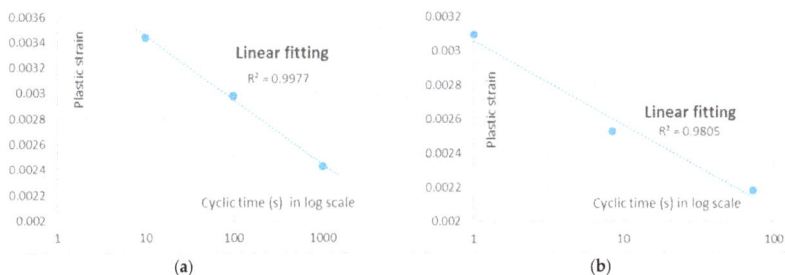

Figure 5. Curve fitting of log of cyclic time vs. strain for experimental data: (**a**) 63Sn37Pb; (**b**) stainless steel 316. Raw data from [21,51] respectively.

When the cycle is reversed, and the load moves into the compression stage, the diffusion is not undone (reversed). This is because changes have occurred at the microstructural level such that it is not the same geometric system as before—the system is inelastic. Consequently, the compression part of the cycle does not completely undo the inelastic strain of the previous stage (during a limited time range imposed for general engineering case). Also, the compression is proposed to undo or at least disturb the flow-limiting effects that arose in the tension stage. Hence the next tension cycle permits further diffusion to occur. We therefore propose that the cyclic time reduces the fatigue capacity per $1 - \left[c_2 \log\left(t_c/t_{ref}\right)\right]$. Although the cycle time provides a limited opportunity for the creep effect to operate, it is still reasonable to assume that the amount of creep that occurs within one part of the cycle follows the logarithmic time-temperature dependency.

Therefore, the diffusion-creep behavior is physically described by the amount of substance (which is logarithmical with temperature) flowing through a specific area during a unit time, thus this gives a logarithmical relation between temperature and time.

4.3. Power-Law Relation between Grain Size and Strain

The derivation of the unified creep-fatigue equation shows that the grain-size component is imposed into creep-related component and is directly extracted from the general creep equation for steady state [32], wherein a power-law relation between creep-related strain and grain size is presented (Equation (9)). This relation appears to accurately represent creep behavior.

Generally, diffusion flux for creep behavior is numerically formulated by Fick's equation (Equation (17)). A microstructural-level-based discussion shown in Section 4.1 shows that diffusion flux is not only related to the time, but also related to the unit area where substance following through (Equation (22)):

$$J = \frac{dD_t}{dA} \propto exp(-1/T) \tag{22}$$

where D_t reflects the amount of substance following during a unit time and A is the area where substance following through. This area can be obtained by the product of number of grains in unit area and average area of grain, wherein the average area has a strong dependency on grain size. Generally, the area of geometry can be numerically related with the key dimension in the form of second power order. In the present work, a complex situation for creep-fatigue condition is presented, wherein creep and fatigue are coupled, and this may cause the intensity of grain-size effect to deviate from the power of two, but the power-law relation should remain. Consequently, based on the general creep mechanism, it is logical that a power-law relation between grain size and strain is presented in the unified formulation.

In addition, bigger grain size is always beneficial for pure creep. This grain-size effect on creep is consistent with the presentation of the unified creep-fatigue equation. According to the validations [25] on the materials of Inconel 718 and GP91 casting steel, the negative exponent to grain size shows the benefit of big grain size on creep resistance.

We propose that the physical explanation for the grain-size dependency is that the diffusion creep phenomenon involves effects at the (irregular) grain boundaries, and to a lesser extent movement within the crystalline structure of the grain. However the latter mechanism becomes stalled once the available dislocations have run their courses. Hence the steady creep loading purges the internal structure of the grain of imperfections. Consequently the larger the grain, the lower the opportunity for diffusion creep to occur. Thus, it is to be expected that plastic strain would be inversely related to grain size, hence m is negative in $\left[A\left(d/d_{ref}\right)^m\right]$. For very small grains the effect becomes disproportionately worse, because a small change in grain size results in a large change in the number of grains in the section. This means more grain boundaries and opportunity for creep. A similar change in size for a large grain has a much smaller effect.

We note that m is approximately -0.5 for both materials considered (Inconel 718 and GP91 casting steel). Since creep damage involves crack propagation along a grain boundary, the decreased grain size results in increased opportunity for crack growth within a given area. For example, one square grain is identified to have four sides, hence four crack-growth potentials, then halving or quartering of grain size gives two more or six more potential directions for crack growth respectively for the same total area. Note that the size of one square grain is a reference condition, based on which the grain is halved or quartered. In this case, the grain size of 1 is not specified as 1 µm, but should rather be considered as a mathematical origin for the scaling effect, hence we refer to this as the pseudo grain size. The relationship between pseudo grain size d' and the number of crack-growth planes is shown in Figure 6. This relationship may be formulated numerically as a power-law relation, with exponent -0.68. In practical situations, the grain boundaries are not flat planes but are instead more irregular. The effect of this is to (a) increase the planar area available, and (b) provide more opportunity for an element of the boundary to be aligned with the preferred crack-growth direction. Hence practical situations are expected to provide greater opportunity for crack growth as the grain size decreases. The effect of this is to change the exponent closer to zero, e.g., if there were two additional planes at each stage then the exponent becomes -0.523.

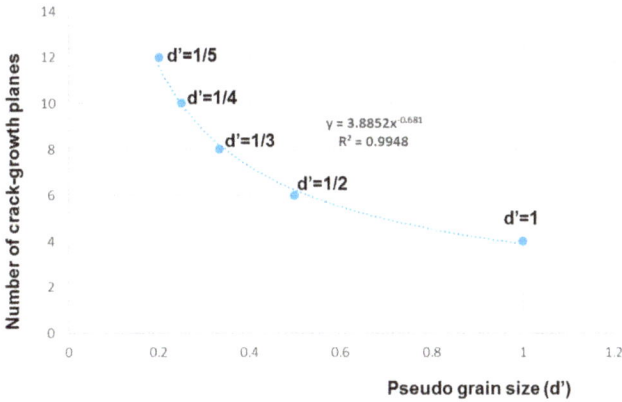

Figure 6. Relationship between pseudo grain size and the number of potential crack-growth planes.

We therefore propose that there are natural reasons for the exponent in the formulation $\left[A \left(d/d_{ref} \right)^m \right]$ to be of the order $m = -0.681$ or larger, and this is compatible with the empirically determined values of $m = -0.5411$ for Inconel 718 and $m = -0.4053$ for GP91 casting steel.

4.4. Power-Law Relation between Life and Strain

The derivation of the unified creep-fatigue equation shows an extension of the Coffin-Manson equation. Therefore, the power-law relation shown in the Coffin-Manson equation is accommodated in the unified formulation. It is acceptable that the power-law relation between reversed loading and number of cycles could well present the process of damage accumulation in fatigue perspective.

In the present work, fracture mechanics is presented as an opening model showing direct apart between two crack surfaces, per [52]. Therefore, the stress intensity factor (K) is shown by Equation (23):

$$K = \sigma \sqrt{\pi a / 2} \tag{23}$$

where σ is the applied stress and a is the crack length. Crack-growth process shows that plastic deformation occurs around the crack tip due to high stress concentration, where a circular plastic zone ahead of the crack tip is formed (Figure 7). According to the definition of the stress intensity and Equation (23), the stress distribution (σ_{ij}) near the crack tip is:

$$\sigma_{ij} = \sigma\sqrt{\frac{a}{2r}}f(\theta) = \frac{K}{2\pi r}f(\theta) \tag{24}$$

where r and θ are polar coordinates. This equation shows that when r tends towards zero, the stress around the crack tip becomes singular. This implies the existence of a plastic zone, with yield stress (σ_y) presented at the boundary of this zone. Then, through letting $\sigma_{ij} = \sigma_y$ and $f(\theta) = 1$, Equation (25) gives the size of plastic zone [52]:

$$r_y = \frac{1}{2\pi}\left(\frac{K}{\sigma_y}\right)^2 \tag{25}$$

where r_y is the radius of the plastic zone.

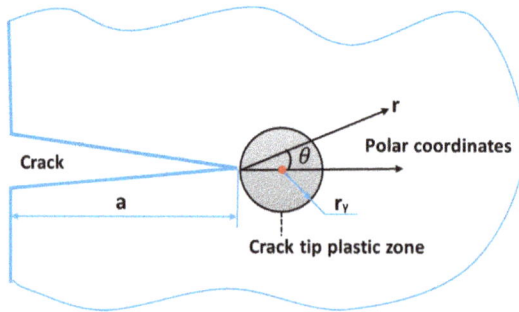

Figure 7. Plastic zone around crack tip.

At the situation with cyclic loading, the stress intensify factor in Equation (25) varies with the change of loading, and then results in expansion or shrinkage of the plastic zone. In this case, the effective stress intensify factor (K_{eff} or ΔK) is introduced and is given as:

$$K_{eff} = K_{max} - K_{min} = \Delta K \tag{26}$$

where ΔK is the difference between maximum (K_{max}) and minimum (K_{min}) stress intensify factors for one cycle. Then, an equivalent radius of the plastic zone ($\overline{r}_y$) can be given by Equation (27):

$$\overline{r}_y = \frac{1}{2\pi}\left(\frac{\Delta K}{\sigma_y}\right)^n \tag{27}$$

This equation shows that the second power order in Equation (25) is reasonably replaced by a general exponent (n) due to the equivalent transformation, which is consistent with FEA result shown by You [53]. Since larger plastic zone gives more crack growth [52], the crack growth in one cycle can be related with ΔK in a power-law relation. This relation (Equation (28)) was initially presented by Paris [3] and then demonstrated amounts of empirical data [54,55].

$$\frac{da}{dN} = C(\Delta K)^m \tag{28}$$

where $\frac{da}{dN}$ gives the increased crack length in one cycle, and C and m are constants. According to Equation (23), effective stress intensify factor also can be expressed as:

$$\Delta K = X\Delta\sigma\sqrt{\pi a} \tag{29}$$

where X is a constant and $\Delta\sigma$ is the stress range. Then, introducing Equation (29) into Equation (28) gives:

$$\frac{da}{dN} = C\left(X\Delta\sigma\sqrt{\pi a}\right)^m \tag{30}$$

Applying the integral operation for life gives [56]:

$$N_f = \frac{2\left(a_c^{\frac{2-m}{2}} - a_i^{\frac{2-m}{2}}\right)}{(2-m)C\left(X\Delta\sigma\sqrt{\pi a}\right)^m} \tag{31}$$

where a_c is the critical crack length at which fracture occurs and a_i is the initial crack length at which crack starts to grow under a given stress range. Significantly, this equation shows a typical power-law relation between fatigue life and applied stress range (applied loading), and thus it is also reasonable to relate plastic strain with fatigue life in a power-law relation.

Therefore, the existence and size of the plastic zone at crack tip are the physical basis for the power-law relation for crack growth. It is reasonable to assume that fatigue damage is accumulated at each cycle by the same form. This results in a power-law relation between applied loading and life, hence explaining the form $N^{-\beta_0}$.

4.5. Numerical Presentation of Creep Effect on Fatigue Capacity

The development of the strain-based unified creep-fatigue equation is based on the concept of "fatigue capacity" [26,27]. Briefly, this concept physically indicates that the full fatigue capacity is gradually consumed by the creep effect, and this process is numerically presented by the form of $(1 - x)$. This form is accommodated in the unified creep-fatigue formulation (Equation (1)).

Equation (9) shown in Section 3.1 indicates that the first term reflects the full fatigue capacity wherein the fatigue behavior under the pure-fatigue condition is presented by the Coffin-Manson equation. In addition, the second term of Equation (9) describes the creep-related effect, where temperature, cyclic time and grain size dependencies are included. This term takes a Coffin-Manson-type formulation, and shows the creep-related damage is accumulated at cyclic loading. The combination of these two terms is numerically presented as the form of "$1 - x$" (Equations (1)) and shows the gradual consumption of full fatigue capacity due to creep effect where the residual fatigue capacity is given.

In addition, the reference condition is introduced to show the threshold between pure fatigue and creep fatigue, and creep effect is dormant below the reference condition. At the reference condition, the full fatigue capacity is presented (the second term of Equation (9) equals to zero) in the form of the Coffin-Manson equation. Therefore, the introduction of the reference condition builds a bridge between pure fatigue and creep fatigue. The ability to describe the pure-fatigue condition is proved on the material of GP91 casting steel, where the ratios of predicted fatigue life (obtained by the degenerated form of the unified model) to experimental result (extracted from [30]) fall within the upper bound (+25%) and the lower bound (−25%) (Figure 8). This implies that the unified formulation provides a good quality prediction of fatigue-life under the pure-fatigue condition, specifically a relatively high correlation between predicted and experimental fatigue life.

Consequently, the negative effect of creep on fatigue capacity is numerically formulated as the form of $(1 - x)$, and the introduction of the reference condition shows the threshold for the activation of creep effect.

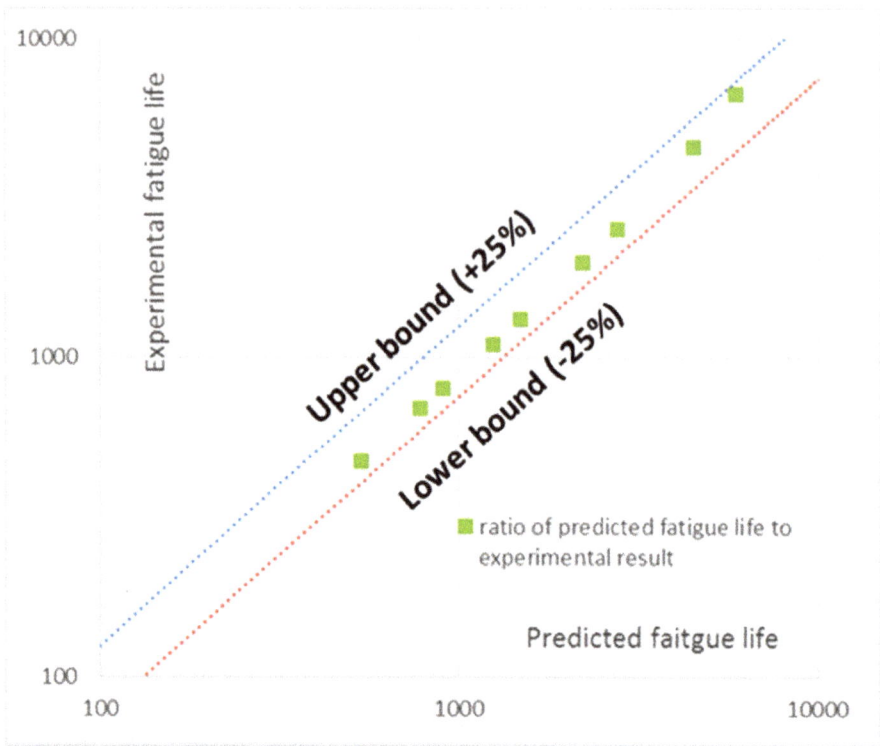

Figure 8. The ratio of predicted fatigue life to experimental result under the pure-fatigue condition, per the unified model, for GP91 casting steel. Raw data from [30].

5. Discussion

5.1. Summary

In summary, we propose the following fundamental mechanisms are at work to determine the creep-fatigue behavior of a material; see Figure 9. In this way we propose that the main structural features of the unified creep-fatigue equation are grounded in deeper physical phenomena.

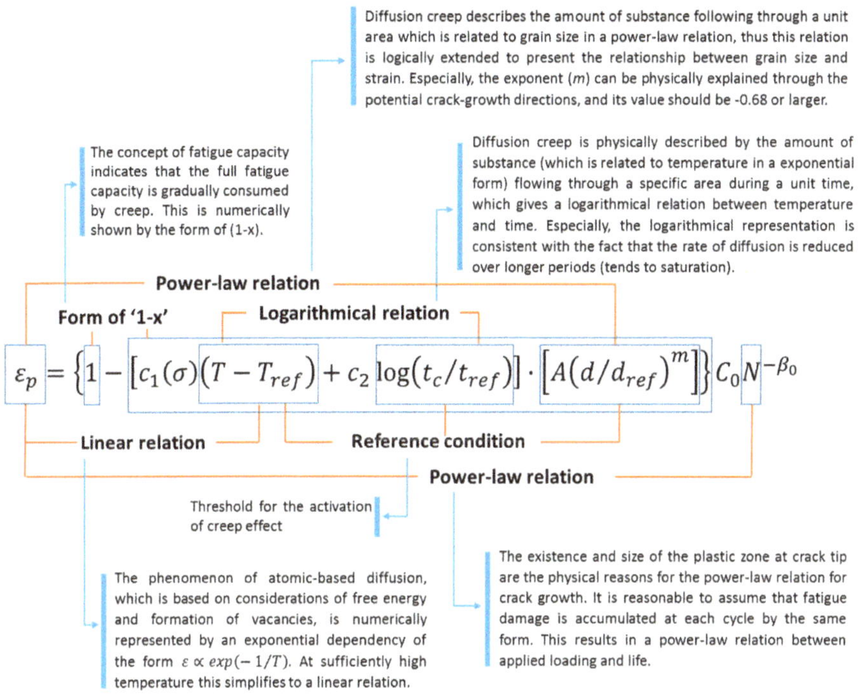

Diffusion creep describes the amount of substance following through a unit area which is related to grain size in a power-law relation, thus this relation is logically extended to present the relationship between grain size and strain. Especially, the exponent (*m*) can be physically explained through the potential crack-growth directions, and its value should be -0.68 or larger.

The concept of fatigue capacity indicates that the full fatigue capacity is gradually consumed by creep. This is numerically shown by the form of (1-x).

Diffusion creep is physically described by the amount of substance (which is related to temperature in a exponential form) flowing through a specific area during a unit time, which gives a logarithmical relation between temperature and time. Especially, the logarithmical representation is consistent with the fact that the rate of diffusion is reduced over longer periods (tends to saturation).

Power-law relation

Form of '1-x' Logarithmical relation

$$\varepsilon_p = \left\{ 1 - \left[c_1(\sigma)\left(T - T_{ref}\right) + c_2 \log\left(t_c/t_{ref}\right) \right] \cdot \left[A\left(d/d_{ref}\right)^m \right] \right\} C_0 N^{-\beta_0}$$

Linear relation Reference condition

Power-law relation

Threshold for the activation of creep effect

The phenomenon of atomic-based diffusion, which is based on considerations of free energy and formation of vacancies, is numerically represented by an exponential dependency of the form $\varepsilon \propto exp(-1/T)$. At sufficiently high temperature this simplifies to a linear relation.

The existence and size of the plastic zone at crack tip are the physical reasons for the power-law relation for crack growth. It is reasonable to assume that fatigue damage is accumulated at each cycle by the same form. This results in a power-law relation between applied loading and life.

Figure 9. Main relationships and fundamental mechanisms in the unified formulation.

5.2. Limitations and Future Work

The discussion shown above indicates that the relationships between different variables in the unified creep-fatigue formulation are consistent with underlying physical mechanisms, thus this creep-fatigue model has the ability to describe creep-fatigue behavior numerically and also in terms of physical meaning. Although the relationships between different parameters were derived from underlying physical phenomena, the numerical values of the coefficients still cannot be predicted with precision. In this case, the coefficients need to be extracted through performing creep-fatigue experiments. This is a limitation for this unified formulation, which implies that the coefficients cannot be predicted without any empirical data of fatigue. This model has an opportunity to be further modified/improved to reduce the dependence on fatigue tests, by exploring for more convenient and economical data collection methods.

Generally, reducing the dependence on fatigue test could be conducted through introducing the material-property-related parameters, such as yield strength, into the coefficients. By this means, the coefficients could be directly evaluated through the material properties, and the fatigue test would be eliminated. This work was initially attempted by Manson [57], who proposed a universal slope formulation for the strain-life relation through introducing the parameter of ductility. This provides a possible method to improve this limitation. In addition, this limitation also could be improved by deeper investigation into physical phenomena of fatigue and creep. For example, the grain-size related coefficients may be related to crystal structure, and the creep-related coefficients may be physically represented through quantitatively investigating the influences of temperature and time on creep-fatigue damage presented by the behaviors of diffusion or thermodynamics.

It is notable that although the new creep-fatigue model is not completely free of the need for empirical data, the method of derivation makes this model fundamentally superior to other existing creep-fatigue models (mentioned in Section 1) because of the good balance between accuracy and economy. The accuracy of fatigue-life prediction has been proved on multiple materials in previous research [25–27]. The good economy of the current model means fewer creep-fatigue data are required to determine the coefficients.

6. Conclusions

Creep-fatigue behavior is normally influenced by temperature, cyclic time and grain size. Generally, the fatigue capacity is gradually consumed by elevated temperature and prolonged cyclic time, and smaller grain size results in better fatigue capacity, but leads to worse creep resistance. These relevant variables are well accommodated in the unified creep-fatigue equation, and the relationships between them are con with underlying physical mechanisms of fatigue and creep. Specifically, the creep-related relationships, including linear relation between temperature and strain, logarithmic relation between temperature and cyclic time and power-low relation between grain size and strain, are extracted from diffusion-creep phenomenon. In addition, crack-growth behavior gives a power-law relationship between life and strain. Finally, based on the concept of fatigue capacity, these physical-mechanism-based relationships are numerically constructed in the form of "$1 - x$", and the reference condition is introduced to present the threshold of creep effect.

The original contribution of this work is that the unified creep-fatigue equation is linked to physical phenomena at a microstructural level. Specifically, the influences of different variables (including temperature, cyclic time and grain size) on creep-fatigue behavior were explored, and the numerical relationships shown in this equation were investigated and explained through proposed deeper physical mechanisms of fatigue and creep. A particular contribution is the proposition of a physical explanation of the grain-size exponent (m) via consideration of crack-growth planes.

Author Contributions: The work was conducted by D.L. and supervised by D.J.P. The investigation of physical-mechanism-based relationships from a microstructural level was conducted by D.L. The explanation of how relevant variables influence creep fatigue was conducted by D.L. and D.J.P. All authors contributed to writing the paper.

Conflicts of Interest: The authors declare no conflict of interest.

References

1. Miner, M.A. Cumulative damage in fatigue. *J. Appl. Mech.* **1945**, *12*, 159–164.
2. Palmgren, A. Die lebensdauer von kugellagern. *Z. Ver. Dtsch. Ing.* **1924**, *68*, 339–341.
3. Paris, P.; Erdogan, F. A critical analysis of crack propagation laws. *J. Basic Eng.* **1963**, *85*, 528–533. [CrossRef]
4. Takahashi, Y. Study on creep-fatigue evaluation procedures for high-chromium steels—Part I: Test results and life prediction based on measured stress relaxation. *Int. J. Press. Vessel. Pip.* **2008**, *85*, 406–422. [CrossRef]
5. Takahashi, Y.; Dogan, B.; Gandy, D. Systematic evaluation of creep-fatigue life prediction methods for various alloys. *J. Press. Vessel. Technol.* **2013**, *135*, 061204. [CrossRef]
6. Payten, W.M.; Dean, D.W.; Snowden, K.U. A strain energy density method for the prediction of creep-fatigue damage in high temperature components. *Mater. Sci. Eng.* **2010**, *527*, 1920–1925. [CrossRef]
7. Neu, R.; Sehitoglu, H. Thermomechanical fatigue, oxidation, and creep: Part II. Life prediction. *Metall. Trans. A* **1989**, *20*, 1769–1783. [CrossRef]
8. Ainsworth, R.; Ruggles, M.; Takahashi, Y. Flaw assessment procedure for high-temperature reactor components. *J. Press. Vessel. Technol.* **1992**, *114*, 166–170. [CrossRef]
9. Cailletaud, G.; Nouailhas, D.; Grattier, J.; Levaillant, C.; Mottot, M.; Tortel, J.; Escavarage, C.; Héliot, J.; Kang, S. A review of creep-fatigue life prediction methods: Identification and extrapolation to long term and low strain cyclic loading. *Nucl. Eng. Des.* **1984**, *83*, 267–278. [CrossRef]
10. June, W. A continuum damage mechanics model for low-cycle fatigue failure of metals. *Eng. Fract. Mech.* **1992**, *41*, 437–441. [CrossRef]
11. Chaboche, J.L. *Une loi Différentielle D'endommagement de Fatigue avec Cumulation non Linéaire*; Office Nationale d'Etudes et de Recherches Aérospatiales: Palaiseau, France, 1974.

12. Metzger, M.; Nieweg, B.; Schweizer, C.; Seifert, T. Lifetime prediction of cast iron materials under combined thermomechanical fatigue and high cycle fatigue loading using a mechanism-based model. *Int. J. Fatigue* **2013**, *53*, 58–66. [CrossRef]

13. Seifert, T.; Riedel, H. Mechanism-based thermomechanical fatigue life prediction of cast iron. Part I: Models. *Int. J. Fatigue* **2010**, *32*, 1358–1367. [CrossRef]

14. Charkaluk, E.; Bignonnet, A.; Constantinescu, A.; Dang Van, K. Fatigue design of structures under thermomechanical loadings. *Fatigue Fract. Eng. Mater. Struct.* **2002**, *25*, 1199–1206. [CrossRef]

15. Constantinescu, A.; Charkaluk, E.; Lederer, G.; Verger, L. A computational approach to thermomechanical fatigue. *Int. J. Fatigue* **2004**, *26*, 805–818. [CrossRef]

16. Basquin, O. *The Exponential Law of Endurance Tests*; ASTM International: West Conshohocken, PA, USA, 1910; pp. 625–630.

17. Coffin, L.F., Jr. *A Study of the Effects of Cyclic Thermal Stresses on a Ductile Metal*; Knolls Atomic Power Lab.: Niskayuna, NY, USA, 1953.

18. Manson, S.S. *Behavior of Materials under Conditions of Thermal Stress*; Lewis Flight Propulsion Lab.: Cleveland, OH, USA, 1954.

19. Coffin, L. Fatigue at high temperature. In *Fatigue at Elevated Temperatures*; ASTM International: West Conshohocken, PA, USA, 1973.

20. Solomon, H. Fatigue of 60/40 solder. *IEEE Trans. Compon. Hybrids Manuf. Technol.* **1986**, *9*, 423–432. [CrossRef]

21. Shi, X.; Pang, H.; Zhou, W.; Wang, Z. Low cycle fatigue analysis of temperature and frequency effects in eutectic solder alloy. *Int. J. Fatigue* **2000**, *22*, 217–228. [CrossRef]

22. Jing, H.; Zhang, Y.; Xu, L.; Zhang, G.; Han, Y.; Wei, J. Low cycle fatigue behavior of a eutectic 80 au/20 sn solder alloy. *Int. J. Fatigue* **2015**, *75*, 100–107. [CrossRef]

23. Engelmaier, W. Fatigue life of leadless chip carrier solder joints during power cycling. *IEEE Trans. Compon. Hybrids Manuf. Technol.* **1983**, 232–237. [CrossRef]

24. Wong, E.; Mai, Y.-W. A unified equation for creep-fatigue. *Int. J. Fatigue* **2014**, *68*, 186–194. [CrossRef]

25. Liu, D.; Pons, D. Development of a unified creep-fatigue equation including heat treatment. *Fatigue Fract. Eng. Mater. Struct.* **2017**. [CrossRef]

26. Liu, D.; Pons, D.; Wong, E.-H. The unified creep-fatigue equation for stainless steel 316. *Metals* **2016**, *6*, 219. [CrossRef]

27. Liu, D.; Pons, D.; Wong, E.-H. Creep-integrated fatigue equation for metals. *Int. J. Fatigue* **2017**, *98*, 167–175. [CrossRef]

28. Manson, S.; Haferd, A. *A Linear Time-Temperature Relation for Extrapolation of Creep and Stress-Rupture Data*; Lewis Flight Propulsion Lab.: Cleveland, OH, USA, 1953.

29. Tabuchi, M.; Hongo, H.; Li, Y.; Watanabe, T.; Takahashi, Y. Evaluation of microstructures and creep damages in the haz of p91 steel weldment. *J. Press. Vessel. Technol.* **2009**, *131*, 021406. [CrossRef]

30. Mroziński, S.; Golański, G. Low cycle fatigue of gx12crmovnbn9–1 cast steel at elevated temperature. *J. Achiev. Mater. Manuf. Eng.* **2011**, *49*, 7–16.

31. Liu, D.; Pons, D. A unified creep-fatigue equation with application to engineering design. *InTechOpen* **2017**, under review.

32. Dowling, N.E. *Mechanical Behavior of Materials: Engineering Methods for Deformation, Fracture, and Fatigue*; Pearson Education (US): London, UK, 2012.

33. Shigley, J.E.; Mischke, C.R. *Mechanical Engineering Design*; McGraw-Hill: New York, NY, USA, 2003.

34. Ashby, M.F.; Shercliff, H.; Cebon, D. *Materials: Engineering, Science, Processing and Design*; Butterworth-Heinemann: Oxford, UK, 2013.

35. Callister, W.D.; Rethwisch, D.G. *Materials Science and Engineering*; John Wiley & Sons: New York, NY, USA, 2011.

36. Finnie, I.; Heller, W.R. *Creep of Engineering Materials*; McGraw-Hill: New York, NY, USA, 1959.

37. Kassner, M.E. *Fundamentals of Creep in Metals and Alloys*; Elsevier: Amsterdam, The Netherlands, 2015.

38. Poirier, J.-P. *Creep of Crystals: High-Temperature Deformation Processes in Metals, Ceramics and Minerals*; Cambridge University Press: Cambridge, UK, 1985.

39. Shewmon, P. *Diffusion in Solids*; Springer: Berlin/Heidelberg, Germany, 2016.

40. Prokoshkina, D.; Esin, V.; Wilde, G.; Divinski, S. Grain boundary width, energy and self-diffusion in nickel: Effect of material purity. *Acta Mater.* **2013**, *61*, 5188–5197. [CrossRef]

41. Dai, C.; Zhang, B.; Xu, J.; Zhang, G. On size effects on fatigue properties of metal foils at micrometer scales. *Mater. Sci. Eng.* **2013**, *575*, 217–222. [CrossRef]

42. Hanlon, T.; Kwon, Y.-N.; Suresh, S. Grain size effects on the fatigue response of nanocrystalline metals. *Scr. Mater.* **2003**, *49*, 675–680. [CrossRef]

43. Cocks, A.; Pontern, A. *Mechanics of Creep Brittle Materials 1*; Springer Science & Business Media: Berlin, Germany, 1989.

44. Ejaz, N.; Qureshi, I.; Rizvi, S. Creep failure of low pressure turbine blade of an aircraft engine. *Eng. Fail. Anal.* **2011**, *18*, 1407–1414. [CrossRef]

45. Král, P.; Dvořák, J.; Kvapilová, M.; Svoboda, M.; Sklenička, V. Creep damage of al and al-sc alloy processed by ecap. In Proceedings of the Acta Metallurgica Slovaca-Conference, Košice, Slovakia, 2012.

46. Hatanaka, K.; Yamada, T. Effect of grain size on low cycle fatigue in low carbon steel. *Bull. JSME* **1981**, *24*, 1692–1699. [CrossRef]

47. Hattori, H.; Kitagawa, M.; Ohtomo, A. Effect of grain size on high temperature low-cycle fatigue properties of inconel 617. *Tetsu-To-Hagane* **1982**, *68*, 2521–2530. [CrossRef]

48. Pieraggi, B.; Uginet, J. *Fatigue and Creep Properties in Relation*; The Minerals, Metals & Materials Society: Pittsburgh, PA, USA, 1994.

49. Thébaud, L.; Villechaise, P.; Cormier, J.; Crozet, C.; Devaux, A.; Béchet, D.; Franchet, J.-M.; Organista, A.; Hamon, F. Relationships between microstructural parameters and time-dependent mechanical properties of a new nickel-based superalloy ad730™. *Metals* **2015**, *5*, 2236–2251. [CrossRef]

50. Fredriksson, H. *Solidification and Crystallization Processing in Metals and Alloys*; John Wiley & Sons: New York, NY, USA, 2012.

51. Kanazawa, K.; Yoshida, S. *Effect of Temperature and Strain Rate on the High Temperature Low-Cycle Fatigue Behavior of Austenitic Stainless Steels*; IAEA: Vienna, Austria, 1975.

52. Weertman, J. Fatigue crack propagation theories. In *Fatigue and Microstructure*; ASM: Metals Park, OH, USA, 1979; pp. 279–306.

53. You, C.; He, B.; Achintha, M.; Reed, P. Numerical modelling of the fatigue crack shape evolution in a shot-peened steam turbine material. *Int. J. Fatigue* **2017**, *104*, 120–135. [CrossRef]

54. Liang, R.; Ji, Y.; Wang, S.; Liu, S. Effect of microstructure on fracture toughness and fatigue crack growth behavior of ti17 alloy. *Metals* **2016**, *6*, 186. [CrossRef]

55. Paggi, M. Crack propagation in honeycomb cellular materials: A computational approach. *Metals* **2012**, *2*, 65–78. [CrossRef]

56. Xiang, Y.; Lu, Z.; Liu, Y. Crack growth-based fatigue life prediction using an equivalent initial flaw model. Part I: Uniaxial loading. *Int. J. Fatigue* **2010**, *32*, 341–349. [CrossRef]

57. Manson, S. A modified universal slopes equation for estimation of fatigue characteristics of metals. *J. Eng. Mater. Technol.* **1988**, *110*, 55.

MDPI AG

St. Alban-Anlage 66

4052 Basel, Switzerland

Tel. +41 61 683 77 34

Fax +41 61 302 89 18

http://www.mdpi.com

Metals Editorial Office

E-mail: metals@mdpi.com

http://www.mdpi.com/journal/metals

www.ingramcontent.com/pod-product-compliance
Lightning Source LLC
Chambersburg PA
CBHW051848210326
41597CB00033B/5813